Lecture Notes i

The Lecture Notes in Physics

The series Lecture Notes in Physics (LNP), founded in 1969, reports new developments in physics research and teaching – quickly and informally, but with a high quality and the explicit aim to summarize and communicate current knowledge in an accessible way. Books published in this series are conceived as bridging material between advanced graduate textbooks and the forefront of research and to serve three purposes:

- to be a compact and modern up-to-date source of reference on a well-defined topic

- to serve as an accessible introduction to the field to postgraduate students and nonspecialist researchers from related areas

- to be a source of advanced teaching material for specialized seminars, courses and schools

Both monographs and multi-author volumes will be considered for publication. Edited volumes should, however, consist of a very limited number of contributions only. Proceedings will not be considered for LNP.

Volumes published in LNP are disseminated both in print and in electronic formats, the electronic archive being available at springerlink.com. The series content is indexed, abstracted and referenced by many abstracting and information services, bibliographic networks, subscription agencies, library networks, and consortia.

Proposals should be sent to a member of the Editorial Board, or directly to the managing editor at Springer:

Christian Caron
Springer Heidelberg
Physics Editorial Department I
Tiergartenstrasse 17
69121 Heidelberg / Germany
christian.caron@springer.com

A.K. Chandra

A. Das

B.K. Chakrabarti (Eds.)

Quantum Quenching, Annealing and Computation

 Springer

Anjan Kumar Chandra
Saha Institute of Nuclear
Physics
Centre for Applied
Mathematics &
Computational Science
1/AF Bidhannagar
Kolkata-700064
India
anjan.chandra@saha.ac.in

Arnab Das
Abdus Salam International
Centre for Theoretical Physics
(ICTP)
Strada Costiera, 11
34014 Trieste
Italy
arnabdas@ictp.it

Bikas K. Chakrabarti
Saha Institute of Nuclear
Physics
Centre for Applied
Mathematics &
Computational Science
1/AF Bidhannagar
Kolkata-700064
India
bikask.chakrabarti@saha.ac.in

Chandra A.K., Das A., Chakrabarti B.K. (Eds.): *Quantum Quenching, Annealing and Computation*, Lect. Notes Phys. 802 (Springer, Berlin Heidelberg 2010), DOI 10.1007/978-3-642-11470-0

Lecture Notes in Physics ISSN 0075-8450   e-ISSN 1616-6361
ISBN 978-3-642-11469-4   e-ISBN 978-3-642-11470-0
DOI 10.1007/978-3-642-11470-0
Springer Heidelberg Dordrecht London New York

Library of Congress Control Number: 2010923000

Cover design: Integra Software Services Pvt. Ltd., Pondicherry

Printed on acid-free paper

Springer is part of Springer Science+Business Media (www.springer.com)

Preface

The process of realizing the ground state of some typical (frustrated) quantum many-body systems, starting from the 'disordered' or excited states, can be formally mapped to the search of solutions for computationally hard problems. The dynamics through the critical point, in between, are therefore extremely crucial. In the context of such computational optimization problems, the dynamics (of rapid quenching or slow annealing), while tuning the appropriate fields or fluctuations, in particular while crossing the quantum critical point, are extremely intriguing and are being investigated these days intensively. Several successful methods and tricks are now well established.

This volume gives a collection of introductory reviews on such developments written by well-known experts. It concentrates on quantum phase transitions and their dynamics as the transition or critical points are crossed. Both the quenching and annealing dynamics are extensively covered.

We hope these timely reviews will inspire the young researchers to join and contribute to this fast-growing, intellectually challenging, as well as technologically demanding field. We are extremely thankful to the contributors for their intensive work and pleasant cooperations. We are also very much indebted to Kausik Das for his help in compiling this book. Finally, we express our gratitude to Johannes Zittartz, Series Editor, LNP, and Christian Caron of physics editorial department of Springer for their encouragement and support.

Trieste and Kolkata, India

Anjan Kumar Chandra
Arnab Das
Bikas K. Chakrabarti

Contents

Chapter 1
Quantum Approach to Classical Thermodynamics and Optimization

R.D. Somma and G. Ortiz

1.1 Introduction

The goal of combinatorial optimization is to find the optimal solution that minimizes a given cost function E. Since the configuration space is typically large, finding such a solution may be a time-consuming task, even when an efficient description of the cost function is provided. Some problems in combinatorial optimization can be then classified according to their computational complexity. This complexity is determined by the number of resources (time and space) needed to find the optimal solution, and the relevant question is how the complexity depends on the problem size (e.g., the dimension of the configuration space, D).

Interestingly, many problems in classical physics can be expressed as combinatorial optimization ones, so we can associate them with the corresponding complexity classes. For example, it was previously shown that computing the ground state energy (or the partition function) of a classical Ising N-spin glass with a random magnetic field in two spatial dimensions is an NP-hard problem [1], i.e., there is no known algorithm that can find the solution with poly(N) resources. After all, the number of possible microscopic configurations of the system ($D = 2^N$) increases exponentially with the system size and, in the worst-case scenario, one has to search in this exponentially large configuration space to find the solution. Still in some cases this complexity can be reduced by using adequate methods that allow one to solve the problem with significantly less resources. This decrease in complexity may be attributed, in some cases, to the presence of conserved quantities (symmetries) that pose constraints and reduce the number of steps needed to find the optimal solution.

It is then desirable to explore different methods to solve combinatorial optimization problems that avoid searching exhaustively over the large configuration

R. D. Somma (✉)
Perimeter Institute for Theoretical Physics, Waterloo, ON N2L 2Y5, Canada,
rsomma@perimeterinstitute.ca

G. Ortiz
Department of Physics, Indiana University, Bloomington, IN 47405, USA, ortizg@indiana.edu

Somma, R. D., Ortiz, G.: *Quantum Approach to Classical Thermodynamics and Optimization*. Lect. Notes Phys. **802**, 1–20 (2010)
DOI 10.1007/978-3-642-11470-0_1

space. Simulated annealing (SA) [2] and quantum annealing (QA) (see [3–6] and references there in) represent possible algorithmic strategies to attack these problems, and can be used to compute properties of classical systems in equilibrium at finite temperatures as well [7, 8]. In physical terms these annealing methods, as applied to optimization problems, are interpreted as follows. First, the cost function is identified with a classical Hamiltonian H that models the interactions of a classical system. As a matrix operator, H is diagonal in the standard (classical) basis. In SA the classical system is put in contact with a thermal bath, initially at temperature $T \gg 1$. Then T is slowly decreased (as a function of time) to zero so that, at the end, the state of the system has high probability of being in the optimal configuration (ground state of H). In QA an ad ψ hoc transverse magnetic field γ is added to H, and the total Hamiltonian can be interpreted as that of a quantum system. The QA process consists of slowly decreasing γ (as a function of time) from a sufficiently large value to zero, while keeping $T = 0$. The annealing procedure is meant to converge to the optimal (ground) state [9].

Both annealing processes can be simulated by a number of computational techniques, each with a corresponding implementation complexity. SA is often simulated classically by using Markov chain Monte Carlo (MCMC) methods. MCMC generates a sequence of states via a Markov process so that, after a number of steps, the probability of finding the optimal solution is sufficiently large. Each configuration is obtained from the previous one by acting with a temperature-dependent transition rule. The latter is determined by the choice of stochastic matrix . In the worst-case scenario, the complexity of SA is dominated by the inverse of the minimum spectral gap of the stochastic matrices, Δ [10]. The most natural device for the implementation of QA would be a quantum computer. In this case, the complexity (or total evolution time) is determined by the adiabatic theorem of quantum mechanics. The *folk* result of the adiabatic theorem gives a complexity dominated by the inverse of the minimum gap of the (quantum) Hamiltonians squared. Since no large quantum computer exists today, many authors attempt to simulate QA using MCMC methods as well (quantum Monte Carlo). A way to implement QA in a classical computer is to associate the quantum system, embedded in d spatial dimensions, with a classical one in $d + 1$ spatial dimensions [11] (see Fig. 1.1 for an example). The complexity of the MCMC method used for QA is not clear and is usually determined on a case-by-case basis by numerical simulations.

One can develop annealing strategies similar to the ones used for optimization problems to study the thermal properties of classical systems in equilibrium [7, 8]. Indeed, optimization problems simply correspond to the $T \to 0$ limit of these more general thermodynamic studies.

In this chapter we present two quantum strategies to compute thermodynamic properties of classical systems in equilibrium at bounded precision. These strategies can also be used for combinatorial optimization by associating the cost function with the Hamiltonian of the classical system. The first strategy is described in the context of QA or adiabatic quantum computation. The goal of this strategy is to prepare a coherent version of the Gibbs state (i.e., the *quantum* Gibbs state (QGS)). A measurement in the standard basis of such a state allows us to sample classical

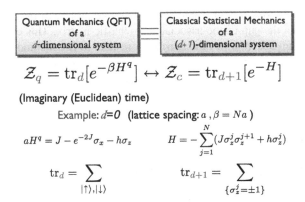

Fig. 1.1 Example of a quantum-to-classical Feynman mapping. In this case, one maps a quantum problem (or quantum field theory) in d space dimensions (H^q) to a classical statistical mechanics problem in $d + 1$ (H). The figure also displays a quantum single-spin example, i.e., $d = 0$

configurations with the desired (equilibrium) Gibbs' distribution. The QGS can be shown to be a non-degenerate eigenstate of a family of quantum Hamiltonians. We then choose a Hamiltonian path to adiabatically prepare the QGS. The adiabatic theorem of quantum mechanics yields a complexity, or total evolution time, of $O[\text{poly}(1/\Delta)]$ for the preparation of the state. Δ is the minimum spectral gap of the quantum Hamiltonians. If Δ coincides with the spectral gap of the stochastic matrices, as it occurs for some paths that are in one-to-one correspondence with thermal evolutions, no quantum speedup (in terms of complexity) with respect to the SA method implemented with MCMC exists. Next we show, using techniques from quantum information (see [12]), a novel way to construct Hamiltonian paths that yield a quadratic improvement of the complexity on the gap of the stochastic matrix.

The second strategy we present is intended to output the partition function of the classical system, at bounded precision, as the inner product or overlap between two quantum states that can be efficiently prepared. This strategy can be interpreted as the quantum counterpart of the classical method that computes the partition function when no importance sampling is used. In the latter, we sample configurations from the uniform distribution and compute the corresponding Boltzmann weight. After sufficiently many samples, the average weight converges to the desired value. Due to the poor resource scaling of this strategy with precision, the disadvantages of this strategy are manyfold. Still, the quantum method yields a quadratic improvement in the error with respect to its classical counterpart.

This chapter is organized as follows. In Sect. 1.2 we describe quantum annealing strategies for optimization. In Sect. 1.2.1 we show a classical-to-quantum mapping and build a family of quantum Hamiltonians whose ground state is the QGS. In Sect. 1.2.2 we determine the resource scaling for two different adiabatic paths that are related to the paths used in the standard SA and QA methods. In Sect. 1.2.3 we explain a possible way to classically simulate adiabatic evolutions using MCMC. In

Sect. 1.2.4 we give a Hamiltonian path where the minimum relevant gap is quadratically bigger than the gap of the associated stochastic matrix. This choice yields a quadratic quantum speedup (on the gap) with respect to SA methods. In Sect. 1.3 we show how a quantum phase estimation algorithm can be used to compute the partition function. Concluding remarks are given in Sect. 1.4.

1.2 Quantum Annealing Strategies for Optimization

1.2.1 Classical-to-Quantum Mapping

We describe and generalize the results of [7, 13–15]. For simplicity, we study classical models on a lattice, where a discrete and finite variable $\sigma_j \in \{\sigma_j^1, \sigma_j^2, \ldots, \sigma_j^{\mathcal{M}}\}$ is defined on each lattice site j. A given spin configuration of the $D = \mathcal{M}^N$ possible ones is $\sigma = [\sigma_1, \ldots, \sigma_N]$, where N is the total number of lattice sites. Here D is the dimensionality of the configuration (search) space and should not be confused with d, the space dimension of the lattice or physical system. An energy functional $E : \sigma \mapsto E_\sigma \in \mathbf{R}$ (cost functional) is defined on the lattice and we assume that E_σ can be computed efficiently for all σ. For example, in the Ising two-level (spin-1/2) model we have $\sigma_\sigma \in \{-1, 1\}$, and the energy functional is determined by $E_\sigma = \sum_{ij} J_{ij} \sigma_i \sigma_j$: Two interacting spins (i, j) contribute J_{ij} $(-J_{ij})$ to the energy if they are in the same (different) state(s). If the system is in equilibrium, the expectation value of a thermodynamic variable A in the canonical ensemble is given by

$$\langle A \rangle_\beta = \frac{1}{\mathcal{Z}_\beta} \sum_\sigma e^{-\beta E_\sigma} A_\sigma, \tag{1.1}$$

where $\mathcal{Z}_\beta = \sum_\sigma e^{-\beta E_\sigma}$ is the partition function, $\beta = (k_B T)^{-1}$ is the inverse temperature, and k_B is the Boltzmann's constant.

Any classical (finite-dimensional) spin model on a lattice can be associated with a (trivial) quantum model, defined on the same lattice, by associating every configuration σ with a quantum state σ. In this basis, the energy functional maps to a diagonal Hamiltonian matrix H, where the diagonal entries are the E_σ's. For classical two-level (spin-1/2) models, H is given by mapping $\sigma_j \rightarrow \sigma_z^j$ in E_σ, where σ_z^j is the Pauli operator acting on the jth site ($D \times D$-dimensional matrix):

$$\sigma_z^j = \mathbb{1} \otimes \mathbb{1} \otimes \cdots \otimes \underbrace{\sigma_z}_{j^{th} \text{ factor}} \otimes \cdots \otimes \mathbb{1} \,, \, \sigma_z = \begin{pmatrix} 1 & 0 \\ 0 & -1 \end{pmatrix}. \tag{1.2}$$

For example, $H = \sum_{ij} J_{ij} \sigma_z^i \sigma_z^j$ in the spin-1/2 Ising model.

Let the N-spin *quantum state* $|\psi_\beta\rangle \propto e^{-\beta H/2} \sum_\sigma |\sigma\rangle$ denote the QGS. It satisfies

$$\langle \hat{A} \rangle = \text{tr}\,[\rho_\beta \hat{A}] = \frac{\langle \psi_\beta | \hat{A} | \psi_\beta \rangle}{\langle \psi_\beta | \psi_\beta \rangle} \equiv \langle A \rangle_\beta, \tag{1.3}$$

where $\rho_\beta = e^{-\beta H}/\mathcal{Z}_\beta$ is the diagonal matrix that represents the classical Gibbs state. The operator $\hat{A}$ is determined by mapping the thermodynamic variable A as described above. Then, $\hat{A}$ is diagonal and $[\hat{A}, H] = 0$. The classical Gibbs state ρ_β can be prepared from $|\psi_\beta\rangle$ by a simple projective measurement in the basis $\{|\sigma\rangle\}_\sigma$.

A natural question in quantum annealing is whether $|\psi_\beta\rangle$ is the unique ground state of some simple Hamiltonian or not. Interestingly, we show now that the QGS is the ground state of a family of quantum Hamiltonians $\{H_\beta^q\}$ defined on the same lattice and space. Each H_β^q can be constructed by applying a similarity transformation to a possible transition matrix M_β of a Markovian process leading to the thermal distribution ρ_β:

$$H_\beta^q = -e^{\beta H/2} M_\beta e^{-\beta H/2}. \tag{1.4}$$

If detailed balance is assumed in M_β, the resulting matrix H_β^q is symmetric and can be regarded as a quantum Hamiltonian. The Markov chain as determined by M_β is assumed to be aperiodic and irreducible. We obtain

$$H_\beta^q |\psi_\beta\rangle = -|\psi_\beta\rangle. \tag{1.5}$$

The ranges of interactions appearing in both, H_β^q and H, are comparable. A finite-T phase transition of a classical system on a lattice in d space dimensions can then be identified with a quantum phase transition of a quantum model in a similar d-dimensional lattice. This follows from the fact that (diagonal) correlation functions coincide in both models. Constructing specific H_β^q's and studying their spectral properties are necessary to build adiabatic algorithms to prepare the QGS and understand their implementation complexity.

Can we systematically realize H_β^q? Let us consider N spin-1/2 systems. A possible H_β^q (up to irrelevant constants) is obtained as follows. The Pauli "spin-flip" operator $\sigma_x^{\,j}$

$$\sigma_x^j = \mathbb{1} \otimes \mathbb{1} \otimes \cdots \otimes \underbrace{\sigma_x}_{j^{th}\ \text{factor}} \otimes \cdots \otimes \mathbb{1}\ ,\ \sigma_x = \begin{pmatrix} 0 & 1 \\ 1 & 0 \end{pmatrix}, \tag{1.6}$$

satisfies $\sigma_x^j e^{-\beta H/2} \sigma_x^j = e^{\beta H_j} e^{-\beta H/2}$, $\forall j \in [1, N]$. The diagonal Hamiltonian H_j only includes the terms in H that contain the operator σ_z^j (i.e., the terms in H that anticommute with σ_x^j). Moreover, $\sigma_x^j \sum_\sigma |\sigma\rangle = \sum_\sigma |\sigma\rangle$ since the spin-flip operator maps the equal superposition state to itself. For example, for a single spin 1/2, $\sigma_x(|0\rangle + |1\rangle) = (|1\rangle + |0\rangle)$. Defining $A_j = \sigma_x^j - e^{\beta H_j}$ we obtain $A_j |\psi_\beta\rangle = 0$. In the basis $\{|\sigma\rangle\}_\sigma$ the off-diagonal elements of A_j are non-negative and, from irreducibility, the coefficients appearing in $|\psi_\beta\rangle$ are positive. The Perron–Frobenius theorem [16] implies that $|\psi_\beta\rangle$ is the unique ground state of the irreducible quantum Hamiltonian $H^q_\beta = -\mathbb{1} - \chi \sum_j A_j$. To satisfy $\|H^q_{\beta \to \infty}\| < \infty$ we need to set $\chi = e^{-\beta p}/N$, with $p \approx \max_j \|H_j\|$.

We emphasize the simplicity of this particular choice: The thermodynamic properties of any finite two-level (spin-1/2) classical system can be obtained by studying the ground state properties of a finite spin-1/2 quantum model, with *classical* (diagonal) interactions determined by β and H (i.e., the classical system) and an external and homogeneous transverse field. Remarkably, this field generates quantum fluctuations that are in one-to-one correspondence with the classical fluctuations at the inverse temperature β. In particular, $H^q_0 = -\mathbb{1} - (\sum_j \sigma^j_x)/N$, thus its ground state has all spins aligned along the external field, i.e., $|\psi_0\rangle \propto \sum_\sigma |\sigma\rangle$. This quantum state can be identified with the completely mixed state (uniform Boltzmann weights) of the classical model for $T \to \infty$. In the low-T limit, we obtain $H^q_{\beta \gg 1} \approx -\mathbb{1} + \chi \sum_j e^{\beta H_j}$, whose expectation value is minimized by the ground state(s) of the classical model; i.e., $|\psi_{\beta \gg 1}\rangle$ is also a lowest energy state of H with high probability.

It is important to remark that the classical-to-quantum mapping described above can be used to study *any* (finite-dimensional) classical system other than Ising-like or $s = 1/2$ models. In other words, the classical-to-quantum mapping proposed is of quite general applicability. For example, in the $d = 1$, $s = 1$ (three-state) Potts model [17], $E_\sigma = J \sum_j \delta(\sigma_j, \sigma_{j+1})$, and the classical Hamiltonian can be written as $H = J \sum_j [\mathbb{1} + S^j_z S^{j+1}_z (1 + 3 S^j_z S^{j+1}_z)/2 - (S^j_z)^2 - (S^{j+1}_z)^2]$, where $S^j_z \in su(2)_j$ is the angular momentum operator along the z-axis

$$S^j_z = \mathbb{1} \otimes \mathbb{1} \otimes \cdots \otimes \underbrace{S_z}_{j\text{th factor}} \otimes \cdots \otimes \mathbb{1} \ , \ S_z = \begin{pmatrix} 1 & 0 & 0 \\ 0 & 0 & 0 \\ 0 & 0 & -1 \end{pmatrix} , \quad (1.7)$$

and determines $H_j = J/2 [S^{j-1}_z S^j_z + S^j_z S^{j+1}_z]$. The other two elements of $su(2)_j$ are $S^j_\pm = S^j_x \pm i S^j_y$. The operator playing the role of *transverse field* is $X_j = \mathbb{1} - (S^j_z)^2 + [(S^j_+)^2 + (S^j_-)^2]/2$ ($X_j \in su(3)_j$ satisfy $[X_j, S^i_z] = 0$, $\forall i \neq j$, and $X_j S^j_z = -S^j_z X_j$, with $X^2_j = \mathbb{1}$). In matrix representation, X has 1's in the anti-diagonal and 0's otherwise, i.e.,

$$X = \begin{pmatrix} 0 & 0 & 1 \\ 0 & 1 & 0 \\ 1 & 0 & 0 \end{pmatrix} . \quad (1.8)$$

An entry in the stochastic matrix represents a transition probability to reach configuration τ from σ in a single Markov step:

$$p_\beta(\tau|\sigma) = \langle \tau|M_\beta|\sigma\rangle , \quad (1.9)$$

and $\sum_\tau p_\beta(\tau|\sigma) = 1$, $\forall \sigma$. For simplicity we write $p_{\sigma\tau} = p_\beta(\tau|\sigma)$. The detailed balance condition reads

$$e^{-\beta E_\sigma} p_{\sigma\tau} = e^{-\beta E_\tau} p_{\tau\sigma}. \tag{1.10}$$

It follows that the entries of H_β^q, as defined in Eq. (1.4), are

$$\langle \sigma | H_\beta^q | \tau \rangle = -\sqrt{p_{\tau\sigma} p_{\sigma\tau}}. \tag{1.11}$$

The Hamiltonians H_β^q are stochastic and their complexity has been studied in [18].

We now study adiabatic state preparation of $|\psi_\beta\rangle$. Later, in Sect. 1.2.5, we show that the QGS associated with one-dimensional classical systems can be efficiently prepared by a sequence of $O(N)$ elementary unitary evolutions.

1.2.2 Adiabatic State Preparation: Rates of Convergence

Let us focus first on the SA algorithm implemented via MCMC. In this case, the annealing process is determined by a choice of an annealing schedule. This schedule is a sequence of inverse temperatures $0 = \beta_1 < \beta_2 < \cdots < \beta_P$ together with transition rules $\{M_{\beta_1}, \ldots, M_{\beta_P}\}$. We assume $\{M_\beta\}$ to be smooth in $\beta \in [0, \beta_P]$, but this requirement could be avoided. The initial distribution is the uniform distribution among all possible classical configurations D. We assume efficient sampling of such a distribution as well as efficient implementation of M_β. If the goal of SA is to find the state that minimizes the expected value of H (i.e., the ground state of H) with high probability, the final temperature can be chosen so that $\beta_P \in O(\log D/\Gamma)$ and $|\beta_i - \beta_{i+1}| \leq c\Delta_i$, for some constant $c > 0$ [10]. Γ is the spectral gap of H, which is the difference between the minimum value of E and the next one. Δ_i is the spectral gap of M_{β_i}, which is the difference between 1 and the second largest eigenvalue of M_{β_i}. We are interested in those cases where $\Gamma \in O(1)$. This occurs, for example, when the cost function is determined by those instances of the classical-spin Ising model where the spin–spin couplings take values in the integers. It follows that the complexity of SA, in terms of the gap, is $O(1/\Delta)$, with $\Delta = \inf_i \Delta_i$.

A quantum annealing version of SA is obtained by defining H_β^q as a *similarity transformation* of the stochastic matrices M_β (see Eq. (1.4) and Sect. 1.2.1). In this context, the quantum version of SA corresponds to a (real-time) quantum evolution with initial quantum state $|\psi_0\rangle \propto \sum_\sigma |\sigma\rangle$, assumed to be efficiently prepared. The system evolves by slowly changing the interaction parameter β, related to the inverse temperature of the classical model, in H_β^q so that it traverses the smooth Hamiltonian path $\{H_\beta^q\}$. If the evolution is adiabatic, the state of the system remains sufficiently *close* to the ground state $|\psi_\beta\rangle$ along the path.

The total evolution time of the above QA algorithm is determined by the adiabatic theorem and strongly depends on the spectral gap Δ_β of H_β^q, which is the difference between the two lowest eigenvalues of H_β^q. For arbitrary precision, such a gap determines a bound on the rate at which β can be increased. It satisfies [7]: $\Delta_\beta \geq 2\sqrt{2\pi N}e^{-(\beta p+1)N} = \bar{\Delta}_\beta$. This bound was determined using the inequalities in [19] and considering that $(-H^q_\beta)^N$ is a strictly positive operator. It is based on the worst-case scenario (i.e., for the most general form of H), so it is expected to be improved depending on the choice of stochastic matrix and thus H^q_β. Considering the adiabatic condition of [20, 21], the rate $\dot{\beta}(t)$ is determined via

$$\frac{\|\partial_\beta|\psi_\beta\rangle\|\|\dot{\beta}(t)\|}{\Delta_{\beta(t)}} \leq \varepsilon, \ 0 \leq t \leq \mathcal{T}, \tag{1.12}$$

where ε is an upper bound to the error probability, $|\psi_\beta\rangle$ is the (normalized) QGS, and $\mathcal{T}$ is the total time of the evolution. One can bind $\|\partial_\beta|\psi_\beta\rangle\|$ independently of the gap as follows:

$$\|\partial_\beta|\psi_\beta\rangle\| = \frac{1}{2}\|(\langle H\rangle_\beta - H)|\psi_\beta\rangle\| \tag{1.13}$$

$$\leq \|H\| = E_M, \tag{1.14}$$

where E_M is the largest $|E_\sigma|$. Equation (1.13) yields a total evolution time $\mathcal{T} \in O[(\beta_P E_M)/(\varepsilon\Delta)]$, with Δ being a lower bound to the minimum gap along the path, that is, $\Delta \leq \inf_\beta \Delta_\beta$. This is a better resource scaling than the one commonly associated with an adiabatic evolution, in which the dependence of $\mathcal{T}$ on the gap Δ is $O(1/\Delta^2)$. We remark that, even if the proof in [20, 21] is not completely rigorous, other adiabatic methods that achieve the $1/\Delta$ scaling in this case exist (see [22]). Integrating Eq. (1.12) gives

$$\beta(t) \approx \frac{\log(\alpha t + 1)}{pN}, \ 0 < t \leq \mathcal{T}, \tag{1.15}$$

where α decreases exponentially with N. That is, if β increases according to Eq. (1.15), the evolved state has high fidelity (i.e., overlap) with the QGS $|\psi_{\beta_P}\rangle$. In the limit $\log t \gg N \gg 1$, we obtain $\beta(t) \in O[\log t/(pN)]$ which agrees with the asymptotic convergence rate obtained in [23] for SA. Such an agreement results from the fact that the energy gap of H^q_β is also the gap of M_β, and the dependence of the complexity on the gap for both, the quantum adiabatic method and the SA method, is given by $1/\Delta$.

In [3] QA has been proposed as an alternative approach to reach the optimal (ground) state of a classical system with Ising-like interactions. Contrary to SA, the time-dependent quantum state in QA does not correspond, in general, to a thermal configuration of the original classical model. In this case, the quantum Hamiltonian is given by $H^q_\gamma = H - \gamma \sum_j \sigma^j_x$, where γ is decreased from a very large value, corresponding to $T \to \infty$, to $\gamma \to 0$, corresponding to $T \to 0$. If γ is slowly (adiabatically) changed, this method also outputs a quantum state sufficiently close to $|\psi_{\beta_P}\rangle$, which has high fidelity with a ground state of H (i.e., it is an optimal configuration with high probability). Numerical and analytical results show that, for certain optimization problems, QA might enable a faster convergence rate to the optimal state than SA [3, 24, 25]. Faster convergence of QA could be attributed to a decrease in the probability of driving the classical system to a local minima. Nevertheless, it has also been observed that in some cases QA performs similarly to SA [26]. Note, however, that one could construct different Hamiltonian paths to approach the optimal state. Each path yields a particular convergence rate that has to be determined on a case-by-case basis.

Standard QA methods can be extended to simulate classical systems at any finite T as follows [7]. [We refer to this finite-temperature extension of the method as *Extended Quantum Annealing* (EQA).] We consider again the example of N spin-1/2 systems. We can construct a quantum Hamiltonian $\tilde{H}_\gamma^q = -\mathbb{1} + \chi \sum_j e^{\beta_P H_j} - \gamma \sum_j \sigma_x^j$, and define $|\psi(\gamma)\rangle$ and $|\psi_m(\gamma)\rangle$ to be its ground and excited states, respectively. Here, γ is adiabatically decreased from a very large value to $\gamma = \chi$. The initial state $\frac{1}{\sqrt{D}} \sum_\sigma |\sigma\rangle$ is then transformed into the desired state $|\psi_{\beta_P}\rangle \equiv |\psi(\gamma = \chi)\rangle$. Notice that, from our unifying viewpoint, QA differs from SA only by the choice of path used to reach the desired state (see Fig. 1.2). To successfully implement this annealing procedure, the rate of change of γ is determined from the adiabatic condition, i.e., by the gap $\Delta(\gamma)$ between the ground and first excited states of $\tilde{H}_\gamma^q$. This gap can be shown to satisfy $\Delta(\gamma) \geq 2\sqrt{2\pi N} e^{-N} (1 + c)^{-N} \gamma^N = \bar{\Delta}(\gamma)$ [7], with $\gamma < c$. For the worst-case scenario, the adiabatic condition [20, 21] yields a rate for γ

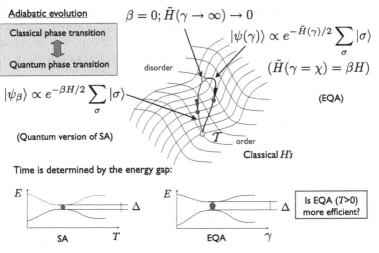

Comparing different Adiabatic Paths

Adiabatic evolution

Classical phase transition

Quantum phase transition

$|\psi_\beta\rangle \propto e^{-\beta H/2} \sum_\sigma |\sigma\rangle$

(Quantum version of SA)

$\beta = 0; \tilde{H}(\gamma \to \infty) \to 0$

$|\psi(\gamma)\rangle \propto e^{-\tilde{H}(\gamma)/2} \sum_\sigma |\sigma\rangle$

disorder

$(\tilde{H}(\gamma = \chi) = \beta H)$

(EQA)

T order

Classical H's

Time is determined by the energy gap:

E — Δ — SA — T

E — Δ — EQA — γ

Is EQA ($T>0$) more efficient?

Fig. 1.2 Different adiabatic paths lead to different algorithms. In this case, simulated annealing (SA) and the extended quantum annealing (EQA) algorithms are illustrated. A thermodynamic phase transition in the classical model corresponds to a quantum phase transition in the quantum model

$$\gamma(t) \in O\left[[(2N-1)(\bar{\alpha}t)]^{-1/(2N-1)} \right], \; 0 < t \leq \mathcal{T}, \tag{1.16}$$

where $\bar{\alpha}$ depends on N, c, and ε, and $\gamma(\mathcal{T}) = \chi$ is determined by $\mathcal{T}$. In the limit $\log t \gg N \gg 1$ and $\gamma(\mathcal{T}) \ll 1$, we obtain

$$\gamma(t) \in O\left[(2N\bar{\alpha}t)^{(-1/2N)} \right]. \tag{1.17}$$

The convergence rate is in agreement with the result obtained in [25]. Note, however, that this annealing schedule does not provide an advantage with respect to SA as γ must be decreased to $\gamma(\mathcal{T}) = \chi$, which is exponentially small in β_P.

1.2.3 Quantum Monte Carlo Methods for Quantum Annealing

The EQA method described above can be classically simulated.[1] For example, if the path-integral Monte Carlo method (PIMC) is chosen to simulate the one-dimensional Ising-like model with nearest-neighbor interactions, $\tilde{H}_\gamma^q$ has to be mapped onto the two-dimensional classical model, with energy functional

$$\bar{E}_\sigma = \frac{\tilde{\beta}}{L} \sum_{k=1}^{L} \sum_{ij} \tilde{J}_{ij}(\beta)\sigma_{ik}\sigma_{jk} + \xi(\beta,t) \sum_{k=1}^{L} \sum_{i=0}^{N} \sigma_{ik}\sigma_{i(k+1)}. \tag{1.18}$$

Here, $\sigma = [\sigma_{11}, \sigma_{21}, \ldots, \sigma_{NL}]$ is one of the 2^{N+L} possible spin configurations, and $\sigma_{ik} = \pm 1$. β is the inverse temperature at which the thermodynamic properties of the one-dimensional classical system are to be computed. The goal of PIMC is to sample with the Gibbs' distribution determined by the squared amplitudes of $|\psi(\gamma)\rangle$ at any γ. The parameter L denotes the number of copies of the system in the extra space dimension (i.e., the Trotter discretization) and satisfies $L \gg 1$. The coupling constants $\tilde{J}_{ij}(\beta)$ are defined via $\chi \sum_j e^{\beta H_j} \equiv \Lambda(\beta) + \sum_{ij} \tilde{J}_{ij}(\beta)\sigma_z^i\sigma_z^j$, with $\tilde{J}_{ij}(\beta \to 0) \to 0$. The coefficient $\tilde{\beta}$ is given by the inverse effective temperature of the quantum system and is not related to β. Therefore, $\tilde{\beta} \gg 1$ and $\tilde{\beta}/L$ is the time-slice of the discretization. The (ferromagnetic) coupling between two adjacent copies is determined by $\xi(t) = \log[\coth(\tilde{\beta}\chi\gamma(t)/L)]/2$, and its magnitude increases as the transverse field $\gamma(t)$ decreases to $\gamma(\mathcal{T}) = \chi$, determined by β. In

---

[1] It is important to emphasize that, as is expected for the simulation of classical systems, the simulations presented in this work do not suffer of the so-called sign(phase) problem since all the coefficients in the expansion of the exact ground state, in terms of the standard basis, are real and positive. This implies the possibility of finding an irreducible quantum Hamiltonian $-H^q$ with non-negative off-diagonal elements such that the quantum state is the highest energy state of $-H^q$. Nevertheless, many optimization problems are known to be NP-hard. It follows that not suffering from the sign problem is not sufficient for a problem to be easy, i.e., to be in the class P.

order to simulate more general classical systems at finite temperature, the interactions appearing in Eq. (1.18) must be modified accordingly.

1.2.4 Quantum Quadratic Speedup

The total evolution time of quantum adiabatic algorithms to prepare the quantum Gibbs state $|\psi_{\beta_P}\rangle$ depends on the minimum gap of the Hamiltonians along the path. So far we showed that there exists a family of quantum Hamiltonians $\{H_\beta^q\}$ whose spectral gap coincides with the relevant spectral gap of the associated stochastic matrix M_β. The natural question is whether we can construct a new family of Hamiltonians $\{I_\beta^q\}$ having $|\psi_\beta\rangle$ as eigenstate and larger spectral gap. The purpose of this section is to construct a possible I_β^q from H_β^q [Eq. (1.4)], where the relevant gap is bounded by $O(\sqrt{\Delta_\beta})$. As before, Δ_β is the relevant gap of H_β^q that coincides with the relevant gap of M_β. This will result in an adiabatic quantum algorithm whose cost dependence on the gap is quadratically improved. The results presented here, for the first time, represent a continuous-time Hamiltonian evolution version of those developed in [12].

To build our I_β^q we need to introduce additional degrees of freedom. Let us denote $\mathcal{H}$ the Hilbert space with orthonormal state basis $\{|\sigma\rangle\}_\sigma$, i.e., the Hilbert space where H_β^q acts on. We then consider two copies of $\mathcal{H}$ and define $\mathcal{I} = \mathcal{H} \otimes \mathcal{H}$. A basis state for $\mathcal{I}$ is the tensor product vector $|\sigma\tau\rangle$, with $|\sigma\rangle$ and $|\tau\rangle$ representing basis states of $\mathcal{H}$. The dimension of the Hilbert space $\mathcal{I}$ is D^2. We consider $|\phi_m(\beta)\rangle$, $1 \leq m \leq D$, to be the eigenstates of H_β^q, so that $|\phi_1(\beta)\rangle = |\psi_\beta\rangle$ is the QGS at β. We have

$$H_\beta^q|\phi_m(\beta)\rangle = e_m(\beta)|\phi_m(\beta)\rangle. \tag{1.19}$$

The eigenvalues $e_m(\beta)$ coincide with the eigenvalues of the stochastic matrix M_β up to a minus sign. We sort out the eigenvalues so that $e_1(\beta) = -1 < e_2(\beta) \leq \cdots \leq e_D(\beta)$. It follows that the spectral gap of H_β^q is $\Delta_\beta = 1 + e_2(\beta)$. With no loss of generality we assume $e_m(\beta) \leq 0$ for all m. If this condition is not naturally satisfied, we can always add a term proportional to the identity in M_β that yields a H_β^q satisfying the desired assumption. We define $\alpha_m(\beta) = \arccos(-e_m(\beta))$, so that $\alpha_1(\beta) = 0$ and $\Delta_\beta = 1 - \cos(\alpha_2(\beta))$. We further define unitaries V and W acting on $\mathcal{I}$ via

$$V|\sigma\ 0\rangle = \sum_\tau \sqrt{p_{\sigma\tau}}\ |\sigma\ \tau\rangle, \tag{1.20}$$

$$W|\tau\ 0\rangle = \sum_\sigma \sqrt{p_{\tau\sigma}}\ |\sigma\ \tau\rangle. \tag{1.21}$$

The state $|0\rangle$ is some simple, efficiently prepared, reference state in $\mathcal{H}$ and the action of V and W on the remaining basis states is irrelevant. Operations V and W represent the quantum walks introduced in [28]. They satisfy

$$\langle \sigma \, 0|V^\dagger W|\tau \, 0\rangle = \sqrt{p_{\sigma\tau}p_{\tau\sigma}} \, . \tag{1.22}$$

In other words, there is a $D \times D$-dimensional block in $V^\dagger W$ which coincides with the elements of $-H_\beta^q$. The block is labeled by the state $|0\rangle$ in the second copy of $\mathcal{H}$ appearing in $\mathcal{I}$. It then follows that

$$\cos(\alpha_m(\beta)) = \langle \phi_m(\beta) \, 0|V^\dagger W|\phi_m(\beta) \, 0\rangle \, , \tag{1.23}$$
$$= \langle \phi_m(\beta) \, 0|W^\dagger V|\phi_m(\beta) \, 0\rangle \, , \tag{1.24}$$

therefore $\alpha_m(\beta)$ can be interpreted as half the *angle* formed between the states $|\phi_m(\beta) \, 0\rangle$ and $V^\dagger W|\phi_m(\beta) \, 0\rangle$ in the corresponding Bloch sphere. Since the Hilbert space dimension has increased, we naturally redefine the QGS to be $|\psi_\beta \, 0\rangle$, so a measurement in the first copy of $\mathcal{H}$ returns state $|\sigma\rangle$ with the desired probability. Thus, $V^\dagger W$ as well as $W^\dagger V$ leave such a QGS invariant:

$$1 = \langle \psi_\beta \, 0|V^\dagger W|\psi_\beta \, 0\rangle = \langle \psi_\beta \, 0|W^\dagger V|\psi_\beta \, 0\rangle \, . \tag{1.25}$$

Let us define the quantum Hamiltonian

$$I_\beta^q = i[V^\dagger WP - PW^\dagger V] \, , \tag{1.26}$$

which acts on $\mathcal{I}$. The operator P is the projector onto the state $|0\rangle$ of the second copy of $\mathcal{H}$ in $\mathcal{I}$. That is,

$$P = \mathbb{1} \otimes |0\rangle\langle 0| \tag{1.27}$$
$$= \sum_\sigma |\sigma \, 0\rangle\langle \sigma \, 0|$$
$$= \sum_{m=1}^{D} |\phi_m(\beta) \, 0\rangle\langle \phi_m(\beta) \, 0| \, ,$$

since both $\{|\sigma\rangle\}_\sigma$ and $\{|\phi_m(\beta)\rangle\}_m$ are complete orthogonal bases for $\mathcal{H}$.

To obtain the spectral properties of the new Hamiltonian, we first note that I_β^q acts irreducibly in the subspaces spanned by $\mathcal{L}_m = \{|\phi_m(\beta) \, 0\rangle, V^\dagger W|\phi_m(\beta) \, 0\rangle\}_m$. This follows from the conditions

$$\delta_{mm'} = \langle \phi_m(\beta) \, 0|\phi_{m'}(\beta) \, 0\rangle \, , \tag{1.28}$$
$$\delta_{mm'} = \langle \phi_m(\beta) \, 0|(V^\dagger W)^\dagger V^\dagger W|\phi_{m'}(\beta) \, 0\rangle \, , \tag{1.29}$$
$$\delta_{mm'} \cos(\alpha_m(\beta)) = \langle \phi_m(\beta) \, 0|V^\dagger W|\phi_{m'}(\beta) \, 0\rangle \, , \tag{1.30}$$
$$\delta_{mm'} \cos(\alpha_m(\beta)) = -\langle \phi_m(\beta)|H_\beta^q|\phi_{m'}(\beta)\rangle \, . \tag{1.31}$$

We then have

$$I_\beta^q |\phi_m(\beta) \, 0\rangle = i \left[V^\dagger W |\phi_m(\beta) \, 0\rangle - \cos{(\alpha_m(\beta))} |\phi_m(\beta) \, 0\rangle \right], \quad (1.32)$$

$$I_\beta^q V^\dagger W |\phi_m(\beta) \, 0\rangle = i \left[\cos{(\alpha_m(\beta))} V^\dagger W |\phi_m(\beta) \, 0\rangle - |\phi_m(\beta) \, 0\rangle \right]. \quad (1.33)$$

In particular [see Eq. (1.25)]

$$I_\beta^q |\psi_\beta \, 0\rangle = 0, \quad (1.34)$$

so the QGS is a zero-eigenvalue eigenstate of I_β^q. Also $I_\beta^q |\xi\rangle = 0$ for all $|\xi\rangle \in \mathcal{I}$ in the subspace orthogonal to all $\mathcal{L}_m$'s. Since each $\mathcal{L}_m$ is a two-dimensional vector space, except for $m = 1$ where it is one-dimensional, the degeneracy of the zero eigenvalue is $D^2 - 2(D-1)$. The remaining eigenvalues of I_β^q, denoted as $\lambda_m^\pm(\beta)$ for $m \in \{2, 3, \cdots, D\}$, are nonzero and are distributed symmetrically around the origin. With no loss of generality one can write

$$V^\dagger W |\phi_m(\beta) \, 0\rangle = \cos{(\alpha_m(\beta))} |\phi_m(\beta) \, 0\rangle + \sin{(\alpha_m(\beta))} |\phi_m(\beta) \, 0\rangle^\perp, \quad (1.35)$$

with $\langle\phi_m(\beta) \, 0|\phi_m(\beta) \, 0\rangle^\perp = 0$. In the basis $\{|\phi_m(\beta) \, 0\rangle, |\phi_m(\beta) \, 0\rangle^\perp\}$, for $m > 1$, the new Hamiltonian takes the form

$$I_\beta^{q,m} = i \begin{pmatrix} 0 & -\sin{(\alpha_m(\beta))} \\ \sin{(\alpha_m(\beta))} & 0 \end{pmatrix}, \quad (1.36)$$

where the index m is used to label the corresponding two-dimensional, invariant subspace. Diagonalization of the above matrix returns the nonzero eigenvalues of I_β^q:

$$\lambda_m^\pm(\beta) = \pm \sin{(\alpha_m(\beta))}, \quad (1.37)$$

for $m \in \{2, \ldots, D\}$.

The adiabatic quantum algorithm in this case starts with the state $|\psi_{\beta=0} \, 0\rangle$, which is assumed to be efficiently prepared, and evolves it slowly according to $I_{\beta(t)}^q$. If the evolution is adiabatic, the evolved state is $|\psi_{\beta(t)} \, 0\rangle$ with high fidelity along the path. The relevant spectral gap of I_β^q is given by the absolute value of the difference between 0 and the next eigenvalue $\lambda_2^+(\beta)$. Let Θ_β denote such a gap. We obtain

$$\begin{aligned}
\Theta_\beta &= \lambda_2^+(\beta) \\
&= \sin{(\alpha_2(\beta))} \\
&= \sqrt{1 - \cos^2{(\alpha_2(\beta))}} \\
&= \sqrt{1 - e_2^2(\beta))} \\
&= \sqrt{(1 - e_2(\beta))(1 + e_2(\beta))} \\
&\geq \sqrt{\Delta_\beta},
\end{aligned} \quad (1.38)$$

where we assumed that $-1 \leq e_2(\beta) \leq 0$. Note that Δ_β is the relevant gap of H_β^q as well as the relevant gap of the stochastic matrix M_β.

Assuming again that the results of the adiabatic theorem in [20, 21] do apply in this case, evolving adiabatically with $I_{\beta(t)}^q$ allows us to prepare the desired QGS $|\Psi_{\beta_p} 0\rangle$ with high fidelity and cost $O(1/\sqrt{\Delta})$. This cost was obtained from Eqs. (1.12) and (1.13) in Sect. 1.2.2, replacing $\Delta_{\beta(t)}$ by $\Theta_{\beta(t)}$. As in Sect. 1.2.2, Δ is a lower bound on the gap Δ_β along the path.

It follows that **the quantum adiabatic algorithm to prepare the quantum Gibbs state by evolving with I_β^q yields a quadratic improvement in the cost (total evolution time) on the gap with respect to the classical SA method implemented via MCMC, which has a cost $O(1/\Delta)$.** This is an important result in those cases where Δ is very small in N (e.g., $\Delta \in O(\exp(-N)))$), which usually occurs in hard instances of the optimization problem.

Some clarifying concepts are in order. Note that the results in [20, 21] for the adiabatic condition assume non-degeneracy of the relevant subspace. This assumption is not satisfied in this case as the zero-eigenvalue subspace is highly degenerate. However, this degeneracy does not pose problems in an adiabatic evolution determined by the Hamiltonian path $\{I_\beta^q\}$, since level crossings between $|\Psi_\beta 0\rangle$ and other orthogonal degenerate subspaces are forbidden due to symmetry reasons. More precisely, let us write $|\xi\rangle$, a zero-eigenvalue eigenstate of I_β^q, orthogonal to $|\Psi_\beta 0\rangle$. We want to prove

$$\langle \xi | \left(\partial_\beta | \Psi_\beta \, 0 \rangle \right) = 0 , \tag{1.39}$$

which is a sufficient condition for the application of the results of the adiabatic theorem, for the non-degenerate case, to our degenerate problem. We can obtain Eq. (1.39) by noticing that

$$\partial_\beta | \Psi_\beta \, 0 \rangle = \sum_{m=1}^{D} c_m | \phi_m(\beta') \, 0 \rangle , \tag{1.40}$$

for any β'. Except for $|\Psi_\beta 0\rangle$, all the remaining zero-eigenvalue eigenstates of I_β^q must be orthogonal to $|\sigma \, 0\rangle$ for all σ. This is because the basis $\{|\phi_m(\beta') \, 0\rangle\}_m$ is complete for the first copy of $\mathcal{H}$ in $\mathcal{I}$. Each state $|\phi_m(\beta') \, 0\rangle$ belongs to the subspace $\mathcal{L}_m$ which is orthogonal to the remaining degenerate states. It follows that

$$\langle \xi | \phi_m(\beta') \, 0 \rangle = 0 , \tag{1.41}$$

which gives the sufficient condition.

Using the techniques just described it is not possible to construct a Hamiltonian J_β^q from I_β^q with a larger gap of $O(\sqrt{\Theta_\beta})$. This follows from the fact that the desired QGS is not the ground state of I_β^q. Additionally, one may ask if the adiabatic quantum evolution with I_β^q can be classically simulated by using techniques such as quantum Monte Carlo and obtain a classical quadratic speedup as well. We point out that in our construction we are interested in adiabatically following an eigenstate of I_β^q other than its ground state, so the method explained in Sect. 1.2.3, replacing H_β^q by I_β^q, cannot be straightforwardly used in this case.

Another important question regards the implementation complexity of evolving with I_β^q for some unit of time t. Remarkably, this implementation complexity is comparable to the implementation complexity of evolving with H_β^q for the same time t as well as implementing the stochastic matrix M_β. Thus the quadratic advantage in implementation complexity was obtained by evaluating the performance of the classical and quantum annealing methods on similar complexity grounds. We refer to [12] for further details in this issue.

1.2.5 Efficient Quantum Gibbs State Preparation of One-Dimensional Systems

We will now show that when the classical system is one-dimensional, $d = 1$; the corresponding QGS can be prepared efficiently. This result is related to the fact that thermodynamic properties of one-dimensional systems in equilibrium can be efficiently computed through the transfer-matrix method. A similar result was pointed out in [29]. In this section we mainly show that the QGS can be easily represented as a (so-called) matrix product state, which can be efficiently prepared using the techniques described in [30].

Consider a one-dimensional N-site classical lattice system with site variables $\sigma_j \in \{\sigma_j^1, \sigma_j^2, \ldots, \sigma_j^\mathcal{M}\}$. For simplicity, assume only nearest-neighbor interactions and periodic boundary conditions (i.e., $\sigma_{N+1} = \sigma_1$). Note that, if longer range interactions exist, we can redefine the site variable by increasing $\mathcal{M}$ so that the only interactions that remain are between nearest neighbors. The cost or energy functional E_σ can then be written as $E_\sigma = \sum_j E_{\sigma_j \sigma_{j+1}}^j$, where the *bond* energies $E_{\sigma_j \sigma_{j+1}}^j$ only depend on σ_j and σ_{j+1}. Up to normalization, we seek to prepare

$$|\psi_\beta\rangle \propto e^{-\beta H/2} \sum_\sigma |\sigma\rangle \tag{1.42}$$

$$= \sum_{\sigma_1,\ldots,\sigma_N} c^{\sigma_1,\ldots,\sigma_N} |\sigma_1 \cdots \sigma_N\rangle, \tag{1.43}$$

with coefficients

$$c^{\sigma_1,\ldots,\sigma_N} = e^{-\beta E_{\sigma_1 \sigma_2}^1/2} \times \cdots \times e^{-\beta E_{\sigma_N \sigma_1}^N/2}. \tag{1.44}$$

To this end, we define the $(\mathcal{M} \times \mathcal{M})$-dimensional transfer matrices

$$T^j = \sum_{\sigma_j,\sigma_{j+1}} e^{-\beta E^j_{\sigma_j\sigma_{j+1}}/2} |\sigma_j\rangle\langle\sigma_{j+1}|, \tag{1.45}$$

to obtain

$$c^{\sigma_1,\dots,\sigma_N} = \langle\sigma_1|T^1|\sigma_2\rangle\langle\sigma_2|T^2|\sigma_3\rangle\cdots\langle\sigma_N|T^N|\sigma_1\rangle. \tag{1.46}$$

At this point one can introduce $\mathcal{M}N$ new ($\mathcal{M} \times \mathcal{M}$)-dimensional matrices

$$T^j_{\bar\sigma_j} = (|\bar\sigma_j\rangle\langle\bar\sigma_j|) \cdot T^j, \tag{1.47}$$

where $\bar\sigma_j \in \{\sigma^1_j,\dots,\sigma^\mu_j,\dots,\sigma^\mathcal{M}_j\}$, and each $|\sigma^\mu_j\rangle\langle\sigma^\mu_j|$ is a ($\mathcal{M} \times \mathcal{M}$)-dimensional diagonal matrix that has a 1 at position μ and 0's elsewhere. Then,

$$c^{\sigma_1,\dots,\sigma_N} = \mathrm{tr}\left[T^1_{\sigma_1}\cdots T^N_{\sigma_N}\right]. \tag{1.48}$$

Quantum states described by Eq. (1.48) are usually defined as *matrix product states* (MPSs) [31]. An important consequence of the matrix product representation is that expectation values of spatially localized operators can be classically computed in time polynomial in $\mathcal{M}$ and linear in N. This computation is carried efficiently when $\mathcal{M} \in O(\mathrm{poly}(N))$, as is the case for one-dimensional systems.

The preparation of an MPS can be sequentially done using $O(N)$ elementary quantum (unitary) gates and by coupling the quantum system to an ancillary system. We refer to [30] for a full description of the method.

1.3 Quantum Algorithms for Numerical Integration with No Importance Sampling

We now explain a quantum method to estimate the partition function $\mathcal{Z}_\beta$ via phase estimation at the quantum metrology limit of precision (see [32]). This method has been previously discussed in [8]. Contrary to some of the quantum annealing methods described in Sect. 1.2, the quantum algorithm we explain in this section does not concern adiabatic evolutions. The classical analogue of this quantum approach can be interpreted as the method that realizes numerical integration by randomly sampling configurations from the uniform distribution. After M samplings, the partition function $\mathcal{Z}_\beta$ for a classical configuration space of dimension D is estimated as

$$\mathcal{Z}_\beta \to \hat{\mathcal{Z}}_\beta = \frac{D}{M} \sum_{n=1}^M e^{-\beta E_{\sigma^n}}, \tag{1.49}$$

where σ^n is the configuration sampled at the nth step. The absolute error in the estimation is upper bounded by

$$\varepsilon = \frac{D\sqrt{\langle e^{-2\beta E}\rangle - \langle e^{-\beta E}\rangle^2}}{\sqrt{M}}, \tag{1.50}$$

where the expectation values are taken over by the probability distribution

$$\Pr(E) = \frac{1}{D}\left(\sum_{\sigma \ s.t. \ E_\sigma = E} 1\right). \tag{1.51}$$

Since the relevant quantity is the relative precision, we have

$$\varepsilon_r = \frac{1}{\sqrt{M}}\sqrt{\frac{\langle e^{-2\beta E}\rangle}{\langle e^{-\beta E}\rangle^2} - 1}. \tag{1.52}$$

In the $\beta \rightarrow 0$ limit (large temperature limit), we have $\langle e^{-\beta E}\rangle \rightarrow 0$ and $\langle e^{-2\beta E}\rangle \rightarrow 0$. The relative uncertainty ε_r approaches to zero in this case, even for sufficiently small M. On the contrary, in the large β limit, the uncertainty is large and small ε_r can be obtained only if $M \in O(D/\varepsilon_r^2)$. To show this, consider the case where the minimum E_0 of E is non-degenerate. With no loss of generality we assume $E_0 = 0$. At large β, $e^{-\beta E} \rightarrow 0$ for $E > E_0$. It follows that $\langle e^{-\beta E}\rangle \rightarrow 1/D$ and $\langle e^{-2\beta E}\rangle \rightarrow 1/D$. Using Eq. (1.52) proves the scaling for M in this limit. The classical method is then very inefficient in the large β limit, where only a few configurations contribute significantly to the partition function. Thus, sampling with a uniform distribution returns the significant configurations with low probabilities. Still, we will show [8] that the quantum version of this classical integration method provides a quadratic improvement on the dependence of the complexity with the relative error ε_r.

The quantum algorithm begins with the preparation of the equal superposition state $\frac{1}{\sqrt{D}}\sum_\sigma |\sigma\rangle$, which is assumed to be prepared efficiently. This is the QGS representing the uniform distribution in the classical case. We then add an extra qubit (spin-1/2), called ancilla a, initialized in the state $|0\rangle$. This ancilla will be *rotated* to the state $|1\rangle$ with amplitude $\sqrt{1 - r_\sigma} = \sqrt{1 - e^{-\beta E_\sigma}}$, controlled in the state $|\sigma\rangle$ of the initial state. To do this, only a single oracle call to a reversible circuit that computes the weight r_σ (with certain bit precision) on input σ is necessary. The unitary operation that acts on a is given by

$$R = \sum_\sigma |\sigma\rangle\langle\sigma|e^{-i\arccos(\sqrt{r_\sigma})\sigma_y^a}, \tag{1.53}$$

where σ_y^a is the Pauli operator acting on the ancilla qubit:

$$\sigma_y^a = \mathbb{1} \otimes \sigma_y, \tag{1.54}$$

$$\sigma_y = \begin{pmatrix} 0 & -i \\ i & 0 \end{pmatrix}. \tag{1.55}$$

Note that we assumed $E_\sigma \geq 0$ so that $r_\sigma \leq 1$. This can be done by adding an irrelevant constant to the energy functional. The state so prepared is

$$|\phi_\beta\rangle = U_\beta |\tilde{0}\, 0_a\rangle = \frac{1}{\sqrt{D}} \sum_\sigma \left[\sqrt{r_\sigma}\, |\sigma\, 0_a\rangle + \sqrt{1 - r_\sigma}\, |\sigma\, 1_a\rangle \right], \qquad (1.56)$$

where U_β is the unitary operator that prepared $|\phi_\beta\rangle$ from some simple initial state $|\tilde{0}\, 0_a\rangle$. Note that if the ancilla were measured in the computational basis and projected onto $|0\rangle$, the state $|\phi_\beta\rangle$ would be projected onto $|\psi_\beta\rangle$, which denotes the QGS described in Sect. 1.2. However, the probability of such event is generally exponentially small in $\log D$ (or system size).

Our quantum algorithm estimates $\mathcal{Z}_\beta$ from the expectation value

$$o = \langle \phi_\beta | P | \phi_\beta \rangle = \frac{\mathcal{Z}_\beta}{D}, \qquad (1.57)$$

where P is the projector onto the state $|0\rangle$ of the ancilla:

$$P = |0_a\rangle\langle 0_a|. \qquad (1.58)$$

This can be done using the techniques developed in [32]. If we seek relative precision ε_r for the estimation of $\mathcal{Z}_\beta$ in the large β limit, the absolute precision at which o has to be estimated is $\varepsilon \in O(\varepsilon_r/D)$. To show this we considered that $\mathcal{Z}_\beta \in O(1)$ in this limit. This is satisfied if the constant removed from E to satisfy $r_\sigma \leq 1$ is not too distant from the minimum value of E.

The methods in [32] achieve this with complexity $O(D/\varepsilon_r)$, which is a quadratic improvement in ε_r with respect to the classical method described above. The idea is to use the phase estimation algorithm [33] to estimate an eigenphase of a unitary operator W that is built upon composition of a reflection over the state $|\phi_\beta\rangle$ followed by a reflection on the state $|0_a\rangle$, that is,

$$W = (\mathbb{1} - 2|0_a\rangle\langle 0_a|)(\mathbb{1} - 2|\phi_\beta\rangle\langle \phi_\beta|). \qquad (1.59)$$

The complexity of implementing W is basically two times the complexity of preparing $|\phi_\beta\rangle$, since W uses each operator U_β and $U_\beta^\dagger$ once. Equivalently, the complexity of implementing W is comparable to the one given by two uses of the circuit that computes r_σ. To resolve the eigenphases of W at precision ε_r/D, only $O(D/\varepsilon_r)$ *computations* of r_σ are necessary.

Note that the complexity dependence on D, in the worst-case scenario, is linear in D. This complexity can be made $O(\sqrt{D})$ with some simple modifications of the method explained above. A commonly used technique for this purpose is *amplitude amplification* [34]. This technique can be also used to coherently prepare the QGS $|\psi_\beta\rangle$ from $|\phi_\beta\rangle$. To show this note that

$$|\phi_\beta\rangle = \alpha_0 |\psi_\beta\, 0_a\rangle + \alpha_1 |\chi_\beta\, 1_a\rangle, \qquad (1.60)$$

with

$$\alpha_0 = \sqrt{\frac{\mathcal{Z}_\beta}{D}}. \tag{1.61}$$

Amplitude amplification can be used to rotate the state $|\phi_\beta\rangle$ to $|\psi_\beta\, 0_a\rangle$ with $O(1/\alpha_0) \in O((D/\mathcal{Z}_\beta)^{1/2})$ uses of U_β, which is equivalent to $O((D/\mathcal{Z}_\beta)^{1/2})$ computations of r_σ. This is done by a sequence of steps, each involving a reflection over $|\phi_\beta\rangle$ followed by a reflection over $|0_a\rangle$. We refer to [34] for further details.

1.4 Conclusions

We presented quantum methods to compute thermal properties of classical systems in equilibrium that can also be used for combinatorial optimization problems. The first method is based on quantum adiabatic evolutions and the idea was to prepare a quantum version of the classical Gibbs' state, the so-called quantum Gibbs' state QGS. A simple projective measurement on this state allows us to sample classical configurations with the desired probability. For the quantum evolution related to the SA classical path, we determined the complexity of the adiabatic algorithm to be $O(1/\Delta)$, where Δ is a bound on the minimum gap of the stochastic matrices (or quantum Hamiltonians) along the path. We later showed that this complexity can be made $O(1/\sqrt{\Delta})$ by using techniques borrowed from quantum information theory, proving a quadratic quantum speedup with respect to SA implemented with MCMC. For other Hamiltonian paths, the complexity is $O(1/\Delta^2)$. We explained how to simulate some of the adiabatic evolutions using the path-integral Monte Carlo method. The implementation complexity of PIMC in this case is not clear.

The second approach we studied is based on the algorithm that samples classical configurations with the uniform probability distribution to estimate the partition function of classical systems. We showed that using known techniques in quantum information, such as phase estimation, a quadratic improvement of the cost dependence on precision can be achieved in the quantum case. We finally showed that amplitude amplification can be used to prepare the quantum Gibbs' state in a time that scales with the square root of the configuration space.

Acknowledgments We are thankful to H. Barnum, C. D. Batista, S. Boixo, and E. Knill for their contribution to some of the results stated in this chapter.

References

1. F. Barahona, J. Phys. A **15**, 3241 (1982).
2. S. Kirkpatrick, C.D. Gelett and M.P. Vecchi, Science **220**, 671 (1983).
3. T. Kadowaki and H. Nishimori, Phys. Rev. E **58**, 5355 (1998)
4. G.E. Santoro et al., Science **295**, 2427 (2002).

5. E. Farhi et al., Science **292**, 472 (2001).
6. A. Das and B. Chakrabarti, *Quantum Annealing and Other Optimization Methods* (Springer, New York, 2005).
7. R.D. Somma, C.D. Batista and G. Ortiz, Phys. Rev. Lett. **99**, 030603 (2007).
8. R.D. Somma, C.D. Batista and G. Ortiz, J. Phys. (Conf. Ser.) **95**, 012020 (2008).
9. A.K. Hartmann and H. Rieger, *Optimization Algorithms in Physics* (Wiley-VCH, Berlin, 2002).
10. D.W. Strook, *An Introduction to Markov Chain Processes* (Springer, Berlin, 2005).
11. A.M. Polyakov, *Gauge Fields and Strings* (Harwood Academic Publishers, Chur, 1993).
12. R.D. Somma, S. Boixo, H. Barnum and E. Knill, Phys. Rev. Lett. **101**, 130504 (2008).
13. C.L. Henley, J. Phys. Cond. Mat. **16**, S891 (2004)
14. C. Castelnovo et al., Ann. Phys. **318**, 316 (2005)
15. F. Verstraete et al., Phys. Rev. Lett. **96**, 220601 (2006).
16. R.A. Horn and C.R. Johnson, *Matrix Analysis* (Cambridge University Press, Cambridge, 1985).
17. R.B. Potts, Proc. Camb. Phil. Soc. **48**, 106 (1952).
18. S. Bravyi, D.P. DiVincenzo, R.I. Oliveira and B.M. Terhal, Quan. Inf. Comp. **8**, 0361 (2008).
19. E. Hopf, J. Math. Mech. **12**, 683 (1963).
20. D. Bohm, *Quantum Theory* (Prentice-Hall, New York, 1951)
21. A. Messiah, *Quantum Mechanics* (Wiley, New York, 1976).
22. S. Boixo, E. Knill and R.D. Somma, e-print arXiv:0903.1652 (2009).
23. S. Geman and D. Geman, IEEE Trans. Pattern. Anal. Mach. Intell. **6**, 721 (1984).
24. S. Suzuki and M. Okada, J. Phys. Soc. Jap. **74**, 1649 (2005).
25. S. Morita and H. Nishimori, J. Phys. A: Math. Gen. **39**, 13903 (2006).
26. E. Farhi, J. Goldstone and S. Gutmann, quant-ph/0201031
27. S. Suzuki, H. Nishimori and M. Suzuki, Phys. Rev. E **75**, 051112 (2007).
28. M. Szegedy, Proc. 45th Annual IEEE Symp. Found. Comp. Sci. (STOC), 32 (2004).
29. M. Van den Nest, W. Dür and H.J. Briegel, Phys. Rev. Lett. **98**, 117207 (2007).
30. C. Schön, E. Solano, F. Verstraete, J.I. Cirac and M.M. Wolf, Phys. Rev. A **75**, 032311 (2007).
31. S. Rommer and S. Ostlund, Phys. Rev. B **55**, 2164 (1997).
32. E. Knill, G. Ortiz and R. Somma, Phys. Rev. A **75**, 012328 (2007).
33. A. Kitaev, e-print quant-ph/9511026 (1995).
34. L.K. Grover, Proc. 28th Annual ACM Symp. Th. Comp. 212 (1996).

Chapter 2
Non-equilibrium Dynamics of Quantum Systems: Order Parameter Evolution, Defect Generation, and Qubit Transfer

S. Mondal, D. Sen, and K. Sengupta

2.1 Introduction

The properties of systems near quantum critical points (QCPs) have been studied extensively in recent years [1, 2]. A QCP is a point across which the symmetry of the ground state of a quantum system changes in a fundamental way; such a point can be accessed by changing some parameter, say λ, in the Hamiltonian governing the system. The change in the ground state across a QCP is mediated by quantum fluctuations. Unlike conventional thermal critical points, thermal fluctuations do not play a crucial role in such transitions. Similar to its thermal counterparts, the low-energy physics near a QCP is associated with a number of critical exponents which characterize the universality class of such a transition. Among these exponents, the dynamical critical exponent z provides the signature of the relative scaling of space and time at the transition and has no counterpart in thermal phase transitions. The other exponent which is going to be important for the purpose of this review is the well-known correlation length exponent ν. These exponents are formally defined as follows. As we approach the critical point at $\lambda = \lambda_c$, the correlation length diverges as $\xi \sim |\lambda - \lambda_c|^{-\nu}$, while the gap between the ground state and first excited state vanishes as $\Delta E \sim \xi^{-z} \sim |\lambda - \lambda_c|^{z\nu}$. Exactly at the critical point $\lambda = \lambda_c$, the energy of the low-lying excitations vanishes at some wave number $\mathbf{k}_0$ as $\omega \sim |\mathbf{k} - \mathbf{k}_0|^z$. The critical exponents are independent of the details of the microscopic Hamiltonian; they depend only on a few parameters such as the dimensionality of the system and

S. Mondal (✉)
Theoretical Physics Department, Indian Association for the Cultivation of Sciences, Jadavpur, Kolkata 700 032, India, `tpsm4@iacs.res.in`

D. Sen
Center for High Energy Physics, Indian Institute of Science, Bangalore 560 012, India, `diptiman@cts.iisc.ernet.in`

K. Sengupta
Theoretical Physics Department, Indian Association for the Cultivation of Sciences, Jadavpur, Kolkata 700 032, India, `tpks@iacs.res.in`

Mondal, S. et al.: *Non-equilibrium Dynamics of Quantum Systems: Order Parameter Evolution, Defect Generation, and Qubit Transfer*. Lect. Notes Phys. **802**, 21–56 (2010)
DOI 10.1007/978-3-642-11470-0_2   © Springer-Verlag Berlin Heidelberg 2010

the symmetry of the order parameter. These features render the low-energy equilibrium physics of a quantum system near a QCP truly universal.

In contrast to this well-understood universality of the equilibrium properties of a system near a QCP, relatively few universal features are known in the non-equilibrium behavior of a quantum system. Initial studies in this field aimed at understanding the near-equilibrium finite temperature dynamics near a quantum critical point using the Boltzman equation approach [3–6]. Such a dynamics is useful in making contact with experiments which are always carried out at finite temperature. Moreover, the excitations near a quantum critical point with a non-zero value of η do not have a simple pole structure like that of the conventional quasiparticle excitations of condensed matter systems; this property makes such a dynamics interesting in its own right.

More recently, significant theoretical [7–16] and experimental [17, 18] endeavors have focussed on out-of-equilibrium dynamics of closed quantum critical systems. On the experimental front, it has been possible, in ultracold atom systems, to gain unprecedented control over the measurement of out-of-equilibrium properties of quantum systems [17, 18]. On the theoretical front, such studies can be broadly classified into two distinct categories. The first type involves a study of the time evolution of a quantum system after a rapid quench through a quantum critical point. Such a study yields information about the order parameter dynamics across a quantum critical point. It turns out that such a dynamics exhibits a universal signature of the quantum critical point crossed during the quench. The second type involves a study of defect production during slow non-adiabatic dynamics through a quantum critical point. Such a defect production mechanism was first pointed out for dynamics through thermal critical points in [19–23]. For a slow enough quenches through quantum critical points, the density of defects produced is known to depend on z and ν which characterize the critical point [24–28].

Quantum communication in spin systems has also been a subject of intense study recently. Following the seminal work in [29], a tremendous amount of theoretical effort has been put in to understand the nature of qubit or entanglement transfer through one- or multi-dimensional spin systems [30–34, Zueco et al. (unpublished)]. One of the major goals of such studies is to characterize the fidelity of the transfer of a qubit across such a spin system. The maximization of both the fidelity and the speed of transfer, in moving a qubit through a spin chain, is an issue of great interest in such studies.

In this chapter, we will review some studies of sudden and slow zero temperature non-equilibrium dynamics of closed quantum systems across critical points. In Sect. 2.2, we consider a sudden quench across a quantum critical point. We study the order parameter dynamics of one-dimensional ultracold atoms in an optical lattice in Sect. 2.2.1 and of the infinite range ferromagnetic Ising model in Sect. 2.2.2. We demonstrate that the dynamics shows universal signatures of the QCP across which the system is quenched. In Sect. 2.3, we discuss defect production for slow non-adiabatic dynamics; typically, we find that the density of defects scales as an inverse power of the quench time τ, where the power depends on the dimensionality d of the system, and the exponents z and ν. In Sect. 2.3.1, we discuss the time

evolution of the system across a quantum critical surface; we find that the defect
scaling exponent in this case depends on the dimensionality of the critical surface.
This is confirmed by a study of defect production in the Kitaev model, which is
an exactly solvable model of spin-1/2's on a honeycomb lattice. In Sect. 2.3.2, we
study the effect of quenching across a QCP in a non-linear way; we find that the
defect scaling exponent also depends on the degree of non-linearity. We illustrate
these ideas by studying two exactly solvable spin-1/2 models in one dimension. In
Sect. 2.3.3, we discuss a number of experimental systems where our results on defect
scaling can possibly be checked. Finally, in Sect. 2.4, we show that non-equilibrium
dynamics, in one- and two-dimensional Heisenberg spin models, can be engineered
to maximize the fidelity and speed of the transfer of qubits.

2.2 Quench Dynamics

2.2.1 Ultracold Atoms in an Optical Lattice

In this section, we shall study a system of ultracold spinless bosons in a one-
dimensional (1D) optical lattice in the presence of a harmonic trap potential [35]. We
will restrict ourselves to the Mott phase of the bosons and will study their response
to a shift in the position of the trap potential. Such a shift acts as an effective "electric
field" for the bosons whose Hamiltonian is given by [35]

$$\mathcal{H} = -t \sum_{ij} \left(b_i^\dagger b_j + b_j^\dagger b_i \right) + \frac{U}{2} \sum_i n_i(n_i - 1) - E \sum_i \mathbf{e} \cdot \mathbf{r}_i n_i, \quad (2.1)$$

where ij represents pairs of nearest-neighbor sites of the optical lattice, $n_i = b_i^\dagger b_i$
is the number operator for the bosons, $\mathbf{r}_i$ are the dimensionless spatial coordinates
of the lattice sites (the lattice spacing is unity), $\mathbf{e}$ is a unit vector in the direction of
the applied electric field, and the effective electric field E (in units of energy) can be
deduced from the shift a of the center of the trap as $E = -a\partial V_{\text{trap}}(x)/\partial x$. In what
follows, we will restrict ourselves to $|U - E|, t \ll E, U$. We note that such a regime
has been achieved in experiments [17, 18].

In the presence of such an electric field, our classical intuition suggests that all the
bosons would gather in the last site of the 1D chain thereby minimizing their energy.
However, this does not happen for two reasons. First, the bosons are interacting and
a state where all the bosons are in a single site leads to a huge interaction energy
cost. But more importantly, even non-interacting bosons (or in the parameter regime
$E \gg Un_0$ for interacting bosons) do not exhibit this behavior. To understand this,
we note that when $U = 0$, $\mathcal{H}$ is simply the Wannier–Stark Hamiltonian whose wave
functions, in the limit of strong electric fields ($t \ll E$), are well-localized Bessel
functions. Thus for $E \gg t$, the bosons remain localized in their respective lattices.
It turns out that for realistic optical lattices where interband energy spacings are
large compared to both U and E, the Zener tunneling time, i.e., the time taken by

the bosons to reach the final ground from this metastable Mott state, is of the order of milliseconds and is larger than the system lifetime [17, 18]. Our strategy will therefore be to start from the parent Mott state of these localized bosons, identify the complete set of states resonantly coupled to this parent state, obtain the effective Hamiltonian within the subspace of these states, and determine its spectrum and correlations. This effective Hamiltonian is expected to describe the low-energy behavior of the system.

The parent Mott state and its resonant dipole excitations are shown in Figs. 2.1 and 2.2 [35]. A dipole here consists of a bound pair of hole at site i and an additional particle at its neighboring $i+1$ site. We note that the dipole excitations cost an energy $U - E$ and hence become energetically favorable when the electric field exceeds the interaction energy. However, once a dipole forms between two adjacent sites, these sites cannot participate in the formation of another dipole since the resultant state lies out of the resonant subspace [35]. This leads to a constraint on the dipole number on any given link ℓ connecting two sites, namely $n_\ell^d \leq 1$. Similar reasoning, elaborated in [35], shows that there can be at most one dipole on two adjacent links: $n_\ell^d n_{\ell+1}^d = 0$. The effective Hamiltonian of these dipoles can be written in terms of the dipole annihilation and creation operators d_ℓ and $d_\ell^\dagger$ as

$$\mathcal{H}_d = -t\sqrt{n_0(n_0 + 1)} \sum_\ell \left(d_\ell + d_\ell^\dagger\right) + (U - E) \sum_\ell d_\ell^\dagger d_\ell. \qquad (2.2)$$

Note that the presence of boson hopping leads to non-conservation of the dipole number since such a hopping can spontaneously create or destroy dipoles on a given link. Also, $\mathcal{H}_d$ needs to be supplemented by the constraint conditions $n_\ell^d \leq 1$ and $n_\ell^d n_{\ell+1}^d = 0$.

The phase diagram of the dipolar system can be easily found by inspecting $\mathcal{H}_d$. For $(U - E)/t = \lambda \to \infty$, the ground state of the system represents a vacuum of dipoles. In contrast, for $\lambda \to -\infty$, the ground state is doubly degenerate because there are two distinct states with maximal dipole number: $(\cdots d_1^\dagger d_3^\dagger d_5^\dagger \cdots)|0\rangle$ and $(\cdots d_2^\dagger d_4^\dagger d_6^\dagger \cdots)|0\rangle$. This immediately suggests the existence of an Ising QCP at some intermediate value of λ, associated with an order parameter $\Delta = \sum_\ell (-1)^\ell d_\ell^\dagger d_\ell$ which is a density wave of dipoles with a period of two lattice spacings. Further analytic evidence for an Ising QCP can be obtained by examining the excitation spectra for the limiting λ regimes and noting their similarity to those on either side of the critical point in the quantum Ising chain [1].

For $\lambda \to \infty$, the lowest excited states are single dipoles: $|\ell\rangle = d_\ell^\dagger |0\rangle$. There are N such states (where N is the number of sites) and, at $\lambda = \infty$, they are all degenerate with energy $U - E$. The degeneracy is lifted at second order in a perturbation theory in $1/\lambda$. By a standard approach using canonical transformations, these corrections can be described by an effective Hamiltonian, $\mathcal{H}_{d,\text{eff}}$, which acts entirely within the subspace of single dipole states. We find that

$$\mathcal{H}_{d,\text{eff}} = (U-E) \sum_{\ell} \left[|\ell\rangle\langle\ell| + \frac{n_0(n_0+1)}{\lambda^2} \left(|\ell\rangle\langle\ell| + |\ell\rangle\langle\ell+1| + |\ell+1\rangle\langle\ell| \right) \right]. \quad (2.3)$$

Notice that, quite remarkably, a local dipole hopping term has appeared in the effective Hamiltonian. The constraints ($n_\ell^d \leq 1$ and $n_\ell^d n_{\ell+1}^d = 0$) played a crucial role in the derivation of Eq. (2.3). Upon considering perturbations to $|\ell\rangle$ from the first term in Eq. (2.2), it initially seems possible to obtain an effective matrix element between any two states $|\ell\rangle$ and $|\ell'\rangle$. However, this connection can generally happen via two possible intermediate states, $|\ell\rangle \rightarrow d_\ell^\dagger d_{\ell'}^\dagger |0\rangle \rightarrow |\ell'\rangle$ and $|\ell\rangle \rightarrow |0\rangle \rightarrow |\ell'\rangle$, and the contributions of the two processes exactly cancel each other for most ℓ, ℓ'. Only when the constraints block the first of these processes is a residual matrix element possible. It is a simple matter to diagonalize $\mathcal{H}_{d,\text{eff}}$ by going to momentum space; we then find a single band of dipole states. The lowest energy dipole state has momentum π: the softening of this state upon reducing λ is then consistent with the appearance of a density wave order with period 2. The higher excited states at large λ consist of multiparticle continua of this band of dipole states, just as in the Ising chain [1]. A related analysis can be carried out for $\lambda \rightarrow -\infty$, and the results are similar to those for the ordered state in the quantum Ising chain [1]. The lowest excited states form a single band of domain walls between the two filled dipole states, and above them are the corresponding multiparticle continua. At an intermediate critical electric field $E_c = U + 1.310t\sqrt{n_0(n_0+1)}$, the system undergoes a quantum phase transition lying in the Ising universality class [35].

Having obtained the equilibrium phase diagram for the model, we now consider the quench dynamics of the dipoles when the value of the electric field is suddenly quenched [7]. We assume that the atoms in the 1D lattice are initially in the ground state $|\Psi_G\rangle$ of the dipole Hamiltonian (2.1) with $E = E_i \ll E_c$. This ground state

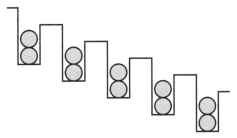

Fig. 2.1 Schematic representation of the parent Mott insulating state with $n_0 = 2$. Each well represents a local minimum of the optical lattice potential – we number these as 1–5 from the *left*. The potential gradient leads to a uniform decrease in the on-site energy of an atom as we move to the *right*. The gray circles are the d_i bosons of Eq. (2.2). The *vertical direction* represents increasing energy: the repulsive interaction energy between the atoms is realized by placing atoms vertically within each well, so that each atom displaces the remaining atoms upward along the energy axis. We have chosen the diameter of the atoms to equal the potential energy drop between neighboring wells – this corresponds to the condition $U = E$. Consequently, *a resonant transition is one in which the top atom in a well moves horizontally to the top of a nearest-neighbor well*; motions either upward or downward are non-resonant

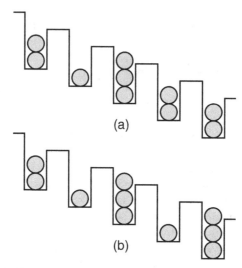

Fig. 2.2 Notation as in Fig. 2.1. (**a**) A dipole on sites 2 and 3; this state is resonantly coupled by an infinitesimal t to the Mott insulator in (**a**) when $E = U$. (**b**) Two dipoles between sites 2 and 3 and between 4 and 5; this state is connected via multiple resonant transitions to the Mott insulator for $E = U$

corresponds to a dipole vacuum. Consider shifting the center of the magnetic trap so that the new potential gradient is E_f. If this change is done suddenly, the system initially remains in the old ground state. The state of the system at time t is therefore given by

$$|\Psi(t)\rangle = \sum_n c_n \exp(-i\varepsilon_n t)|n\rangle, \tag{2.4}$$

where $|n\rangle$ denotes the complete set of energy eigenstates of the Hamiltonian $\mathcal{H}_d$ with $E = E_f$, $\varepsilon_n = \langle n|\mathcal{H}_d[E_f]|n\rangle$ is the energy eigenvalue corresponding to state $|n\rangle$, and $c_n = \langle n|\Psi(t = 0)\rangle = \langle n|\Psi_G\rangle$ denotes the overlap of the old ground state with the state $|n\rangle$. (We have set $\hbar = 1$.) Notice that the state $|\Psi(t)\rangle$ is no longer the ground state of the new Hamiltonian. Furthermore, in the absence of any dissipative mechanism, which is the case for ultracold atoms in optical lattices, $|\Psi(t)\rangle$ will never reach the ground state of the new Hamiltonian. Rather, in general, we expect the system to thermalize at long enough times, so that the correlations are similar to those of $H_{1D}[E_f]$ at some finite temperature.

We are now in a position to study the dynamics of the Ising density wave order parameter

$$O = \frac{1}{N}\langle\Psi|\Delta|\Psi\rangle, \tag{2.5}$$

where N is the number of sites. The time evolution of O is given by

$$O(t) = \frac{1}{N} \sum_{m,n} c_m c_n \cos\left[(E_m - E_n)\,t\right] \langle m| \sum_\ell (-1)^\ell d_\ell^\dagger d_\ell |n\rangle. \quad (2.6)$$

Equation (2.6) is solved numerically using exact diagonalization to obtain the eigenstates and eigenvalues of the Hamiltonian $H_{1D}[E_f]$. Before resorting to numerics, it is useful to discuss the behavior of $O(t)$ qualitatively. We note that if E_f is close to E_i, the old ground state will have a large overlap with the new one, i.e., $c_m \sim \delta_{m1}$. Hence in this case we expect $O(t)$ to have small oscillations about $O(t = 0)$. On the other hand, if $E_f \gg E_c$, the two ground states will have very little overlap, and we again expect $O(t)$ to have a small oscillation amplitude. This situation is in stark contrast with the adiabatic turning on of the potential gradient, where the systems always remain in the ground state of the new Hamiltonian $H_{1D}[E_f]$ and therefore has a maximal value of $\langle O \rangle$ for $E_f \gg E_c$. In between, for $E_f \sim E_c$, the ground state $|\Psi\rangle$ has a finite overlap with many states $|m\rangle$, and hence we expect $O(t)$ to display significant oscillations. Furthermore, if the symmetry between the two Ising ordered states is broken slightly (as is the case in our studies below), the time-averaged value of $O(t)$ will be non-zero.

This qualitative discussion is supported by numerical calculations for finite size systems with size $N = 9, 11, 13$. For numerical computations with finite systems, we choose systems with an odd number of sites and open boundary conditions, so that dipole formation on odd sites is favored, thus breaking the Z_2 symmetry. The results are shown in Figs. 2.3, 2.4, 2.5, and 2.6.

Figure 2.3 shows the oscillations of the order parameter $O(t)$ for different values of E_f for $N = 13$. In agreement with our qualitative expectations, the oscillations have maximum amplitude when $E_f/t \approx 40$ is near the critical value $E_c/t = 41.85$. For either $E_f \ll E_c$ or $E_f \gg E_c$, the oscillations have a small amplitude around $O(t = 0)$. Furthermore, it is only for $E_f \approx E_c$ that the time-averaged value of $O(t)$ is appreciable. Figure 2.4 shows the system size dependence of the time evolution for $E_f = U = 40t$. We find that the oscillations remain visible as we go to higher system

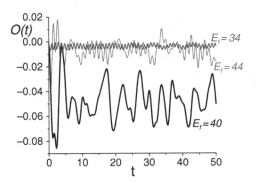

Fig. 2.3 Evolution of the Ising order parameter in Eq. (2.5) under the Hamiltonian $H_{1D}[E_f]$ for $n_0 = 1$. The initial state is the ground state of $H_{1D}[E_i]$. All the plots in this section have $U = 40$, $t = 1$, and $E_i = 32$, and consequently the equilibrium QCP is at $E_c = 41.85$

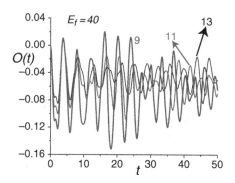

Fig. 2.4 System size (N) dependence of the results of Fig. 2.3 for $E_f = 40$. The *curves* are labeled by the value of N

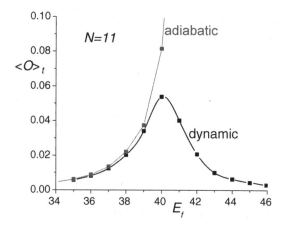

Fig. 2.5 The *curve* labeled 'dynamic' is the long-time limit $\langle O \rangle_t$ of the Ising order parameter in Eq. (2.6) as a function of E_f (for $N = 11$), with other parameters as in Fig. 2.1. This long-time limit can be obtained simply by setting $m = n$ in Eq. (2.6). For comparison, in the curve labeled 'adiabatic,' we show the expectation value of the Ising order O in the ground state of $H_{1D}[E_f]$; such an order would be observed if the value of E was changed adiabatically. Note that the dynamic curve has its maximal value near (but not exactly at) the equilibrium QCP $E_c = 41.85$, where the system is able to respond most easily to the change in the value of E; this dynamic curve is our theory of the 'resonant' response in the experiments of [17, 18] discussed in Sect. 2.1. In contrast, the adiabatic result *increases monotonically* with E_f into the $E > E_c$ phase where the Ising symmetry is spontaneously broken

sizes, although they do weaken somewhat. More significantly, the time-averaged value of $O(t)$ remains non-zero and has a weaker decrease with system size. In Fig. 2.5, we plot the long-time limit of the Ising order parameter, $\langle O \rangle_t$, as a function of E_f and compare it with O_{ad}, the value of the order parameter when E reaches E_f adiabatically, and the wave function is that of the ground state at $E = E_f$. We find that $\langle O \rangle_t$ stays close to O_{ad} as long as there is a large overlap between the old and the new ground states. However, as we approach the adiabatic phase transition

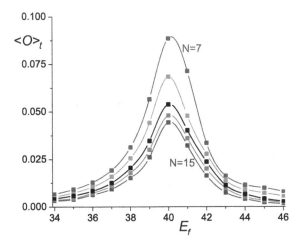

Fig. 2.6 Size dependence of the 'dynamic' results in Fig. 2.5. The sizes range from $N = 7$ to $N = 15$ (as labeled), with the intermediate values $N = 9, 11, 13$: $\langle O \rangle_t$ decreases monotonically with N

point, this overlap decreases and $\langle O \rangle_t$ cannot follow O_{ad} anymore. The deviation of $\langle O \rangle_t$ is therefore a signature that the system is now in a different phase for the new value of the electric field. The 'dynamic' curve in Fig. 2.5 shows that the Mott insulator has a resonantly strong response to an electric field $E \sim U$ induced by the proximity of a QCP.

We comment briefly on the nature of the thermodynamic limit, $N \to \infty$, for the results in Figs. 2.2 and 2.3. For O_{ad} it is clear that there is a non-zero limit only for $E > E_c$, when it equals the order parameter of the spontaneously broken Ising symmetry. If we assume that the system thermalizes at long times for the dynamic case, then $\langle O \rangle_t$ corresponds to the expectation value of the equilibrium order parameter in $H_{1D}[E_f]$ at some finite temperature. In one dimension, it is not possible to break a discrete symmetry at finite temperatures, and so the thermodynamic limit of the order parameter must always vanish. By this reasoning, we expect $\langle O \rangle_t$ to also vanish in the thermodynamic limit. This is consistent with the results in Fig. 2.6, where we show the N dependence of the long-time limit $\langle O \rangle_t$. Our data are at present not extensive enough to definitely characterize the dependence of $\langle O \rangle_t$ on N.

2.2.2 Infinite Range Ising Model in a Transverse Field

The analysis of the quench dynamics of 1D ultracold atoms does not permit an analytical description of the long-time value of the order parameter. In particular, the system size dependence of the peak height of $\langle O \rangle_t$ is not easy to understand analytically in this model. For this purpose, we now consider a simple model system, the infinite range ferromagnetic spin-1/2 Ising model in a transverse field, and study its quench dynamics due to a sudden variation of the transverse field. The model Hamiltonian is given by [13]

$$H = -\frac{J}{N} \sum_{i<j} S_i^z S_j^z - \Gamma \sum_i S_i^x , \qquad (2.7)$$

where $S_i^a = \sigma_i^a/2$, $a = x, y, z$, denote the components of the spin-1/2 operator represented by the standard Pauli spin matrices σ^a. Here we assume that $J \geq 0$ (ferromagnetic Ising interaction). This Hamiltonian is invariant under the Z_2 symmetry $S_i^x \rightarrow S_i^x$, $S_i^y \rightarrow -S_i^y$, and $S_i^z \rightarrow -S_i^z$. (The Z_2 symmetry would not be present if there was a longitudinal magnetic field coupling to $\sum_i S_i^z$.) We take $\Gamma \geq 0$ without loss of generality since we can always resort to the unitary transformation $S_i^x \rightarrow -S_i^x$, $S_i^y \rightarrow -S_i^y$, and $S_i^z \rightarrow S_i^z$, which flips the sign of Γ but leaves J unchanged. Equation (2.7) can be written as

$$H = -\frac{J}{2N} (S_{\text{tot}}^z)^2 - \Gamma S_{\text{tot}}^x , \qquad (2.8)$$

where

$$S_{\text{tot}}^z = \sum_i S_i^z, \quad S_{\text{tot}}^x = \sum_i S_i^x, \qquad (2.9)$$

and we have dropped a constant $(J/2N) \sum_i (S_i^z)^2 = J/8$ from the Hamiltonian in Eq. (2.8). This model has been studied extensively, particularly from the point of view of quantum entanglement [36–38]. Note that this model differs from the one studied in [39, 40], where the spins were taken to be living on two sub-lattices, with Ising interactions only between spins on different sub-lattices.

We begin with a mean field analysis of the thermodynamics of the model described by Eq. (2.7). Denoting the mean field value $m = \sum_i \langle S_i^z \rangle /N$, the Hamiltonian governing any one of the spins is given by

$$h = -Jm S_{\text{tot}}^z - \Gamma S_{\text{tot}}^x . \qquad (2.10)$$

This is a two-state problem whose partition function can be found at any temperature T. If $\beta = 1/(k_B T)$, we find that m must satisfy the self-consistent equation

$$m = \frac{Jm}{2\sqrt{\Gamma^2 + J^2 m^2}} \tanh\left(\frac{\beta\sqrt{\Gamma^2 + J^2 m^2}}{2}\right) . \qquad (2.11)$$

This always has the trivial solution $m = 0$. In the limit of zero temperature, there is a non-trivial solution if $\Gamma < J/2$, with $|m| = (1/2)\sqrt{1 - 4\Gamma^2/J^2}$; the energy gap in that case is given by $J/2$. If $\Gamma > J/2$, we have $m = 0$; and the gap is given by $\Gamma - J/2$. Hence there is a zero temperature phase transition at $\Gamma_c = J/2$. The Z_2 symmetry mentioned after Eq. (2.7) is spontaneously broken and $< S_i^z >$ becomes non-zero when one crosses from the paramagnetic phase at $\Gamma > J/2$ into the ferromagnetic phase $\Gamma < J/2$.

In the plane of $(k_B T/J, \Gamma/J)$, there is a ferromagnetic (FM) region in which the solution with $m \neq 0$ has a lower free energy (the Z_2 symmetry is broken) and a paramagnetic (PM) region in which $m = 0$. The boundary between the two is obtained by taking the limit $m \to 0$ in Eq. (2.11). This gives $2\Gamma/J = \tanh(\beta\Gamma/2)$, i.e.,

$$\frac{k_B T}{J} = \frac{\Gamma}{J} \left[\ln \left(\frac{1 + 2\Gamma/J}{1 - 2\Gamma/J} \right) \right]^{-1}. \tag{2.12}$$

The mean field phase diagram is shown in Fig. 2.7. We note that the exact excitation spectrum of this model can also be obtained analytically [13].

Having obtained the phase diagram, we now study the quench dynamics across the QCP. To begin with, we study the dynamics of the equal-time order parameter correlation function (EOC) (defined as $\langle (S_{tot}^z)^2 \rangle / S^2)$ by changing the transverse field Γ from an initial value $\Gamma_i \gg \Gamma_c$ to a final value Γ_f suddenly, so that the ground state of the system has no time to change during the quench. In this case, just after the quench, the ground state of the system can be expressed, in terms of the eigenstates $|n\rangle$ of the new Hamiltonian $\mathcal{H}_f = -(J/4S)(S_{tot}^z)^2 - \Gamma_f S_{tot}^x$ as

$$|\psi\rangle = \sum_n c_n |n\rangle, \tag{2.13}$$

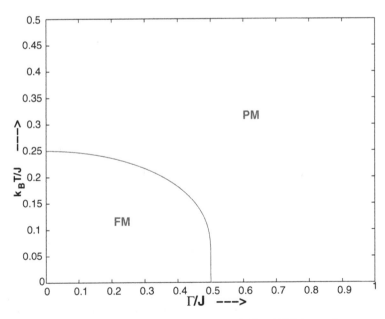

Fig. 2.7 Phase diagram of the model in mean field theory. FM and PM denote ferromagnetic and paramagnetic regions, respectively

where c_n denotes the overlap of the eigenstate $|n\rangle$ with the old ground state $|\psi\rangle$. As the state of the system evolves, it is given at time t by

$$|\psi(t)\rangle = \sum_n c_n e^{-iE_n t} |n\rangle, \qquad (2.14)$$

where $E_n = \langle n| \mathcal{H}_f |n\rangle$ are the energy eigenvalues of the Hamiltonian $\mathcal{H}_f$. The EOC can thus be written as

$$\langle \psi(t)| (S_{\text{tot}}^z)^2 / S^2 |\psi(t)\rangle = \sum_{m,n} c_n c_m \cos\left[(E_n - E_m) t\right] \langle m| (S_{\text{tot}}^z)^2 / S^2 |n\rangle. \quad (2.15)$$

Equation (2.15) can be solved numerically to obtain the time evolution of the EOC. We note that, similar to the case of the dipole model discussed in Sect. 2.2.1, we expect the amplitude of oscillations to be maximum when Γ_f is near Γ_c. This is verified in Fig. 2.8. Here, we have quenched the transverse fields to $\Gamma_f/J = 0.9, 0.01$, and 0.4 starting from $\Gamma_i/J = 2.0$. The oscillation amplitudes of the EOC for $S = 100$, as shown in Fig. 2.8, are small for $\Gamma_f = 0.9$ and 0.01, whereas it is substantially larger for $\Gamma_f = 0.4$.

Next, to understand the dynamics of the EOC in a little more detail, we study its long-time-averaged value given by

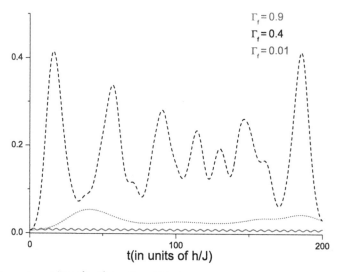

Fig. 2.8 Dynamics of $\langle (S_{\text{tot}}^z)^2 \rangle / S^2$ for $S = 100$ after quenching the transverse field to different values Γ_f/J from an initial field $\Gamma_i/J = 2$. The oscillation amplitudes are small, as seen from the *solid (red)* and *dotted (blue) curves* corresponding to $\Gamma_f/J = 0.9$ and 0.01, respectively, far away from the critical point $\Gamma_c/J = 0.5$. The oscillation is large in the ordered phase near the critical point as seen from the *dashed (black) curve* $\Gamma_f/J = 0.4$

$$O = \lim_{T \to \infty} \left\langle \left\langle (S^z_{\text{tot}})^2(t) \right\rangle \right\rangle_T / S^2$$

$$= \frac{1}{S^2} \sum_n c_n^2 \langle n | (S^z_{\text{tot}})^2 | n \rangle \tag{2.16}$$

for different Γ_f. Note that the long-time average depends on the product of the overlap of the state $|n\rangle$ with the old ground state and the expectation of $(S^z_{\text{tot}})^2$ in that state. From our earlier discussion in Sect. 2.2.1, we therefore expect O to have a peak somewhere near the critical point where such an overlap is maximized. This is verified by explicit numerical computation of Eq. (2.16) in Fig. 2.9 for several values of S and $\Gamma_i/J = 2$. We find that O peaks around $\Gamma_f/J = 0.25$, and the peak height decreases slowly with increasing S.

To understand the position and the system size dependence of the peak in O, we now look at the thermodynamic (large system size) limit; in the present model, this is also the large S and therefore classical limit. With this observation, we study the classical equations of motion for $\mathbf{S} = S(\cos\phi\sin\theta, \sin\phi\sin\theta, \cos\theta)$ for $\Gamma = \Gamma_f$. In the present model, S is a constant. Thus in the classical limit, we need to study the equations of motion for θ and ϕ. To this end, we note that the classical Lagrangian can be written in terms of θ and ϕ as [41]

$$L = -S\,[1 - \cos\theta]\,\frac{d\phi}{dt} - \mathcal{H}[\theta, \phi]. \tag{2.17}$$

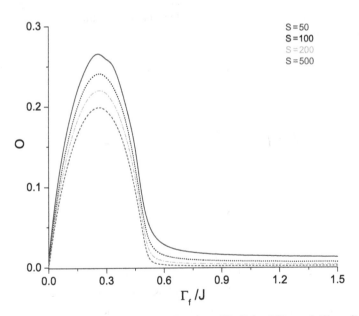

Fig. 2.9 Plot of the long-time average O as a function of Γ_f/J for different S. The *solid (blue)*, *dotted (black)*, *dash-dotted (green)*, and the *dashed (red) lines* represent, respectively, the results for $S = 50$, $S = 100$, $S = 200$, and $S = 500$. O peaks around $\Gamma_f/J = 0.25$, and the peak value decreases with increasing S. We have chosen $\Gamma_i/J = 2$ for all the plots

This gives the equations of motion

$$\frac{d\theta}{dt} = \Gamma_f \sin\phi,$$

$$\frac{d\phi}{dt} = -\frac{J}{2} \cos\theta + \Gamma_f \cot\theta \cos\phi. \qquad (2.18)$$

Equation (2.18) has to be supplemented with the initial condition that $S_{tot}^x = S$ at $t = 0$. The condition $S_{tot}^x = S$ corresponds to $\theta = \pi/2$, $\phi = 0$ which is also a fixed point of Eq. (2.18). Therefore we shall start from an initial condition which is very close to the fixed point: $\theta = \pi/2 - \varepsilon$, $\phi = \varepsilon$, where ε is an arbitrarily small constant. Further, since the motion occurs on a constant energy surface after the quench has taken place, we have

$$\Gamma_f = \frac{J}{4} \cos^2\theta + \Gamma_f \sin\theta \cos\phi. \qquad (2.19)$$

Using Eqs. (2.18) and (2.19), we get an equation of motion for θ in closed form,

$$\frac{d\theta}{dt} = \frac{\sqrt{\Gamma_f^2 \sin^2(\theta) - \left[\Gamma_f - \frac{J}{4}\cos^2\theta\right]^2}}{\sin\theta} \equiv f(\theta). \qquad (2.20)$$

It can be seen that the motion of θ is oscillatory and has classical turning points at $\theta_1 = \sin^{-1}\left(\left|1 - 4\Gamma_f/J\right|\right)$ and $\theta_2 = \pi/2$. One can now obtain $\langle(S_{tot}^z)^2\rangle_T = \langle\cos^2\theta\rangle_T$ from Eq. (2.20),

$$\langle\cos^2\theta\rangle_T = \mathcal{N}/\mathcal{D}, \qquad (2.21)$$

where

$$\mathcal{N} = \int_{\theta_1}^{\theta_2} d\theta \, \frac{\cos^2\theta}{f(\theta)} = 4\sqrt{8\Gamma_f(J - 2\Gamma_f)/J}, \qquad (2.22)$$

and

$$\mathcal{D} = \int_{\theta_1}^{\theta_2} d\theta \, \frac{1}{f(\theta)}. \qquad (2.23)$$

When trying to evaluate $\mathcal{D}$, we find that the integral has an endpoint singularity at θ_2; this can be regulated by a cutoff η so that $\theta_2 = \pi/2 - \eta$. With this regularization, $\mathcal{D} = -J\ln(\eta)/\sqrt{\Gamma_f(J - 2\Gamma_f)}/2$. The cutoff used here has a physical meaning and is not arbitrary. To see this, note that the angles (θ, ϕ) define the surface of a unit sphere of area 4π. This surface, for a system with spin S, is also the phase space which has $2S + 1$ quantum mechanical states. For large S, the area of the surface occupied by each quantum mechanical state is therefore $4\pi/(2S + 1) \simeq 2\pi/S$. In other words, each quantum mechanical state will have a linear dimension of order

$1/\sqrt{S}$; this is how close we can get to a given point on the surface of the sphere. Note that this closeness is determined purely by quantum fluctuation and vanishes for $S \to \infty$. Thus η, which is also a measure of how close to the point $\theta = \pi/2$ we can get, must be of the order of $1/\sqrt{S}$; this determines the system size dependence of $\langle\cos^2 \theta\rangle_T$. Using Eq. (2.21), we finally get

$$\langle\cos^2 \theta\rangle_T = \frac{16\Gamma_f \left(J - 2\Gamma_f\right)}{J^2 \ln(S)}. \tag{2.24}$$

Equation (2.24) is one of the main results of this section. It demonstrates that the long-time average of the EOC must be peaked at $\Gamma_f/J = 0.25$ which agrees perfectly with the exact quantum mechanical numerical analysis leading to Fig. 2.9. Moreover, it provides an analytical understanding of the S (and hence system size) dependence of the peak values of Γ_f/J. A plot of the peak height of O as a function of $1/\ln(S)$ indeed fits a straight line, as shown in Fig. 2.10. So we conclude that the peak in O vanishes logarithmically with the system size S. Such a slow variation with S shows that it might be experimentally possible to observe an experimental signature of a QCP for a possible realization of this model with ultracold atoms where $N \sim 10^5$–10^6 [13].

The results obtained in this section can be tested in two kinds of experimental systems. One class of systems are those with long-range dipole–dipole interactions such as KH_2PO_4 or $Dy(C_2H_5SO_4)_39H_2O$ [42] which exhibit order–disorder transitions driven by tunneling fields. The other class of systems are two-component Bose–Einstein condensates where the inter-species interaction is strong compared to the intra-species interaction; the relative strengths of these interactions can be

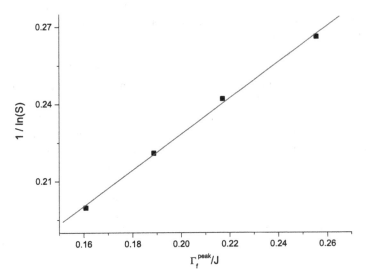

Fig. 2.10 Plot of the maximum peak height $O_{\max}$ of the long-time average of the EOC as a function of $1/\ln(S)$. The *straight line* shows the linear fit

changed by tuning the system to be near a Feshbach resonance as discussed for the
^{41}K $-^{87}$ Rb system in [43–47]. The quench dynamics that we have discussed can be
realized by applying a radio frequency pulse to the system and suddenly changing
the frequency of the pulse.

We end this section with the observation that the resonant response of the order
parameter during quench dynamics has been found in two very disparate models (1D
ultracold atoms and infinite-dimensional Ising ferromagnet) and therefore seems to
be a universal signature of the QCP through which the system passes during its
evolution.

2.3 Non-adiabatic Dynamics

In recent years, there have been extensive studies of what happens when a parameter λ in the Hamiltonian of a quantum system is varied in time slowly (non-
adiabatically) so as to take the system through a QCP. A quantum phase transition
is necessarily accompanied by diverging length and timescales, or, equivalently, a
vanishing energy gap between the ground state and the first excited state [1]. A consequence of this is that the system fails to be in the adiabatic limit when it crosses
a critical point, namely, when λ is varied across the QCP located at $\lambda = \lambda_c$ at a
finite rate given by $1/\tau$ (where τ will be called the quench time), the system fails
to follow the instantaneous ground state in a finite region around λ_c. As a result,
defects are produced [19, 20, 48, 49]. For a slow quench (for instance, for values
of τ much larger than the inverse bandwidth) which takes the system across a QCP
in a linear way, it is well known that the density of defects n scales as a power of
the quench time, $n \sim 1/\tau^{d\nu/(z\nu+1)}$, where ν and z are, respectively, the correlation
length and the dynamical critical exponents characterizing the critical point [24, 50].
A theoretical study of a quench dynamics requires a knowledge of the excited states
of the system. Hence, early studies of the quench problem were mostly restricted to
quantum phase transitions in exactly solvable models such as the 1D Ising model
in a transverse field [7, 51–54], the 1D XY spin-1/2 model [55, 56], quantum spin
chains [57–59], the Bose–Hubbard model [60], the Falicov–Kimball model [61],
and 1D spinless fermionic chains [62]. Experimentally, trapped ultracold atoms in
optical lattices provide possibilities of realization of many of the above-mentioned
systems [63–65]. Experimental studies of defect production due to quenching of the
magnetic field in a spin-1 Bose condensate have also been undertaken [66].

A class of models in which the above power law scaling can be derived easily
is one in which, due to the existence of a mapping to a system of non-interacting
fermions, the system decomposes into a product of two-level systems. For instance,
this occurs in the 1D XY spin-1/2 model and in the two-dimensional (2D) spin-1/2
Kitaev model. In both these cases, it turns out that the Hamiltonian is given by a
sum of terms of the form

$$H_{\mathbf{k}} = \alpha(\mathbf{k})(c_{\mathbf{k}}^{\dagger}c_{\mathbf{k}} + c_{-\mathbf{k}}^{\dagger}c_{\mathbf{k}}) + \Delta^*(\mathbf{k})c_{\mathbf{k}}^{\dagger}c_{-\mathbf{k}}^{\dagger} + \Delta(\mathbf{k})c_{-\mathbf{k}}c_{\mathbf{k}}, \tag{2.25}$$

where $\mathbf{k}$ runs over *half* the Brillouin zone (BZ) and $\alpha(\mathbf{k})$ is real. This Hamiltonian acts on a space spanned by four states, namely the empty state $|0\rangle$, two one-fermion states $c_{\mathbf{k}}^{\dagger}|0\rangle = |\mathbf{k}\rangle$ and $c_{-\mathbf{k}}^{\dagger}|0\rangle = |-\mathbf{k}\rangle$, and a two-fermion state $c_{\mathbf{k}}^{\dagger}c_{-\mathbf{k}}^{\dagger}|0\rangle = |\mathbf{k}, -\mathbf{k}\rangle$. Both the one-particle states are eigenstates of $H_{\mathbf{k}}$ with the same eigenvalue $\alpha(\mathbf{k})$. On the other hand, the states $|0\rangle$ and $|\mathbf{k}, -\mathbf{k}\rangle$ are governed by a 2×2 Hamiltonian given by

$$h_{\mathbf{k}} = \begin{pmatrix} 0 & \Delta(\mathbf{k}) \\ \Delta^{*}(\mathbf{k}) & 2\alpha(\mathbf{k}) \end{pmatrix}. \tag{2.26}$$

The eigenvalues of this are given by $\alpha(\mathbf{k}) \pm \sqrt{\alpha^{2}(\mathbf{k}) + |\Delta(\mathbf{k})|^{2}}$; since the lower eigenvalue is less than $\alpha(\mathbf{k})$, the ground state lies within the subspace of the states $|0\rangle$ and $|\mathbf{k}, -\mathbf{k}\rangle$. We observe that the Hamiltonian in Eq. (2.25) does not mix the states $|0\rangle$ and $|\mathbf{k}, -\mathbf{k}\rangle$ with the states $|\mathbf{k}\rangle$ and $|-\mathbf{k}\rangle$, *even if* $\alpha(\mathbf{k})$ and $\Delta(\mathbf{k})$ change with time. Hence, if we start at time $t \to -\infty$ with a linear superposition of $|0\rangle$ and $|\mathbf{k}, -\mathbf{k}\rangle$, we will end at $t \to \infty$ with a superposition of the same two states. In that case, it is sufficient to restrict our attention to the Hamiltonian for a two-level system given in (2.26). We can rewrite that as

$$h_{\mathbf{k}} = \alpha(\mathbf{k}) I_{2} - \alpha(\mathbf{k}) \sigma_{\mathbf{k}}^{3} + \Delta(\mathbf{k}) \sigma_{\mathbf{k}}^{+} + \Delta^{*}(\mathbf{k}) \sigma_{\mathbf{k}}^{-}, \tag{2.27}$$

where I_{2} is the identity matrix, and σ^{3} and $\sigma^{\pm} = (\sigma^{1} \pm i\sigma^{2})/2$ denote the Pauli matrices. We can ignore the term $\alpha(\mathbf{k})I_{2}$ in Eq. (2.27) since this only affects the wave function by a time-dependent phase factor.

Let us now consider what happens if $\alpha(\mathbf{k})$ varies linearly with time. Then the total Hamiltonian H_{d} for all the two-level systems can be written as

$$H_{d} = \sum_{\mathbf{k}} h_{\mathbf{k}}, \quad \text{where} \quad h_{\mathbf{k}} = \frac{t}{\tau} \varepsilon(\mathbf{k}) \sigma_{\mathbf{k}}^{3} + \Delta(\mathbf{k}) \sigma_{\mathbf{k}}^{+} + \Delta^{*}(\mathbf{k}) \sigma_{\mathbf{k}}^{-}, \tag{2.28}$$

where d is the number of spatial dimensions and the sum over $\mathbf{k}$ runs over half the BZ. If $\varepsilon(\mathbf{k}) > 0$, the ground state is given by $|0\rangle$ as $t \to -\infty$ and by $|\mathbf{k}, -\mathbf{k}\rangle$ as $t \to \infty$. If we begin with the state $|0\rangle$ at $t = -\infty$ and evolve the system using the time-dependent Schrödinger equation, we end at $t = \infty$ in a state which is a superposition of states $|0\rangle$ and $|\mathbf{k}, -\mathbf{k}\rangle$ with probabilities $p_{\mathbf{k}}$ and $1 - p_{\mathbf{k}}$, where $p_{\mathbf{k}}$ is given by the Landau–Zener expression [67, 68]

$$p_{\mathbf{k}} = e^{-\pi\tau|\Delta(\mathbf{k})|^{2}/\varepsilon(\mathbf{k})}. \tag{2.29}$$

Note that $p_{\mathbf{k}} \to 1$ for $\tau \to 0$ (sudden quench) and $\to 0$ for $\tau \to \infty$ (adiabatic quench), as expected. Let us now assume that the gap function $\Delta(\mathbf{k})$ vanishes at some point $\mathbf{k}_{0}$ in the BZ as $|\Delta(\mathbf{k})| \simeq a_{0}|\mathbf{k} - \mathbf{k}_{0}|^{z}$, while $\varepsilon(\mathbf{k}_{0}) = b_{0}$ is finite; this corresponds to a QCP with an arbitrary value of z, but with $zv = 1$. The density of defects n in the final state is given by the density of fermions $n = \int d^{d}k p_{\mathbf{k}}$. In the adiabatic limit $\tau \to \infty$, this is given by

$$\int_{\mathbf{k} \sim \mathbf{k}_0} d^d k \, e^{-\pi \tau a_0^2 |\mathbf{k} - \mathbf{k}_0|^{2z}/b_0} \sim 1/\tau^{d/(2z)}, \qquad (2.30)$$

which is the expected result for $zv = 1$.

It is useful to note that the derivation of the scaling law in Eq. (2.30) does not require a knowledge of the precise functional form given in Eq. (2.29). It is enough to know that $p_{\mathbf{k}}$ must be a function of the form $f(\tau |\Delta(\mathbf{k})|^2 / \varepsilon(\mathbf{k}))$, where $f(x) \to 1$ for $\tau \to 0$ and $\to 0$ for $\tau \to \infty$; these limiting values follow from general properties of the time-dependent Schrödinger equation. The argument of the function f can be derived by considering the equation $i \partial \psi_{\mathbf{k}}(t) / \partial t = h_{\mathbf{k}} \psi_{\mathbf{k}}(t)$, performing a phase re-definition to change $\Delta(\mathbf{k})$ to $|\Delta(\mathbf{k})|$, multiplying both sides by $\sqrt{\tau/\varepsilon(\mathbf{k})}$ and rescaling t to $t\sqrt{\tau/\varepsilon(\mathbf{k})}$. This effectively converts $h_{\mathbf{k}}$ to the form to $\sigma_{\mathbf{k}}^3 + |\Delta(\mathbf{k})| \sqrt{\tau/\varepsilon(\mathbf{k})}(\sigma_{\mathbf{k}}^+ + \sigma_{\mathbf{k}}^-)$; hence the probability $p_{\mathbf{k}}$ of starting in the ground state $\begin{pmatrix} 1 \\ 0 \end{pmatrix}$ at $t = -\infty$ and ending in the same state (which is the excited state) at $t = \infty$ must be a function of $\tau |\Delta(\mathbf{k})|^2 / \varepsilon(\mathbf{k})$.

We can generalize the above results to a QCP with arbitrary values of z and v. We consider a generic time-dependent Hamiltonian $H(t) \equiv H[\lambda(t)]$, whose states are labeled by $|\mathbf{k}\rangle$, and $|0\rangle$ denotes the ground state. If there is a second-order phase transition, the basis states change continuously with time during this evolution and can be written as $|\psi(t)\rangle = \sum_{\mathbf{k}} a_{\mathbf{k}}(t) |\mathbf{k}[\lambda(t)]\rangle$. The defect density can be obtained in terms of the coefficients $a_{\mathbf{k}}(t)$ as $n = \sum_{\mathbf{k} \neq 0} |a_{\mathbf{k}}(t \to \infty)|^2$; hence one gets [24]

$$n \sim \int d^d k \left| \int_{-\infty}^{\infty} d\lambda \, \langle \mathbf{k} | \frac{d}{d\lambda} | 0 \rangle \, e^{i\tau \int^\lambda d\lambda' \delta \omega_{\mathbf{k}}(\lambda')} \right|^2, \qquad (2.31)$$

where $\delta \omega_{\mathbf{k}}(\lambda) = \omega_{\mathbf{k}}(\lambda) - \omega_0(\lambda)$ are the instantaneous excitation energies. Following [24], we note that near a QCP, $\delta \omega_{\mathbf{k}}(\lambda) = \Delta F(\Delta / |\mathbf{k}|^z)$, where Δ is the energy gap, z is the dynamical critical exponent, and $F(x) \sim 1/x$ for large x; we have assumed here that the gap vanishes as $\mathbf{k} = 0$. Also, since the quench term vanishes at the critical point as $\Delta \sim |\lambda|^{zv}$, one can write $\delta \omega_{\mathbf{k}}(\lambda) = |\lambda|^{zv} \tilde{F}(|\lambda|^{zv}/|\mathbf{k}|^z)$, where $\tilde{F}(x) \sim 1/x$ for large x. Further, one has $\langle \mathbf{k} | \frac{d}{d\Delta} | 0 \rangle = |\mathbf{k}|^{-z} G(\Delta/|\mathbf{k}|^z)$ near a critical point where $G(0)$ is a constant. This allows us to write $\langle \mathbf{k} | \frac{d}{d\lambda} | 0 \rangle = \lambda^{zv-1} |\mathbf{k}|^{-z} G'(\lambda^{zv}/|\mathbf{k}|^z)$, where $G'(0)$ is a constant [1, 24]. Substituting these in Eq. (2.31) and changing the integration variables to $\eta = \tau^{v/(zv+1)} |\mathbf{k}|$ and $\xi = |\mathbf{k}|^{-1/v} \lambda$, we find that

$$n \sim \tau^{-dv/(zv+1)}. \qquad (2.32)$$

We will now discuss two major extensions of the above results: (i) what happens if the system is taken across a $d - m$ dimensional quantum critical surface instead of a QCP [25, 26] and (ii) what happens if the quenching across a QCP is non-linear in time [27, 28, 69, 70]. We will show that in both cases, the defect density still scales as a power of the quench time, but the power is not equal to the universal value $dv/(zv + 1)$ mentioned above; rather it depends on other parameters such as m or the degree of non-linearity.

2.3.1 Quenching Across a Critical Surface

When a quench takes a quantum system across a critical surface rather than a critical point, the density of defects scales in a different way with the quench time. To give a simple argument, consider a d-dimensional model with $z = v = 1$ which is described by the Hamiltonian given in Eq. (2.28). Suppose that a quench takes the system through a critical surface of $d - m$ dimensions. The defect density for a sufficiently slow quench is then given by [67, 68] $n \sim \int_{BZ} d^d k e^{-\pi \tau f(\mathbf{k})}$, where $f(\mathbf{k}) = |\Delta(\mathbf{k})|^2/|\varepsilon(\mathbf{k})|$ vanishes on the $d - m$ dimensional critical surface. We can then write

$$n \sim \int_{BZ} d^d k \exp \left[-\pi \tau \sum_{\alpha,\beta=1}^{m} g_{\alpha\beta} k_\alpha k_\beta \right] \sim 1/\tau^{m/2}, \qquad (2.33)$$

where α, β denote one of the m directions orthogonal to the critical surface and $g_{\alpha\beta} = [\partial^2 f(\mathbf{k})/\partial k_\alpha \partial k_\beta]_{\mathbf{k}\in\text{critical surface}}$. Note that this result depends only on the property that $f(\mathbf{k})$ vanishes on a $d - m$ dimensional surface and not on the precise form of $f(\mathbf{k})$. For general values of z and v, we note that the Landau–Zener type of scaling argument yields $\Delta \sim 1/\tau^{dv/(zv+1)}$, where Δ is the energy gap [24]. When one crosses a $d - m$ dimensional critical surface during the quench, the available phase space Ω for defect production scales as $\Omega \sim k^m \sim \Delta^{m/z} \sim 1/\tau^{mv/(zv+1)}$; this leads to $n \sim 1/\tau^{mv/(zv+1)}$. For a quench through a critical point where $m = d$, we retrieve the results of [24].

To give an example of a quench across a critical line, let us consider a model which was proposed recently by Kitaev. This is a 2D spin-1/2 model on a honeycomb lattice as shown in Fig. 2.11; the Hamiltonian is given by [71]

$$H_K = \sum_{j+l=\text{even}} (J_1 \sigma_{j,l}^x \sigma_{j+1,l}^x + J_2 \sigma_{j-1,l}^y \sigma_{j,l}^y + J_3 \sigma_{j,l}^z \sigma_{j,l+1}^z), \qquad (2.34)$$

where j and l denote the column and row indices of the honeycomb lattice. This model has been studied extensively and it exhibits several interesting features [72–77]. It provides a rare example of a 2D model which can be exactly solved using a Jordan–Wigner transformation [71, 72, 76, 77]. It has been shown in [71] that the presence of magnetic field, which induces a gap in the 2D gapless phase, leads to non-Abelian statistics of the low-lying excitations of the model; these excitations can be viewed as robust qubits in a quantum computer [78].

The Jordan–Wigner transformation of the Kitaev model to a model of non-interacting fermions works as follows. One can write

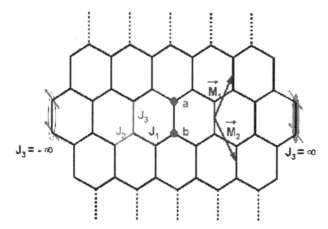

Fig. 2.11 Schematic representation of the Kitaev model on a honeycomb lattice showing the bonds J_1, J_2, and J_3. Schematic pictures of the ground states, which correspond to pairs of spins on vertical bonds locked parallel (antiparallel) to each other in the limit of large negative (positive) J_3, are shown at one bond on the *left (right)* edge, respectively. $\mathbf{M}_1$ and $\mathbf{M}_2$ are spanning vectors of the lattice, and a and b represent inequivalent sites

$$a_{jl} = \left(\prod_{i=-\infty}^{j-1} \sigma_{il}^z \right) \sigma_{jl}^y \text{ for even } j+l,$$

$$b_{jl} = \left(\prod_{i=-\infty}^{j-1} \sigma_{il}^z \right) \sigma_{jl}^x \text{ for odd } j+l, \tag{2.35}$$

where the a_{jl} and b_{jl} are Majorana fermion operators (and hence Hermitian) obeying the anticommutation relations $\{a_{jl}, a_{j'l'}\} = \{b_{jl}, b_{j'l'}\} = \delta_{jj'}\delta_{ll'}$. This transformation maps the spin Hamiltonian in Eq. (2.34) to a fermionic Hamiltonian given by

$$H_K = i \sum_{\mathbf{n}} [J_1 \, b_{\mathbf{n}} a_{\mathbf{n}-\mathbf{M}_1} + J_2 \, b_{\mathbf{n}} a_{\mathbf{n}+\mathbf{M}_2} + J_3 D_{\mathbf{n}} \, b_{\mathbf{n}} a_{\mathbf{n}}], \tag{2.36}$$

where $\mathbf{n} = \sqrt{3}\hat{i} \, n_1 + (\frac{\sqrt{3}}{2}\hat{i} + \frac{3}{2}\hat{j}) \, n_2$ denote the midpoints of the vertical bonds. Here n_1, n_2 run over all integers so that the vectors $\mathbf{n}$ form a triangular lattice whose vertices lie at the centers of the vertical bonds of the underlying honeycomb lattice; the Majorana fermions $a_{\mathbf{n}}$ and $b_{\mathbf{n}}$ sit at the top and bottom sites, respectively, of the bond labeled $\mathbf{n}$. The vectors $\mathbf{M}_1 = \frac{\sqrt{3}}{2}\hat{i} + \frac{3}{2}\hat{j}$ and $\mathbf{M}_2 = \frac{\sqrt{3}}{2}\hat{i} - \frac{3}{2}\hat{j}$ are spanning vectors for the reciprocal lattice, and $D_{\mathbf{n}}$ can take the values ± 1 independently for each $\mathbf{n}$. The crucial point that makes the solution of Kitaev model feasible is that $D_{\mathbf{n}}$ commutes with H_K, so that all the eigenstates of H_K can be labeled by a specific set of values of $D_{\mathbf{n}}$. It has been shown that for any value of the parameters J_i, the ground state of the model always corresponds to $D_{\mathbf{n}} = 1$ on all the bonds. Since $D_{\mathbf{n}}$

is a constant of motion, the dynamics of the model starting from any ground state never takes the system outside the manifold of states with $D_{\mathbf{n}} = 1$.

For $D_{\mathbf{n}} = 1$, it is straightforward to diagonalize H_K in momentum space. We define Fourier transforms of the Majorana operators $a_{\mathbf{n}}$ as

$$a_{\mathbf{n}} = \sqrt{\frac{4}{N}} \sum_{\mathbf{k}} [a_{\mathbf{k}} \, e^{i\mathbf{k}\cdot\mathbf{n}} + a_{\mathbf{k}}^{\dagger} \, e^{-i\mathbf{k}\cdot\mathbf{n}}],$$

$$b_{\mathbf{n}} = \sqrt{\frac{4}{N}} \sum_{\mathbf{k}} [b_{\mathbf{k}} \, e^{i\mathbf{k}\cdot\mathbf{n}} + b_{\mathbf{k}}^{\dagger} \, e^{-i\mathbf{k}\cdot\mathbf{n}}], \tag{2.37}$$

where N is the number of sites (assumed to be even, so that the number of unit cells $N/2$ is an integer) and the sum over $\mathbf{k}$ extends over half the Brillouin zone of the honeycomb lattice. We have the anticommutation relations $\{a_{\mathbf{k}}, a_{\mathbf{k}'}^{\dagger}\} = \delta_{\mathbf{k},\mathbf{k}'}$, $\{a_{\mathbf{k}}, a_{\mathbf{k}'}\} = 0$, and similarly for $b_{\mathbf{k}}$ and $b_{\mathbf{k}}^{\dagger}$. We then obtain $H_K = \sum_{\mathbf{k}} \psi_{\mathbf{k}}^{\dagger} h_{\mathbf{k}} \psi_{\mathbf{k}}$, where $\psi_{\mathbf{k}}^{\dagger} = (a_{\mathbf{k}}^{\dagger}, b_{\mathbf{k}}^{\dagger})$, and $h_{\mathbf{k}}$ can be expressed in terms of Pauli matrices $\sigma^{1,2,3}$ as

$$h_{\mathbf{k}} = 2\,[J_1 \sin(\mathbf{k}\cdot\mathbf{M}_1) - J_2 \sin(\mathbf{k}\cdot\mathbf{M}_2)]\,\sigma^1$$
$$+ 2\,[J_3 + J_1 \cos(\mathbf{k}\cdot\mathbf{M}_1) + J_2 \cos(\mathbf{k}\cdot\mathbf{M}_2)]\,\sigma^2. \tag{2.38}$$

The energy spectrum of H_K consists of two bands with energies

$$E_{\mathbf{k}}^{\pm} = \pm 2\,[(J_1 \sin(\mathbf{k}\cdot\mathbf{M}_1) - J_2 \sin(\mathbf{k}\cdot\mathbf{M}_2))^2$$
$$+ (J_3 + J_1 \cos(\mathbf{k}\cdot\mathbf{M}_1) + J_2 \cos(\mathbf{k}\cdot\mathbf{M}_2))^2]^{1/2}. \tag{2.39}$$

We note that for $|J_1 - J_2| \le J_3 \le J_1 + J_2$, these bands touch each other so that the energy gap $\Delta_{\mathbf{k}} = E_{\mathbf{k}}^{+} - E_{\mathbf{k}}^{-}$ vanishes for special values of $\mathbf{k}$ leading to a gapless phase of the model [71, 72, 74, 76].

We will now quench $J_3(t) = Jt/\tau$ at a fixed rate $1/\tau$, from $-\infty$ to ∞, keeping J, J_1, and J_2 fixed at some non-zero values; we have introduced the quantity J to fix the scale of energy. We note that the ground states of H_K corresponding to $J_3 \to -\infty(\infty)$ are gapped and have $\sigma_{j,l}^z \sigma_{j,l+1}^z = 1(-1)$ for all lattice sites (j, l). To study the state of the system after the quench, we first note that after an unitary transformation $U = \exp(-i\sigma_1\pi/4)$, one can write $H_K = \sum_{\mathbf{k}} \psi_{\mathbf{k}}'^{\dagger} h_{\mathbf{k}}' \psi_{\mathbf{k}}'$, where $h_{\mathbf{k}}' = U h_{\mathbf{k}} U^{\dagger}$ is given by

$$h_{\mathbf{k}}' = 2\,[J_1 \sin(\mathbf{k}\cdot\mathbf{M}_1) - J_2 \sin(\mathbf{k}\cdot\mathbf{M}_2)]\,\sigma^1$$
$$+ 2\,[J_3(t) + J_1 \cos(\mathbf{k}\cdot\mathbf{M}_1) + J_2 \cos(\mathbf{k}\cdot\mathbf{M}_2)]\,\sigma^3. \tag{2.40}$$

Hence the off-diagonal elements of $h_{\mathbf{k}}'$ remain time-independent, and the problem of quench dynamics reduces to a Landau–Zener problem for each $\mathbf{k}$. The defect density can then be computed following a standard prescription [67, 68]

$$n = \frac{1}{A} \int_{\mathbf{k}} d^2\mathbf{k}\, p_{\mathbf{k}},$$

$$p_{\mathbf{k}} = e^{-2\pi\tau\,[J_1 \sin(\mathbf{k}\cdot\mathbf{M}_1) - J_2 \sin(\mathbf{k}\cdot\mathbf{M}_2)]^2/J}, \tag{2.41}$$

where $A = 4\pi^2/(3\sqrt{3})$ denotes the area of half the Brillouin zone over which the integration is carried out. Since the integrand in Eq. (2.41) is an even function of $\mathbf{k}$, one can extend the region of integration over the full Brillouin zone. This region can be chosen to be a rhombus with vertices lying at $(k_x, k_y) = (\pm 2\pi/\sqrt{3}, 0)$ and $(0, \pm 2\pi/3)$. Introducing two independent integration variables v_1, v_2, each with a range $0 \le v_1, v_2 \le 1$, one finds that

$$k_x = 2\pi \frac{v_1 + v_2 - 1}{\sqrt{3}}, \quad k_y = 2\pi \frac{v_2 - v_1}{3}. \tag{2.42}$$

Such a substitution covers the rhombus uniformly and facilitates the numerical integration necessary for computing n.

A plot of n as a function of the quench time $J\tau$ and $\alpha = \tan^{-1}(J_2/J_1)$ (we have taken $J_{1[2]} = J\cos\alpha[\sin\alpha]$) is shown in Fig. 2.12. We note that the density of defects produced is maximum when $J_1 = J_2$. This is due to the fact that the length of the gapless line through which the system passes during the quench is maximum at this point. This allows the system to remain in the non-adiabatic state for the maximum time during the quench, leading to the maximum density of defects. For $J_1/J_3 > 2J_2/J_3$, the system does not pass through a gapless phase during the quench, and the defect production is exponentially suppressed.

For sufficiently slow quench $2\pi J\tau \gg 1$, $p_{\mathbf{k}}$ is exponentially small for all values of $\mathbf{k}$ except in the region near the line

$$J_1 \sin(\mathbf{k} \cdot \mathbf{M}_1) - J_2 \sin(\mathbf{k} \cdot \mathbf{M}_2) = 0, \tag{2.43}$$

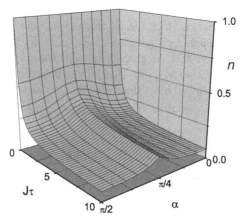

Fig. 2.12 Plot of defect density n as a function of the quench time $J\tau$ and $\alpha = \tan^{-1}(J_2/J_1)$. The density of defects is maximum at $J_1 = J_2$

and the contribution to the momentum integral in Eq. (2.41) comes from values of $\mathbf{k}$ close to this line of zeroes. We note that the line of zeroes where $p_\mathbf{k} = 1$ precisely corresponds to the zeroes of the energy gap $\Delta_\mathbf{k}$ as J_3 is varied for a fixed J_2 and J_1. Thus the system becomes non-adiabatic when it passes through the intermediate gapless phase in the interval $|J_1 - J_2| \leq J_3(t) \leq J_1 + J_2$. It is then easy to see, by expanding $p_\mathbf{k}$ about this line that in the limit of slow quench, the defect density scales as $n \sim 1/\sqrt{\tau}$. We thus see that the scaling of the defect density with the quench rate when the system passes through a critical *line* in momentum space is different from the situation where the quench takes the system through a critical *point*. The Kitaev model is an example of a system in which $d = 2$, $m = 1$, and $z = \nu = 1$; this gives rise to a defect density scaling as $1/\sqrt{\tau}$.

Before ending this section, it is interesting to consider another aspect of the system after the quench. Since the time evolution of the system is unitary, it will always be in a pure state. However, for each value of k, the wave function is given by $\sqrt{1 - p_\mathbf{k}}\psi_{2\mathbf{k}}e^{-iE_{2k}t} + \sqrt{p_\mathbf{k}}\psi_{1\mathbf{k}}e^{-iE_{1k}t}$, where E_{1k} $(E_{2k}) = \infty$ $(-\infty)$. As a result, the final density matrix of the system will have off-diagonal terms involving $\psi_{2\mathbf{k}}^*\psi_{1\mathbf{k}}$ and $\psi_{1\mathbf{k}}^*\psi_{2\mathbf{k}}$ which vary extremely rapidly with time; their effects on physical quantities will therefore average to zero. Hence, for each momentum $\mathbf{k}$, the final density matrix $\rho_\mathbf{k}$ is effectively diagonal like that of a mixed state [55], where the diagonal entries are time-independent as $t \to \infty$ and are given by $1 - p_k$ and p_k. Such a density matrix is associated with an entropy which we will now calculate. The density matrix of the entire system takes the product form $\rho = \otimes \rho_\mathbf{k}$. The von Neumann entropy density corresponding to this state is given by

$$s = -\frac{1}{A} \int d^2\mathbf{k}\,[\,(1 - p_\mathbf{k})\ln(1 - p_\mathbf{k}) + p_\mathbf{k}\ln p_\mathbf{k}\,], \qquad (2.44)$$

where the integral again goes half the Brillouin zone. Let us now consider the dependence of this quantity on the quench time τ [56]. If τ is very small, the system stays in its initial state and $p_\mathbf{k}$ will be close to 1 for all values of $\mathbf{k}$; for the same reason, $\langle O_0 \rangle$ will remain close to 1. If τ is very large, the system makes a transition to the final ground state for all momentum except near the line described in Eq. (2.43). Hence $p_\mathbf{k}$ will be close to 0 for all $\mathbf{k}$ except near that line, and $\langle O_0 \rangle$ will be close to -1. In both these cases, the entropy density will be small. We therefore expect that there will be an intermediate region of values of τ in which s will show a maximum and $\langle O_0 \rangle$ will show a crossover from -1 to 1. A plot of s as a function of $J\tau$ and α shown in Fig. 2.13 confirms this expectation. We find that the entropy reaches a maximum for an intermediate value of $J\tau$ where $\langle O_0 \rangle$ crosses over from -1 to 1 for all values of α.

2.3.2 Non-linear Quenching Across a Critical Point

Let us now consider what happens if we start with a Hamiltonian similar to the one given in Eq. (2.28), except that

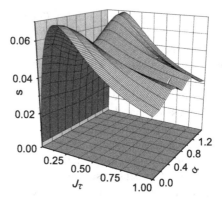

Fig. 2.13 Plot of the entropy density s as a function of $J\tau$ and $\alpha = \tan^{-1}(J_2/J_1)$. The entropy density peaks when $\langle O_0 \rangle$ crosses from -1 to 1

$$h_\mathbf{k}(t) = (\lambda(t) + b(\mathbf{k}))\sigma_\mathbf{k}^3 + \Delta(\mathbf{k})\,\sigma_\mathbf{k}^+ + \Delta^*(\mathbf{k})\,\sigma_\mathbf{k}^-, \qquad (2.45)$$

where $\lambda(t) = \lambda_0 |t/\tau|^\alpha \mathrm{sign}(t)$ is the quench parameter; $\alpha = 1$ corresponds to a linear quench. The instantaneous energies of the Hamiltonian in Eq. (2.45) are given by

$$E(\mathbf{k}) = \pm\sqrt{(\lambda(t) + b(\mathbf{k}))^2 + |\Delta(\mathbf{k})|^2}. \qquad (2.46)$$

These energy levels touch each other at $t = t_0$ and $\mathbf{k} = \mathbf{k}_0$, so that $|\Delta(\mathbf{k})| \sim |\mathbf{k} - \mathbf{k}_0|$ and $|t_0| = \tau |b(\mathbf{k}_0)/\lambda_0|^{1/\alpha} = \tau g^{1/\alpha}$, where $g = |b(\mathbf{k}_0)/\lambda_0|$ is a non-universal model-dependent parameter. At this point the energy levels cross and we have a QCP with $z = \nu = 1$. Note that the critical point is reached at $t = 0$ only if $b(\mathbf{k}_0)$ vanishes.

Let us first consider the case where $b(\mathbf{k}_0) = 0$ so that the system passes through the critical point at $t = 0$. In what follows, we shall assume that $|\Delta(\mathbf{k})| \sim |\mathbf{k} - \mathbf{k}_0|$ and $b(\mathbf{k}) \sim |\mathbf{k} - \mathbf{k}_0|^{z_1}$ at the critical point, where $z_1 \geq 1$ so that $E \sim |\mathbf{k} - \mathbf{k}_0|$ and $z = 1$. In the rest of the analysis, we will scale $t \rightarrow t\lambda_0$, $\tau \rightarrow \tau\lambda_0$, $\Delta(\mathbf{k}) \rightarrow \Delta(\mathbf{k})/\lambda_0$, and $b(\mathbf{k}) \rightarrow b(\mathbf{k})/\lambda_0$.

To obtain the probability $p_\mathbf{k}$ of ending in the excited state at $t = \infty$, we study the time evolution of the system governed by the Schrödinger equation $i\partial\psi_\mathbf{k}/\partial t = h_\mathbf{k}\psi_\mathbf{k}$. This leads to the equations

$$i\dot{c}_{1\mathbf{k}} = (|t/\tau|^\alpha \mathrm{sign}(t) + b(\mathbf{k}))\,c_{1\mathbf{k}} + \Delta(\mathbf{k})\,c_{2\mathbf{k}},$$
$$i\dot{c}_{2\mathbf{k}} = -(|t/\tau|^\alpha \mathrm{sign}(t) + b(\mathbf{k}))\,c_{2\mathbf{k}} + \Delta^*(\mathbf{k})\,c_{1\mathbf{k}}, \qquad (2.47)$$

where $\dot{c}_{1\mathbf{k}(2\mathbf{k})} \equiv \partial_t c_{1\mathbf{k}(2\mathbf{k})}$. To solve these equations, we define

$$c'_{1\mathbf{k}} = c_{1\mathbf{k}}\,e^{i\int^t dt'(|t'/\tau|^\alpha \mathrm{sign}(t') + b(\mathbf{k}))},$$
$$c'_{2\mathbf{k}} = c_{2\mathbf{k}}\,e^{-i\int^t dt'(|t'/\tau|^\alpha \mathrm{sign}(t') + b(\mathbf{k}))}. \qquad (2.48)$$

Substituting Eq. (2.48) in Eq. (2.47) and eliminating c'_{2k} from the resulting equations, we get

$$\ddot{c}'_{1k} - 2i\left[|t/\tau|^\alpha \text{sign}(t) + b(\mathbf{k})\right]\dot{c}'_{1k} + |\Delta(\mathbf{k})|^2 c'_{1k} = 0. \tag{2.49}$$

Now we scale $t \to t\tau^{\alpha/(\alpha+1)}$ so that Eq. (2.49) becomes

$$\ddot{c}'_{1k} - 2i\left[|t|^\alpha \text{sign}(t) + b(\mathbf{k})\tau^{\alpha/(\alpha+1)}\right]\dot{c}'_{1k} + |\Delta(\mathbf{k})|^2\tau^{2\alpha/(\alpha+1)} c'_{1k} = 0. \tag{2.50}$$

From Eq. (2.50) we note that since c_{1k} and c'_{1k} differ only by a phase factor, $p_\mathbf{k}$ must be of the form

$$p_\mathbf{k} = f[b(\mathbf{k})\tau^{\frac{\alpha}{\alpha+1}}, |\Delta(\mathbf{k})|^2\tau^{\frac{2\alpha}{\alpha+1}}], \tag{2.51}$$

where f is a function whose analytical form is not known for $\alpha \neq 1$. Nevertheless, we note that for a slow quench (large τ), $p_\mathbf{k}$ becomes appreciable only when the instantaneous energy gap, as obtained from Eq. (2.46), becomes small at some point of time during the quench. Consequently, f must vanish when either of its arguments are large: $f(\infty, a) = f(a, \infty) = 0$ for any value of a. Thus for a slow quench (large τ), the defect density n is given by

$$n \sim \int_{BZ} d^d k\, f\left[b(\mathbf{k})\tau^{\frac{\alpha}{\alpha+1}}, |\Delta(\mathbf{k})|^2\tau^{\frac{2\alpha}{\alpha+1}}\right] \tag{2.52}$$

and receives its main contribution from values of f near $\mathbf{k} = \mathbf{k}_0$ where both $b(\mathbf{k})$ and $\Delta(\mathbf{k})$ vanish. Thus one obtains, after extending the range of the integration to ∞,

$$n \sim \int d^d k\, f\left[|\mathbf{k} - \mathbf{k}_0|^{z_1}\tau^{\frac{\alpha}{\alpha+1}}; |\mathbf{k} - \mathbf{k}_0|^2\tau^{\frac{2\alpha}{\alpha+1}}\right]. \tag{2.53}$$

Now scaling $\mathbf{k} \to (\mathbf{k} - \mathbf{k}_0)\tau^{\alpha/(\alpha+1)}$, we find that

$$n \sim \tau^{-\frac{d\alpha}{\alpha+1}} \int d^d k\, f(|\mathbf{k}|^{z_1}\tau^{\alpha(1-z_1)/(\alpha+1)}; |\mathbf{k}|^2)$$
$$\sim \tau^{-\frac{d\alpha}{\alpha+1}} \int d^d k\, f(0; |\mathbf{k}|^2) \sim \tau^{-\frac{d\alpha}{\alpha+1}}, \tag{2.54}$$

where, in arriving at the last line, we have used $z_1 > 1$ and $\tau \to \infty$. (If $z_1 = 1$, the integral in the first line is independent of τ, so the scaling argument still holds.) Note that for $\alpha = 1$, Eq. (2.54) reduces to its counterpart for a linear quench [24]. It turns out that the case $z_1 < 1$ deserves a detailed discussion which is given in [27, 28].

Next we generalize our results to a critical point with arbitrary values of z and ν. We use arguments similar to those given in the discussion around Eq. (2.31), namely

$$n \sim \int d^d k \left| \int_{-\infty}^{\infty} d\lambda \, \langle \mathbf{k} | \frac{d}{d\lambda} | 0 \rangle \, e^{i\tau \int^\lambda d\lambda' \delta\omega_\mathbf{k}(\lambda')} \right|^2. \tag{2.55}$$

In the present case, the quench term vanishes at the critical point as $\Delta \sim |\lambda|^{\alpha z \nu}$ for a non-linear quench, and we can write

$$\delta\omega_\mathbf{k}(\lambda) = |\lambda|^{\alpha z \nu} \tilde{F}(|\lambda|^{\alpha z \nu}/|\mathbf{k} - \mathbf{k}_0|^z), \tag{2.56}$$

where $\tilde{F}(x) \sim 1/x$ for large x. Further, $\langle \mathbf{k}| \frac{d}{d\Delta} |0 \rangle = |\mathbf{k} - \mathbf{k}_0|^{-z} G(\Delta/|\mathbf{k} - \mathbf{k}_0|^z)$ near a critical point, where $G(0)$ is a constant. This allows us to write

$$\langle \mathbf{k}| \frac{d}{d\lambda} |0 \rangle = \frac{\lambda^{\alpha z \nu - 1}}{|\mathbf{k} - \mathbf{k}_0|^z} G'(\lambda^{\alpha z \nu}/|\mathbf{k} - \mathbf{k}_0|^z), \tag{2.57}$$

where $G'(0)$ is a constant [1, 24]. Substituting Eqs. (2.56) and (2.57) in Eq. (2.55) and changing the integration variables to $\eta = \tau^{\alpha\nu/(\alpha z\nu+1)}|\mathbf{k} - \mathbf{k}_0|$ and $\xi = |\mathbf{k} - \mathbf{k}_0|^{-1/(\alpha\nu)}\lambda$, we find that

$$n \sim \tau^{-\alpha d\nu/(\alpha z\nu+1)}. \tag{2.58}$$

Next we consider the case where the quench term does not vanish at the QCP for $\mathbf{k} = \mathbf{k}_0$. We again consider the Hamiltonian $h_\mathbf{k}(t)$ in Eq. (2.45), but now assume that the critical point is reached at $t = t_0 \neq 0$. This renders our previous scaling argument invalid since $\Delta(\mathbf{k}_0) = 0$ but $b(\mathbf{k}_0) \neq 0$. In this situation, $|t_0/\tau| = g^{1/\alpha}$ so that the energy gap ΔE may vanish at the critical point for $\mathbf{k} = \mathbf{k}_0$. We now note that the most important contribution to the defect production comes from times near t_0 and from wave numbers near k_0. Hence we expand the diagonal terms in $h_\mathbf{k}(t)$ about $t = t_0$ and $\mathbf{k} = \mathbf{k}_0$ to obtain

$$H = \sum_\mathbf{k} \left[\left\{ \alpha g^{(\alpha-1)/\alpha} \left(\frac{t - t_0}{\tau} \right) + b'(\delta\mathbf{k}) \right\} \sigma_\mathbf{k}^3 + \Delta(\mathbf{k})\sigma_\mathbf{k}^+ + \Delta^*(\mathbf{k})\sigma_\mathbf{k}^- \right], \tag{2.59}$$

where $b'(\delta\mathbf{k})$ represents all the terms in the expansion of $b(\mathbf{k})$ about $\mathbf{k} = \mathbf{k}_0$, and we have neglected all terms

$$R_n = (\alpha - n + 1)(\alpha - n + 2) \cdots (\alpha) \, g^{(\alpha-n)/\alpha}|(t - t_0)/\tau|^n \text{sign}(t)/n! \tag{2.60}$$

for $n > 1$ in the expansion of $\lambda(t)$ about t_0. We shall justify neglecting these higher order terms shortly.

Equation (2.59) describes a linear quench of the system with $\tau_{\text{eff}}(\alpha) = \tau/(\alpha g^{(\alpha-1)/\alpha})$. Hence one can use the well-known results of Landau–Zener dynamics [67, 68] to write an expression for the defect density,

$$n \sim \int_{\text{BZ}} d^d k \, p_\mathbf{k} \sim \int_{\text{BZ}} d^d k \, \exp[-\pi|\Delta(\mathbf{k})|^2 \tau_{\text{eff}}(\alpha)]. \tag{2.61}$$

For a slow quench, the contribution to n comes from $\mathbf{k}$ near $\mathbf{k}_0$; hence

$$n \sim \tau_{\text{eff}}(\alpha)^{-d/2} = \left(\alpha g^{(\alpha-1)/\alpha}/\tau\right)^{d/2}. \qquad (2.62)$$

Note that for the special case $\alpha = 1$, we recover the familiar result $n \sim \tau^{-d/2}$, and the dependence of n on the non-universal constant g vanishes. Also, since the quench is effectively linear, we can use the results of [24] to find the scaling of the defect density when the critical point at $t = t_0$ is characterized by arbitrary z and ν,

$$n \sim \left(\alpha g^{(\alpha-1)/\alpha}/\tau\right)^{\nu d/(z\nu+1)}. \qquad (2.63)$$

Next we justify neglecting the higher order terms R_n. We note that significant contributions to n come at times t when the instantaneous energy levels of H in Eq. (2.59) for a given $\mathbf{k}$ are close to each other, i.e., $(t - t_0)/\tau \sim \Delta(\mathbf{k})$. Also, for a slow quench, the contribution to the defect density is substantial only when $p_\mathbf{k}$ is significant, namely when $|\Delta(\mathbf{k})|^2 \sim 1/\tau_{\text{eff}}(\alpha)$. Using these arguments, we see that

$$R_n/R_{n-1} = (\alpha - n + 1)g^{-1/\alpha}(t - t_0)/(n\tau) \sim (\alpha - n + 1)/(n\sqrt{\tau}). \qquad (2.64)$$

Thus we find that all higher order terms $R_{n>1}$, which were neglected in arriving at Eq. (2.62), are unimportant in the limit of slow quench (large τ).

The scaling relations for the defect density n given by Eqs. (2.58) and (2.63) represent the central results of this section. For such power law quenches, unlike their linear counterpart, n depends crucially on whether or not the quench term vanishes at the critical point. For quenches that do not vanish at the critical point, n scales with the same exponent as that of a linear quench but is characterized by a modified non-universal effective rate $\tau_{\text{eff}}(\alpha)$. If, however, the quench term vanishes at the critical point, we find that n scales with a novel α-dependent exponent $\alpha d\nu/(\alpha z\nu + 1)$. For $\alpha = 1$, $\tau_{\text{eff}}(\alpha) = \tau$ and $\alpha d\nu/(\alpha z\nu + 1) = d\nu/(z\nu + 1)$; hence both Eqs. (2.58) and (2.63) reproduce the well-known defect production law for linear quenches as a special case [24]. We note that the scaling of n will show a crossover between the expressions given in Eqs. (2.58) and (2.63) near some value of $\tau = \tau_0$ which can be found by equating these two expressions; this yields $\tau_0 \sim |b(\mathbf{k}_0)|^{-z\nu-1/\alpha}$. For $\alpha > 1$, the scaling law will thus be given by Eqs. (2.58) and (2.63) for $\tau \ll (\gg)\tau_0$. We also note here that the results of this section assumes that the system passes from one gapped phase to another through a critical point and do not apply to quenches which take a system along a critical line [25, 26, 79–81].

To illustrate the form of defect scaling for a non-linear quench, let us consider the 1D spin-1/2 Kitaev model which is governed by the Hamiltonian

$$H = \sum_{i\in\text{even}} \left(J_1 S_i^x S_{i+1}^x + J_2 S_i^y S_{i-1}^y\right), \qquad (2.65)$$

where $S_i^a = \sigma_i^a/2$. Using the standard Jordan–Wigner transformation, this can be mapped to a Hamiltonian of non-interacting fermions

$$H = \sum_{\mathbf{k}} \psi_k^\dagger \, h_k \, \psi_k,$$

where

$$h_k = -2 \, (J_- \sin k \, \tau_3 + J_+ \cos k \, \tau_2). \tag{2.66}$$

Here $J_\pm = J_1 \pm J_2$ and $\psi_k = (c_1(k), c_2(k))$ are the fermionic fields. We now perform a quench by keeping J_+ fixed and varying the parameter J_- with time as $J_-(t) = J|t/\tau|^\alpha \mathrm{sign}(t)$. We then pass through a QCP at $t = 0$ at the wave number $k = \pi/2$. From Eq. (2.58) we expect the defect density to go as $n \sim \tau^{-\alpha/(\alpha+1)}$ since $\nu = z = 1$ for this system. To check this prediction, we numerically solve the Schrödinger equation $i\partial\psi(k,t)/\partial t = h_k(t)\psi(k,t)$ and compute the defect density $n = \int_0^\pi (dk/\pi) \, p_k$ as a function of the quench rate τ for different α, with fixed $J_+/J = 1$. A plot of $\ln(n)$ vs $\ln(\tau)$ for different values of α is shown in Fig. 2.14. The slopes of these lines, as can be seen from Fig. 2.14, changes from -0.67 toward -1 as α increases from 2 toward larger values. This behavior is consistent with the prediction of Eq. (2.58). The slopes of these lines show excellent agreement with Eq. (2.58) as shown in the inset of Fig. 2.14.

To illustrate what happens if the QCP is crossed at a time t which is different from 0, we consider the 1D Ising model in a transverse magnetic field described by

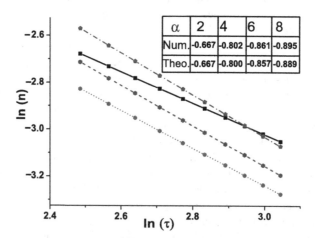

α	2	4	6	8
Num.	-0.667	-0.802	-0.861	-0.895
Theo.	-0.667	-0.800	-0.857	-0.889

Fig. 2.14 (Color online) Plot of $\ln(n)$ vs $\ln(\tau)$ for the 1D Kitaev model for $\alpha = 2$ (*black solid line*), $\alpha = 4$ (*red dotted line*), $\alpha = 6$ (*blue dashed line*), and $\alpha = 8$ (*green dash–dotted line*). The slopes of these lines agree reasonably with the predicted theoretical values $-\alpha/(\alpha + 1)$ as shown in the table

$$H_{\text{Ising}} = -J \left(\sum_i S_i^z S_{i+1}^z + g \sum_i S_i^x \right), \tag{2.67}$$

where J is the strength of the nearest-neighbor interaction and $g = h/J$ is the dimensionless transverse field. In what follows, we shall quench the transverse field as $g(t) = |t/\tau|^\alpha \text{sign}(t)$ and compute the density of the resultant defects.

We begin by mapping H_{Ising} to a system of free fermions via the Jordan–Wigner transformation

$$H = -J \sum_k [(g - \cos k) \sigma_k^3 + \sin k\, \sigma_k^1]. \tag{2.68}$$

If the field g is varied with time as $g(t) = g_0 |t/\tau|^\alpha \text{sign}(t)$, the system will go through two QCPs at $g = 1$ and -1. The energy gap vanishes at these QCPs at $k = k_0 = 0$ and π. As a result, defects are produced in non-adiabatic regions near these points. For this model, the QCP is at $t = t_0 \neq 0$ and $z = \nu = 1$. Hence, $\tau_{\text{eff}} = \tau/\alpha$ for both the QCPs. From Eq. (2.63), therefore, we expect the defect density produced in this system to be given by $n \sim (\tau/\alpha)^{-1/2}$.

To verify this, we numerically solve the Schrödinger equation $i\partial \psi_k(t)/\partial t = h_k(t)\psi_k(t)$ and obtain the probability p_k for the system to be in the excited state. Finally, integrating over all k within the Brillouin zone, we obtain the defect density n for different values of $\alpha > 1$ with fixed τ. The plot of n as a function of α for $\tau = 10$, 15, and 20 is shown in Fig. 2.15. A fit to these curves gives the values of the exponents to be 0.506 ± 0.006, 0.504 ± 0.004, and 0.505 ± 0.002 for $\tau = 10$, 15, and 20, respectively, which are remarkably close to the theoretical value $1/2$. The systematic positive deviation of the exponents from the theoretical value $1/2$ comes from the contribution of the higher order terms neglected in the derivation of Eqs. (2.62) and (2.63). We note that the region of validity of our linear expansion, as can be seen from Fig. 2.15, grows with τ which is in accordance with the result in Eq. (2.64).

2.3.3 Experimental Realizations

The validity of our results can be checked in a variety of experimental systems. We first observe that all our results have been obtained at zero temperature with the assumption that the system does not relax significantly during the quenching process and until the defect density has been measured. This might seem rather restrictive. We note, however, that systems of ultracold atoms in optical or magnetic traps and/or optical lattices can easily satisfy the required criteria since they have a very long relaxation time which often gets close to the system lifetime [63]. We will briefly list some possible experiments here. First, there has been a proposal for realizing the Kitaev model using an optical lattice [64, 65]. In such a realization, all the couplings can be independently tuned using separate microwave radiations. In the proposed

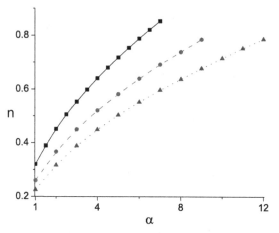

Fig. 2.15 (Color online) Variation of the defect density n with the quench exponent α for representative values of $\tau = 10$ (*black solid line*), $\tau = 15$ (*red dashed line*), and $\tau = 20$ (*blue dotted line*). A polynomial fit of the form $n = a\alpha^b$ yields exponents which are very close to the theoretical result $1/2$ for all values of τ

experiment, one needs to keep $J_3 = 0$ and vary $J_{1(2)} = J(1 \pm |t/\tau|^\alpha \mathrm{sign}(t))/2$, so that J_+ remains constant while J_- varies in time. The variation of the defect density, which in the experimental setup would correspond to the bosons being in the wrong spin state, would then show the theoretically predicted power law behavior in Eq. (2.58). Second, a similar quench experiment can be carried out with spin-1 bosons in a magnetic field described by an effective Hamiltonian $H_{\mathrm{eff}} = c_2 n_0 \langle \mathbf{S} \rangle^2 + c_1 B^2 \langle S_z^2 \rangle$ [66], where $c_2 < 0$ and n_0 is the boson density. Such a system undergoes a quantum phase transition from a ferromagnetic state to a polar condensate at $B^* = \sqrt{|c_2| n_0 / c_1}$. A quench of the magnetic field $B^2 = B_0^2 |t/\tau|^\alpha$ would thus lead to a scaling of the defect density with an effective rate $\tau_{\mathrm{eff}}(\alpha) = \tau/(\alpha g^{(\alpha-1)/\alpha})$, where $g = |c_2| n_0 / c_1$. A measurement of the dependence of the defect density n on α should therefore serve as a test of the prediction in Eq. (2.63). Finally, spin gap dimer compounds such as $BaCuSi_2O_6$ are known to undergo a singlet–triplet quantum phase transition of the Bose–Einstein condensation type at $B_c \simeq 23\,\mathrm{T}$; the critical exponents for this are given by $z = 2$ and $\nu = 2/d$. Experimentally, the exponent ν appears to be $2/3$ above a temperature window of 0.65–$0.9\,\mathrm{K}$ [82] and 1 below that temperature window due to a dimensional reduction from $d = 3$ to $d = 2$ [83]. Thus a non-linear quench of the magnetic field through its critical value $B = B_c + B_0 |t/\tau|^\alpha \mathrm{sign}(t)$ should lead to a scaling of the defects $n \sim \tau^{-6\alpha/(4\alpha+3)}$ for $d = 3$, $\nu = 2/3$, and $n \sim \tau^{-2\alpha/(2\alpha+1)}$ for $d = 2$, $\nu = 1$. It would be interesting to see if the defect scaling exponent depends on the temperature range in the same way as the exponent ν. In the experiment, the defect density would correspond to residual singlets in the final state which can be computed by measuring the total magnetization of the system immediately after the quench. We note that for these dimer systems, it would be necessary to take special care to achieve the criterion of long relaxation time mentioned earlier.

2.4 Quantum Communication

In this section, we demonstrate that a properly engineered non-adiabatic dynamics may lead to larger fidelity and higher speed for the transfer of a qubit through a system. For this purpose, we begin with a Heisenberg spin-1/2 chain described by a generic time-dependent Hamiltonian

$$H = -J_0(t) \sum_{ij} (S_i^x S_j^x + S_i^y S_j^y) + \Delta(t) \sum_{ij} S_i^z S_j^z + B(t) \sum_i S_i^z. \qquad (2.69)$$

We assume that the spin system is on a ring with N sites. We start with the initial ground state being ferromagnetic and denote this state by $|G\rangle$. At the start of the procedure of qubit transfer, we put a state $\cos(\theta/2)|\uparrow\rangle + \sin(\theta/2)\exp(i\phi)|\downarrow\rangle$ at the rth site of the chain. Thus the initial state of the system at the start of the evolution is [29]

$$|\psi_{in}\rangle = \cos(\theta/2)|G\rangle + \sin(\theta/2)e^{i\phi}|r\rangle, \qquad (2.70)$$

where $|r\rangle$ denotes the state of the spin chain with one flipped spin at the site r. We now consider the evolution of this state under a time-dependent Hamiltonian H. The specific form of the interaction need not be specified at the moment. Since the total spin is a conserved quantity ($[\sum_i S_i^z, H] = 0$), the state of the system at time t becomes

$$|\psi(t)\rangle = \cos(\theta/2)|G\rangle + \sin(\theta/2)e^{i\phi} \sum_n f_{nr}(t)|n\rangle,$$

where

$$f_{nr}(t) = \langle n|e^{-i\int^t H(t')dt'}|r\rangle. \qquad (2.71)$$

Since the idea of communication through the chain involves performing measurement on the state at site sth site, we would like to compute the reduced density matrix of this site at time t. To this end, we write the wave function

$$|\psi(t)\rangle = \cos(\theta/2)|G\rangle + \sin(\theta/2)e^{i\phi} \sum_{n\neq s} f_{nr}(t)|n\rangle + \sin(\theta/2)e^{i\phi}f_{sr}(t)|s\rangle, \qquad (2.72)$$

where the first line of the last equation is the contribution from all terms to $|\psi(t)\rangle$ where the spin in the sth site is $\uparrow$. Note that for normalization of the wave function, one needs

$$\cos^2(\theta/2) + \sin^2(\theta/2) \sum_{n\neq s} |f_{nr}(t)|^2 = 1 - |f_{sr}(t)|^2 \sin^2(\theta/2). \qquad (2.73)$$

Using Eqs. (2.72) and (2.73), one find that the reduced density matrix for the sth site of the system is

$$\rho_s(t) = (1 - |f_{sr}(t)|^2 \sin^2(\theta/2)) \, | \uparrow \rangle \langle \uparrow | + |f_{sr}(t)|^2 \sin^2(\theta/2) \, | \downarrow \rangle \langle \downarrow |$$
$$+ \frac{\sin(\theta)}{2} \left(e^{i\phi} f_{sr}(t) | \downarrow \rangle \langle \uparrow | + e^{-i\phi} f_{sr}^*(t) | \uparrow \rangle \langle \downarrow | \right). \tag{2.74}$$

The fidelity of the state transfer at the given time t is thus defined as [29]

$$F(t) = \frac{1}{4\pi} \int d\Omega \langle \psi_{in} | \rho_s(t) | \psi_{in} \rangle$$
$$= \frac{1}{2} + \frac{|f_{sr}(t)|^2}{6} + \frac{Re[f_{rs}(t)]}{3}, \tag{2.75}$$

where the integration is over the Bloch sphere involving θ and ϕ. Thus to obtain the fidelity of a state transfer we need to obtain the matrix elements $f_{sr}(t)$. To do this, we note that since the Hamiltonian in Eq. (2.69) conserves the z-component of the spin, an arbitrary time-dependent dynamics always restricts the system to lie within the subspace of one flipped spin. This allows us to write the wave function after an evolution through a time t to be

$$|\phi(t)\rangle = \sum_n c_n(t)|n\rangle = \sum_k c_k(t)|k\rangle, \tag{2.76}$$

where the real space basis $|n\rangle$ and the wave number space basis $|k\rangle$ are related by $|n\rangle = \sum_k \exp(-ikn)|k\rangle$ for a chain with a periodic boundary condition. The Schrödinger equation for $|\phi(t)\rangle$ now leads to the following equation for $c_k(t)$

$$i\dot{c}_k(t) = \left(2J(t)\cos(k) + \frac{1}{4}[\Delta(t) + 2B(t)] \right) c_k(t), \tag{2.77}$$

where we have neglected factors of $1/N$ (N being the chain length which approaches infinity in the thermodynamic limit) in the expression for $\beta(t)$. These equations are to be solved with the boundary condition $c_n(t = 0) = \delta_{nr}$. This equation has a straightforward solution

$$c_k(t) = e^{-i(2\alpha(t)\cos(k) + \beta(t))},$$

where

$$\alpha(t) = \int^t J(t')dt' \quad \text{and} \quad \beta(t) = \int^t \frac{1}{4}[\Delta(t') + 2B(t')]dt'. \tag{2.78}$$

Using Eq. (2.78), one gets

$$f_{sr}(t) = \langle s | e^{-i\int^t H(t')dt'} | r \rangle = \langle s | \phi(t) \rangle$$
$$= \sum_k e^{-i[k(r-s) + 2\alpha(t)\cos(k) + \beta(t)]}. \tag{2.79}$$

For an infinite chain, the momentum sum can be converted to an integral and exactly evaluated to yield

$$f_{sr}(t) = J_{r-s}(2\alpha(t)) \, e^{-i\beta(t)}. \tag{2.80}$$

From this result, we note the following points. First, we need to choose a time when we shall perform a measurement on the state. This time, t_0, is chosen so as to maximize the fidelity of the state transfer. In the present model, this occurs at the time t_0 when the argument $2\alpha(t_0)$ of the Bessel function approximately equals $r - s$. This suggests that one can reach the maximum fidelity (i.e., maximum $|f_{sr}(t)|$ and maximum $Re[f_{sr}(t)]$ which requires a separate adjustment of the phase factor) for a given separation $r - s$ at a much shorter time for a suitable non-adiabatic dynamics. This ensures faster communication through the channel. Note that by choosing an appropriate form of $J(t)$, the communication can be made exponentially faster compared to adiabatic dynamics since we may ramp up the effective instantaneous velocity so that a given separation $r - s$ is reached at a much shorter time. Second, the non-adiabatic dynamics gives us an additional handle on the phase and hence the real part of $f_{sr}(t)$. Thus one can adjust the phase using a user-chosen classical control parameter (such as frequency in the case of AC dynamics) to obtain maximum fidelity for a given $|f_{sr}(t)|$. Finally, it is straightforward to generalize the derivation of $f_{sr}(t)$ to higher dimensions. The result for a 2D system is

$$f_{sr} = J_{r_x-s_x}(2\alpha(t)) \, J_{r_y-s_y}(2\mu\alpha(t)) \, e^{-i\beta(t)}, \tag{2.81}$$

where μ is an anisotropy parameter which signifies the relative strengths of couplings of the S_x and S_y terms in the two orthogonal spatial directions. For $\mu = 1$, i.e., the isotropic case, we find that the fidelity is maximized when propagation takes place along the diagonal. But in general, the angle of maximum propagation is a function of μ and this can, in general, also be controlled. A similar analysis can be easily extended to higher dimensions; however, as can be seen from Eq. (2.81), the fidelity of the qubit transfer using this method rapidly decays with increasing dimensions. Thus we find, via a simple analysis of a Heisenberg spin model with a time-dependent Hamiltonian, that both the fidelity and the speed of quantum communication may be improved by using suitable non-equilibrium dynamics. We have also shown that such a procedure can lead to direction-specific state transfer in higher dimensional spin systems. Since engineering such time-dependent Hamiltonians have become an experimental reality, this might, in principle, provide a realizable way for faster communication of qubits in future experiments.

2.5 Discussion

To summarize, we first discussed the response of a system of interacting bosons in a 1D optical lattice to a sudden change in a harmonic trap potential. The system can be mapped to a system of dipoles described by an Ising order parameter. After the sudden shift, the order parameter oscillates in time; the amplitude of oscillations depends on the initial and final trap potentials. We then considered an infinite range ferromagnetic Ising model in a transverse magnetic field; we studied what happens after the field is changed suddenly. Once again, the variation of the order parameter (the magnetization in this problem) with time depends in an interesting way on the initial and final fields and the system size.

Next, we considered what happens when a system is taken across a quantum critical point or a critical surface in a non-adiabatic way which is governed by a quench time τ. This leads to the production of defects; the density of defects scales as an inverse power of τ, where the power depends on the dimensionalities of the system and the critical surface and the critical exponents z and ν. This was illustrated by considering the Kitaev model which is an exactly solvable spin-1/2 model defined on the honeycomb lattice; this can be solved by mapping it to a system of non-interacting Majorana fermions using a Jordan–Wigner transformation. We then considered the effect of taking a system across a QCP in a non-linear manner at time $t = 0$; the non-linearity is parametrized by an exponent α. We found that two different things happen depending on whether the system passes through the QCP at $t = 0$ or at a non-zero value of t. In the former case, the power appearing in the scaling of the defect density with τ also depends on α; in the latter case, the power is the same as in a linear quench (corresponding to $\alpha = 1$), but the effective quench time τ_{eff} depends on α. These ideas are illustrated by considering two models in 1D, namely a 1D version of the Kitaev model and the Ising model in a transverse magnetic field; both of these can be solved by mapping them to systems of non-interacting fermions by a Jordan–Wigner transformation. We then discussed some experimental systems where our results for the defect scaling can be checked.

Finally, we used some Heisenberg spin-1/2 models in one and two dimensions to discuss how a qubit can be transferred across the system. In particular, we examined how the speed and fidelity of the transfer can be maximized by choosing the couplings in the Hamiltonian appropriately.

Before ending, we would like to mention two possible extensions of the work discussed here. First, it would be interesting to study whether the defects studied in Sect. 2.3 have any non-trivial topology associated with them. If they are not topological, it would be interesting to find other ways of changing the parameters in the Hamiltonian in order to produce defects which do have a topological character. It is known that topology can affect defect production in a profound way [84]. Second, the defect density discussed in Sect. 2.3 follows from the density matrix of a single site obtained by integrating out all the other sites of the system. It is interesting to compute the two-site density matrix and use that to obtain various measures of two-site entanglement. This has been studied recently for both a sudden quench [85, 86] and a slow quench [87] through a QCP. Other kinds of entanglement produced due by a quench have also been studied [88, 89].

Acknowledgments We thank Amit Dutta and Anatoli Polkovnikov for stimulating discussions.

References

1. S. Sachdev, *Quantum Phase Transitions* (Cambridge University Press, Cambridge, 1999).
2. S.K. Ma, *Modern Theory of Critical Phenomena* (Addison-Wesley, New York, 1996).
3. K. Damle and S. Sachdev, Phys. Rev. B **56**, 8714 (1997).
4. S. Sachdev and K. Damle, Phys. Rev. Lett. **78**, 943 (1997).
5. S.A. Hartnoll, P.K. Kovtun, M. Müller and S. Sachdev, Phys. Rev. B **76**, 144502 (2007).
6. A. Del Maestro, B. Rosenow, N. Shah and S. Sachdev, Phys. Rev. B **77**, 180501(R) (2008).
7. K. Sengupta, S. Powell and S. Sachdev, Phys. Rev. A **69**, 053616 (2004).
8. R.A. Barankov and L.S. Levitov, Phys. Rev. Lett. **96**, 230403 (2006).
9. A.A. Burkov, M.D. Lukin and E. Demler, Phys. Rev. Lett. **98**, 200404 (2007).
10. R.W. Cherng, V. Gritsev, D.M. Stamper-Kurn and E. Demler, Phys. Rev. Lett. **100**, 180404 (2008).
11. E. Altman, A. Polkovnikov, E. Demler, B. Halperin and M.D. Lukin, Phys. Rev. Lett. **95**, 020402 (2005).
12. V. Gritsev, E. Demler, M. Lukin and A. Polkovnikov, Phys. Rev. Lett. **99**, 200404 (2007).
13. A. Das, K. Sengupta, D. Sen and B.K. Chakrabarti, Phys. Rev. B **74**, 144423 (2006).
14. M. Rigol, A. Muramatsu and M. Olshanii, Phys. Rev. A **74**, 053616 (2006).
15. M. Rigol, Phys. Rev. Lett. **103**, 100403 (2009).
16. I. Klich, C. Lannert and G. Refael, Phys. Rev. Lett. **99**, 205303 (2007).
17. M. Greiner, O. Mandel, T. Esslinger, T.W. Hänsch and I. Bloch, Nature **415**, 39 (2002).
18. For a review, see I. Bloch, Nat. Phys. **1**, 23 (2005).
19. T.W.B. Kibble, J. Phys. A **9**, 1387 (1976).
20. W.H. Zurek, Nature **317**, 505 (1985).
21. N.D. Antunes, L.M.A. Bettencourt and W.H. Zurek, Phys. Rev. Lett. **82**, 2824 (1999).
22. J. Dziarmaga, P. Laguna and W.H. Zurek, Phys. Rev. Lett. **82**, 4749 (1999).
23. J.R. Anglin and W.H. Zurek, Phys. Rev. Lett. **83**, 1707 (1999).
24. A. Polkovnikov, Phys. Rev. B **72**, 161201(R) (2005).
25. K. Sengupta, D. Sen and S. Mondal, Phys. Rev. Lett. **100**, (2008) 077204.
26. S. Mondal, D. Sen and K. Sengupta, Phys. Rev. B **78**, (2008) 045101.
27. D. Sen, K. Sengupta and S. Mondal, Phys. Rev. Lett. **101**, (2008) 016806.
28. S. Mondal, K. Sengupta and D. Sen, Phys. Rev. B **79**, (2009) 045128.
29. S. Bose, Phys. Rev. Lett. **91**, 207901 (2003).
30. M. Christandl, N. Datta, A. Ekert and A.J. Landahl, Phys. Rev. Lett. **92**, 187902 (2004).
31. C. Albanese, M. Christandl, N. Datta and A. Ekert, Phys. Rev. Lett. **93**, 230502 (2004).
32. M. Christandl, N. Datta, T. Dorlas, A. Ekert, A. Kay and A. Landahl, Phys. Rev. A **71**, 032312 (2005).
33. A. Kay, Phys. Rev. Lett. **98**, 010501 (2007).
34. F. Galve, D. Zueco, S. Kohler, E. Lutz and P. Hänggi, Phys. Rev. A **79**, 032332 (2009).
35. S. Sachdev, K. Sengupta and S. M. Girvin, Phys. Rev. B **66**, 075128 (2002).
36. S. Dusuel and J. Vidal, Phys. Rev. Lett. **93**, 237204 (2004).
37. S. Dusuel and J. Vidal, Phys. Rev. B **71**, 224420 (2005).
38. J. Vidal, G. Palacios and J. Aslangul, Phys. Rev. A **70**, 062304 (2004).
39. B.K. Chakrabarti and J.-I. Inoue, Ind. J. Phys. **80**, 609 (2006).
40. B.K. Chakrabarti, A. Das and J.-I. Inoue, Euro. Phys. J. B **51**, 321 (2006).
41. E. Fradkin, *Field Theories of Condensed Matter Systems* (Addison-Wesley, Reading, 1991).
42. B.K. Chakrabarti, A. Dutta and P. Sen, *Quantum Ising Phases and Transitions in Transverse Ising Models* (Springer, Heidelberg, 1996).
43. D. Gordon and C.M. Savage, Phys. Rev. A **59**, 4623 (1999).
44. A. Micheli, D. Jaksch, J.I. Cirac and P. Zoller, Phys. Rev. A **67**, 013607 (2003).
45. A.P. Hines, R.H. McKenzie and G.J. Milburn, Phys. Rev. A **67**, 013609 (2004).
46. A. Simoni, F. Ferlaino, G. Roati, G. Modugno and M. Inguscio, Phys. Rev. Lett **90**, 163202 (2003).

47. J. Cirac, M. Lewenstein, K. Molmer and P. Zoller, Phys. Rev. A **57**, 1208 (1998).
48. B. Damski, Phys. Rev. Lett. **95**, 035701 (2005).
49. W.H. Zurek, U. Dorner and P. Zoller, Phys. Rev. Lett. **95**, 105701 (2005).
50. A. Polkovnikov and V. Gritsev, Nat. Phys. **4**, 477 (2008).
51. J. Dziarmaga, Phys. Rev. Lett. **95**, 245701 (2005).
52. J. Dziarmaga, Phys. Rev. B **74**, 064416 (2006).
53. P. Calabrese and J. Cardy, J. Stat. Mech: Theory Expt P04010 (2005).
54. P. Calabrese and Phys. Rev. Lett. **96**, 136801 (2006).
55. R.W. Cherng and L. Levitov, Phys. Rev. A **73**, 043614 (2006).
56. V. Mukherjee, U. Divakaran, A. Dutta and D. Sen, Phys. Rev. B **76**, 174303 (2007).
57. B. Damski and W.H. Zurek, Phys. Rev. A **73** 063405 (2006).
58. T. Caneva, R. Fazio and G.E. Santoro, Phys. Rev. B **76**, 144427 (2007).
59. F.M. Cucchietti, B. Damski, J. Dziarmaga and W.H. Zurek, Phys. Rev. A **75**, 023603 (2007).
60. C. Kollath, A.M. Lauchli and E. Altman, Phys. Rev. Lett. **98**, 180601 (2007).
61. M. Eckstein and M. Kollar, Phys. Rev. Lett. **100**, 120404 (2008).
62. S.R. Manmana, S. Wessel, R.M. Noack and A. Muramatsu, Phys. Rev. Lett. **98**, 210405 (2007).
63. For a review, see I. Bloch, J. Dalibard and W.Zwerger, Rev. Mod. Phys. **80**, 885 (2008).
64. L.-M. Duan, E. Demler and M.D. Lukin, Phys. Rev. Lett. **91**, 090402 (2003).
65. A. Micheli, G.K. Brennen and P. Zoller, Nature Physics **2**, 341 (2006).
66. L.E. Sadler, J.M. Higbie, S.R. Leslie, M. Vengalattore and D.M. Stamper-Kurn, Nature **443**, 312 (2006).
67. See for example, L. Landau and E.M. Lifshitz, *Quantum Mechanics: Non-relativistic Theory*, Second Edition (Pergamon Press, Oxford, 1965).
68. C. Zener, Proc. Roy. Soc. London, Ser. A **137**, 696 (1932).
69. R. Barankov and A. Polkovnikov, Phys. Rev. Lett. **101**, 076801 (2008).
70. C. De Grandi, R.A. Barankov and A. Polkovnikov, Phys. Rev. Lett. **101**, 230402 (2008).
71. A. Kitaev, Ann. Phys. **321**, 2 (2006).
72. X.-Y. Feng, G.-M. Zhang and T. Xiang, Phys. Rev. Lett. **98**, 087204 (2007).
73. G. Baskaran, S. Mandal and R. Shankar, Phys. Rev. Lett. **98**, 247201 (2007).
74. D.-H. Lee, G.-M. Zhang and T. Xiang, Phys. Rev. Lett. **99**, 196805 (2007).
75. K.P. Schmidt, S. Dusuel and J. Vidal, Phys. Rev. Lett. **100**, 057208 (2008).
76. H.-D. Chen and Z. Nussinov, J. Phys. A **41**, 075001 (2008).
77. Z. Nussinov and G. Ortiz, Phys. Rev. B **77**, 064302 (2008).
78. A. Kitaev, Ann. Phys. **303**, 2 (2003).
79. F. Pellegrini, S. Montangero, G.E. Santoro and R. Fazio, Phys. Rev. B **77**, 140404(R) (2008).
80. U. Divakaran, A. Dutta and D. Sen, Phys. Rev. B **78**, 144301 (2008).
81. U. Divakaran, V. Mukherjee, A. Dutta and D. Sen, J. Stat. Mech. P02007 (2009).
82. S.E. Sebastian, P.A. Sharma, M. Jaime, N. Harrison, V. Correa, L. Balicas, N. Kawashima, C.D. Batista and I.R. Fisher, Phys. Rev. B **72**, 100404(R) (2005).
83. S.E. Sebastian, N. Harrison, C.D. Batista, L. Balicas, M. Jaime, P.A. Sharma, N. Kawashima and I.R. Fisher, Nature **441**, 617 (2006).
84. A. Bermudez, D. Patanè, L. Amico and M.A. Martin-Delgado, Phys. Rev. Lett. **102**, 135702 (2009).
85. A. Sen(De), U. Sen and M. Lewenstein, Phys. Rev. A **72**, 052319 (2005).
86. H. Wichterich and S. Bose, Phys. Rev. A **79**, 060302(R) (2009).
87. K. Sengupta and D. Sen, Phys. Rev. A **80**, 032304 (2009).
88. L. Cincio, J. Dziarmaga, M.M. Rams and W.H. Zurek, Phys. Rev. A **75**, 052321 (2007).
89. S. Deng, L. Viola and G. Ortiz, Proceedings of the 14th International Conference on Recent Progress in Many-Body Theories, Series on Advances in Many-Body Theory, vol. 11 (World Scientific, 2008), pp. 387–397.

Chapter 3
Defect Production Due to Quenching Through a Multicritical Point and Along a Gapless Line

U. Divakaran, V. Mukherjee, A. Dutta, and D. Sen

3.1 Introduction

The exciting physics of quantum phase transitions has been explored extensively in the last few years [1, 2]. The non-equilibrium dynamics of a quantum system when quenched very fast [3, 4] or slowly across a quantum critical point [5–7] has attracted the attention of several groups recently. The possibility of experimental realizations of quantum dynamics in spin-1 Bose condensates [8] and atoms trapped in optical lattices [9–12] has led to an upsurge in studies of related theoretical models [3–7, 13–44].

In this review, we concentrate on the dynamics of quantum spin chains swept across a quantum critical or multicritical point or along a gapless line by a slow (adiabatic) variation of a parameter appearing in the Hamiltonian of the system. Our aim is to find the scaling form of the density of defects (which, in our case, is the density of wrongly oriented spins) in the final state which is reached after the system is prepared in an initial ground state and then slowly quenched through a quantum critical point. The dynamics in the vicinity of a quantum critical point is necessarily non-adiabatic due to the divergence of the relaxation time of the underlying quantum system which forces the system to be infinitely sluggish; thus the system fails to respond to a change in a parameter of the Hamiltonian no matter how slow that rate of change may be!

U. Divakaran (✉)
Department of Physics, Indian Institute of Technology Kanpur, Kanpur 208 016, India
udiva@iitk.ac.in

V. Mukherjee
Department of Physics, Indian Institute of Technology Kanpur, Kanpur 208 016, India
victor@iitk.ac.in

A. Dutta
Department of Physics, Indian Institute of Technology Kanpur, Kanpur 208 016, India
dutta@iitk.ac.in

D. Sen
Center for High Energy Physics, Indian Institute of Science, Bangalore 560 012, India
diptiman@cts.iisc.ernet.in

Divakarn, U. et al.: *Defect Production Due to Quenching Through a Multicritical Point and Along a Gapless Line*. Lect. Notes Phys. **802**, 57–73 (2010)
DOI 10.1007/978-3-642-11470-0_3

We first recall the Kibble–Zurek argument [45, 46] which predicts a scaling form for the defect density following a slow quench through a quantum critical point. We assume that a parameter g of the Hamiltonian is varied in a linear fashion such that $g - g_c \sim t/\tau$, where $g = g_c$ denotes the value of g at the quantum critical point and τ is the quenching time. Our interest is in the adiabatic limit, $\tau \to \infty$. The energy gap of the quantum Hamiltonian vanishes at the critical point as $(g - g_c)^{\nu z}$ whereas the relaxation time ξ_τ, which is inverse of the gap, diverges at the critical point. It is clear that non-adiabaticity becomes important at a time $\hat{t}$ when the characteristic timescale of the quantum system (i.e., the relaxation time) is of the order of the inverse of the rate of change of the Hamiltonian; this yields

$$\hat{t} \sim \xi_\tau(\hat{t}) \to \hat{t} \sim (g - g_c)^{-\nu z} \sim \left(\frac{\hat{t}}{\tau}\right)^{-\nu z}. \tag{3.1}$$

This leads to a characteristic length scale ξ given by

$$\xi \sim \tau^{-\nu/(\nu z+1)}, \tag{3.2}$$

where ξ is the correlation length or healing length. The healing length typically denotes the length over which a single defect is present. The density of defects in a d-dimensional system scales as $1/\xi^d$ which leads to the Kibble–Zurek scaling form for the density of defects n given by

$$n \sim \tau^{-d\nu/(\nu z+1)}. \tag{3.3}$$

The interesting aspect of the Kibble–Zurek prediction is that the scaling form of the defect density in the final state of a driven quantum system varies in a power-law fashion with the rate of quenching $1/\tau$, and the exponent of the power law depends on the spatial dimension d and the static quantum critical exponents ν and z.

The Kibble–Zurek scaling has been verified in various exactly solvable spin models and systems of interacting bosons [5–7, 19, 21, 25, 26]; it has been generalized to quenching through a multicritical point [42], across a gapless phase [29, 30, 34], and along a gapless line [37, 39] and to systems with quenched disorder [22, 23], white noise [27], infinite-range interactions [38], and edge states [41]. Studies have also been made to estimate the defect density for quenching with a non-linear form [31–33], an oscillatory variation of an applied magnetic field [43], or under a reversal of the magnetic field [44]. In Chap. 2 [47], the quenching dynamics through a gapless phase and quenching with a power-law form of the change of a parameter as well as the possibility of experimental realizations have been discussed in detail. It should be mentioned that in addition to the density of defects in the final state, the degree of non-adiabaticity can also be quantified by looking at various quantities like residual energy [13, 48] and fidelity [5, 6].

3.2 A Spin Model: Transverse and Anisotropic Quenching

In this chapter, we will focus mainly on the one-dimensional anisotropic XY spin-1/2 chain in a transverse field [49–52] which is represented by the Hamiltonian

$$H = - \sum_n [J_x \sigma_n^x \sigma_{n+1}^x + J_y \sigma_n^y \sigma_{n+1}^y + h \sigma_n^z], \tag{3.4}$$

where σ's are the usual Pauli matrices (we will set Planck's constant $\hbar = 1$). The spectrum of this Hamiltonian can be found exactly by first mapping the σ matrices to Jordan–Wigner fermions c_n as [50–52]

$$c_n = \sigma_n^- \exp\left(i\pi \sum_{j=1}^{n-1} \sigma_j^+ \sigma_j^-\right),$$

$$\sigma_n^z = 2c_n^\dagger c_n - 1, \tag{3.5}$$

where $\sigma_n^\pm = \sigma_n^x \pm i\sigma_n^y$ are spin raising and lowering operators, respectively. These Jordan–Wigner fermion operators follow the usual anticommutation rules

$$\{c_m^\dagger, c_n\} = \delta_{mn}; \quad \{c_m, c_n\} = 0 = \{c_m^\dagger, c_n^\dagger\}. \tag{3.6}$$

Applying a Fourier transformation to the Jordan–Wigner fermions along with the periodic boundary conditions, the Hamiltonian in (3.4) can be rewritten as

$$H = - \sum_{k>0} \{ [(J_x + J_y)\cos k + h] (c_k^\dagger c_k + c_{-k}^\dagger c_{-k})$$

$$+ i(J_x - J_y)\sin k (c_k^\dagger c_{-k}^\dagger - c_{-k} c_k\}, \tag{3.7}$$

where k lies in the range $[0, \pi]$. The Hamiltonian is quadratic in fermion operators and hence exactly solvable using an appropriate Bogoliubov transformation [50–52]. Using the basis vectors $|0\rangle$ (where no fermions are present) and $|k, -k\rangle = c_k^\dagger c_{-k}^\dagger |0\rangle$, we can recast the Hamiltonian for wave number k in a 2×2 matrix form

$$H_k = \begin{bmatrix} h + (J_x + J_y)\cos k & i(J_x - J_y)\sin k \\ -i(J_x - J_y)\sin k & -h - (J_x + J_y)\cos k \end{bmatrix}. \tag{3.8}$$

The spectrum for this is given by

$$\varepsilon_k = [h^2 + J_x^2 + J_y^2 + 2h(J_x + J_y)\cos k + 2J_x J_y \cos 2k]^{1/2}. \tag{3.9}$$

This Hamiltonian has a very rich phase diagram as shown in Fig. 3.1. The vanishing of the energy gap for $h = \pm(J_x + J_y)$ for the critical modes at $k = \pi$ and 0, respectively, signals quantum phase transitions from a ferromagnetically ordered phase to

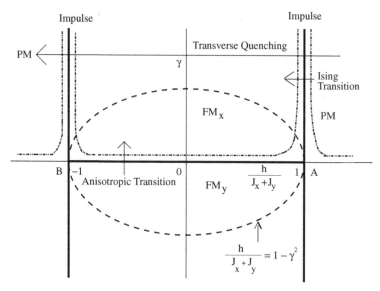

Fig. 3.1 The phase diagram of the anisotropic XY model in a transverse field in the $h/(J_x + J_y) - \gamma$ plane, where $\gamma \equiv (J_x - J_y)/(J_x + J_y)$. The *vertical bold lines* given by $h/(J_x + J_y) = \pm 1$ denote the Ising transitions. The system is also gapless on the *horizontal bold line* $\gamma = 0$ for $|h| < J_x + J_y$. FM$_x$ (FM$_y$) is a long-range ordered phase with ferromagnetic ordering in the $x(y)$-direction. The *thick dashed line* marks the boundary between the commensurate and incommensurate ferromagnetic phases. The *thin dotted lines* indicate the adiabatic and impulse regions when the field h is quenched from $-\infty$ to ∞. The two points with coordinates $\gamma = 0$ and $h/(J_x + J_y) = \pm 1$ denoted by A and B are the multicritical points

a paramagnetic phase with critical exponents $\nu = z = 1$ [49]. On the other hand, the vanishing energy gap at $J_x = J_y$ and $h < J_x + J_y$ denotes the anisotropic transition which marks the boundary between the two ferromagnetic phases with the critical exponents ν and z being identical to the Ising transition. The meeting points of these two transition lines at $h = \pm(J_x + J_y)$ and $J_x = J_y$ are multicritical points.

Let us initiate our discussions with two types of quenching schemes:

(i) quenching the magnetic field h as t/τ which we call transverse quenching [19, 21], and
(ii) the quenching of the interaction J_x as t/τ which is referred to as anisotropic quenching [25].

In the process of anisotropic quenching, the system can be made to cross the multicritical points A and B shown in the phase diagram. The quenching through a multicritical point is a cardinal point of our discussion. We shall also discuss the quenching scheme where the anisotropy parameter $\gamma = J_x - J_y$ is quenched in a linear fashion keeping the system always on the gapless line $h = J_x + J_y$. It will be shown that in either case, we arrive at new scaling behaviors for the defect density which cannot be obtained by a simple fine tuning of the Kibble–Zurek scaling form $n \sim 1/\tau^{\nu d/\nu z+1}$.

The reduction of the general Hamiltonian to a direct product of 2×2 matrices as given in (3.8) facilitates the application of the Landau–Zener (LZ) transition formula [53, 54] to estimate the non-adiabatic transition probability on passing through a quantum critical point. The general LZ Hamiltonian given by

$$H = \varepsilon_1|1\rangle\langle 1| + \varepsilon_2|2\rangle\langle 2| + \Delta \left(|1\rangle\langle 2| + |2\rangle\langle 1|\right) \tag{3.10}$$

closely resembles the reduced 2×2 spin Hamiltonian given in Eq. (3.8). In the LZ Hamiltonian, we set $\varepsilon_1 - \varepsilon_2 = t/\tau$. The two energy levels $\pm\sqrt{\varepsilon^2 + \Delta^2}$, where $\varepsilon_1 = -\varepsilon_2 = \varepsilon$, approach each other with a minimum gap 2Δ at $t = 0$ as shown in Fig. 3.2. The system is prepared in its initial ground state $|1\rangle$ at time $t \to -\infty$ and should reach the final ground state $|2\rangle$ at $t \to +\infty$ if the dynamics is adiabatic throughout. A general state during the time evolution can be written in the form $|\psi(t)\rangle = C_1(t)|1\rangle + C_2(t)|2\rangle$ with the initial condition $|C_1(t \to -\infty)|^2 = 1$. The LZ probability for the non-adiabatic transition is given by [48, 53, 54]

$$p = |C_1(t \to \infty)|^2 = e^{-2\pi \Delta^2/|\frac{\partial}{\partial t}(\varepsilon_1 - \varepsilon_2)|}. \tag{3.11}$$

It is to be noted that the above formula is valid for a linear variation of the bare levels ε_1 and ε_2 and also when the off-diagonal term does not include any time dependence.

Let us now discuss the transverse quenching of the XY spin chain discussed above. The phase diagram in (1.1) shows that when the transverse field is quenched from $-\infty$ to $+\infty$, the system is swept across the two Ising transition lines at $h = \pm(J_x + J_y)$. Referring to the Hamiltonian in (3.8), it is clear that when h is quenched as $h \sim t/\tau$ [21], the dynamics of the spin chain effectively reduces

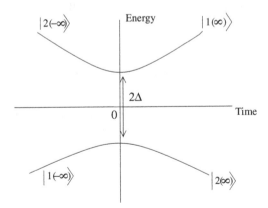

Fig. 3.2 The two levels correspond to the energy levels varying with time; 2Δ is the minimum gap between these two levels, and $|1\rangle$ and $|2\rangle$ are the eigenstates in the asymptotic limits $t \to \pm\infty$. The ground state changes its characteristics from $|1\rangle$ at $t \to -\infty$ to $|2\rangle$ at $t \to +\infty$

to a LZ problem in the two-dimensional reduced Hilbert space spanned by the basis vectors $|0\rangle$ and $|k, -k\rangle$, with the initial ground state being $|0\rangle$. Here, the diagonal and off-diagonal terms are denoted by $\varepsilon_k^{\pm} = \pm[h + (J_x + J_y)\cos k]$ and $\Delta = i(J_x - J_y)\sin k$, respectively. Denoting a general state vector $\psi_k(t)$ at an instant t as $|\psi_k(t)\rangle = C_{1,k}|0\rangle + C_{2,k}|k, -k\rangle$ with $|C_{1,k}(t \to -\infty)|^2 = 1$, the non-adiabatic transition probability for the mode k is directly obtained using the LZ transition formula

$$p_k = |C_{1,k}(t \to \infty)|^2 = e^{-\pi \tau |J_x - J_y|^2 \sin^2 k}. \tag{3.12}$$

The density of defects (i.e., density of wrongly oriented spins) in the final state is obtained by integrating p_k over the Brillouin zone [21],

$$n = \frac{1}{2\pi} \int_{-\pi}^{\pi} p_k dk \approx \frac{1}{\pi \sqrt{\tau(J_x - J_y)}}. \tag{3.13}$$

In the adiabatic limit $\tau \to \infty$, the transition probability is non-zero only for modes close to the critical modes $k = 0$ and π; this allows us to extend the limits of integration from $-\infty$ to $+\infty$. Note that the critical exponents $\nu = z = 1$ for the Ising transition and $d = 1$, hence the $1/\sqrt{\tau}$ scaling of the defect density is consistent with the Kibble–Zurek prediction.

On the other hand, if we look at the anisotropic quenching $J_x \sim t/\tau$, the off-diagonal terms of the Hamiltonian in (3.8) pick up a time dependence and hence a direct application of the LZ transition probability is not possible. To overcome this problem, we rewrite the Hamiltonian in the basis of the initial and final eigenstates when $J_x \to -\infty$ and $J_x \to \infty$. The eigenstates of the Hamiltonian in these limits are given by

$$|e_{1k}\rangle = \sin(k/2)|0\rangle + i\cos(k/2)|k, -k\rangle$$

and

$$|e_{2k}\rangle = \cos(k/2)|0\rangle - i\sin(k/2)|k, -k\rangle,$$

with eigenvalues $\lambda_1 = t/\tau$ and $\lambda_2 = -t/\tau$, respectively; the system is in the state $|e_{1k}\rangle$ initially. A general state vector can be expressed as a linear combination of $|e_{1k}\rangle$ and $|e_{2k}\rangle$,

$$|\psi_k(t)\rangle = C_{1k}(t)|e_{1k}\rangle + C_{2k}(t)|e_{2k}\rangle. \tag{3.14}$$

The initial condition in the anisotropic case is $C_{1k}(-\infty) = 1$ and $C_{2k}(-\infty) = 0$. The unitary transformation to rewrite the Hamiltonian in the $|e_{1k}\rangle$ and $|e_{2k}\rangle$ basis is given by $H_k'(t) = U^{\dagger} H_k(t) U$, where

$$U = \begin{bmatrix} \cos(k/2) & \sin(k/2) \\ -i\sin(k/2) & i\cos(k/2) \end{bmatrix},$$

and the new Hamiltonian is

$$H'_k(t) = -[h + (J_x + J_y)\cos k]\, I_2$$
$$+ \begin{bmatrix} J_x + J_y\cos 2k + h\cos k & J_y\sin 2k + h\sin k \\ J_y\sin 2k + h\sin k & -J_x - J_y\cos 2k - h\cos k \end{bmatrix}. \quad (3.15)$$

By virtue of the unitary transformation, the time dependence is entirely shifted to the diagonal terms which make it possible to apply the LZ transition formula. Evaluating the probability of a non-adiabatic transition for the mode k and integrating over the Brillouin zone, it can be shown that the density of defects in the final state following the anisotropic quenching is given by [25]

$$n \sim \frac{4J_y}{\pi\sqrt{\tau}(4J_y^2 - h^2)} \quad \text{for} \quad h < 2J_y,$$

$$\sim \frac{h}{\pi\sqrt{\tau}(h^2 - 4J_y^2)} \quad \text{for} \quad h > 2J_y. \quad (3.16)$$

The scaling behavior of the density of defects with τ is shown in Fig. 3.3.

Although Eqs. (3.16) satisfy the Kibble–Zurek scaling, the density of defects diverges for $h = 2J_y$, i.e., in a passage through the multicritical points. This necessitates a generalization of the Kibble–Zurek prediction for quenching through a multicritical point. In the next section, we propose a generalized scaling for the density of defects valid for quenching through a critical point as well as a multicritical point.

In Fig. 3.4, we present the variation of the von Neumann entropy density of the final state defined by [21]

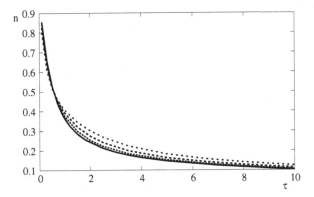

Fig. 3.3 Plot of kink density n versus τ as obtained for $h = 0.2, 0.4, 0.6, 0.8$ (from *bottom* to *top* in the large τ region), with $J_y = 1$. For large τ, n increases with increasing h, whereas for small τ, it decreases with increasing h

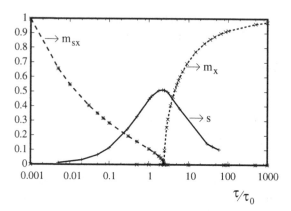

Fig. 3.4 Plot of von Neumann entropy density s, staggered magnetization m_{sx}, and magnetization m_x as a function of τ/τ_0, for $J_y = 1$ and $h = 0.2$

$$s = -\int_0^\pi \frac{dk}{\pi} \left[p_k \ln(p_k) + (1 - p_k) \ln(1 - p_k) \right], \qquad (3.17)$$

following the anisotropic quenching with the rate of quenching. Even though the final state is a pure state as a result of a unitary dynamics, it can also be viewed locally, or on a coarse-grained wave vector scale, as a decohered (mixed) state [21]. In the limit of $\tau \to 0$, the system does not get enough time to evolve; hence it largely retains its initial antiferromagnetic order which results in a low local entropy density. On the other hand, for slow quenching ($\tau \to \infty$), the system evolves adiabatically, always remaining close to its instantaneous ground state; this results in a final state with local ferromagnetic ordering, and hence again with a low local entropy density. We observe that the entropy density shows a maximum at a characteristic timescale τ_0 where the magnetic ordering of the final state also changes from antiferromagnetic to ferromagnetic.

Let us briefly mention a few other interesting variants of the transverse quenching scheme. In a repeated quenching dynamics of the same model, the transverse field $h(= t/\tau)$ is quenched repeatedly between $-\infty$ and $+\infty$ [36]. We refer to a single passage from $h \to -\infty$ to $h \to +\infty$ or the other way around as a half-period of quenching. An even number of half-periods corresponds to the transverse field being brought back to its initial value of $-\infty$, whereas, in the case of an odd number of half-periods, the dynamics is stopped at $h \to +\infty$. The probability of a non-adiabatic transition at the end of l half-periods can be shown to follow the recursion relation

$$p_k(l) = (1 - e^{-2\pi\gamma}) - (1 - 2e^{-2\pi\gamma}) \left[1 - p_k(l - 1) \right], \qquad (3.18)$$

eventually yielding the simplified form of

$$p_k(l) = \frac{1}{2} - \frac{(1 - 2e^{-2\pi\gamma})^l}{2}. \tag{3.19}$$

For large τ, the density of defects is generally found to vary as $1/\sqrt{\tau}$ for any number of half-periods. On the other hand, for small τ, it shows an increase in kink density for even values of l. However, the magnitude is found to depend on the number of half-periods of quenching. For two successive half-periods, the defect density is found to decrease in comparison to a single half-period, suggesting the existence of a corrective mechanism in the reverse path. The entropy density increases monotonously with the number of half-periods and shows qualitatively the same behavior for any number of half-periods. For a large number of repetitions ($l \to \infty$), the defect density saturates to the value $1/2$ while the local entropy density saturates to $\ln 2$ (Fig. 3.5).

The effects of interference on the quenching dynamics of Hamiltonian in (3.8) when the transverse field $h(t)$ varies sinusoidally with time as $h = h_0 \cos \omega t$, with $|t| \leq \pi/\omega$ has also been studied in a recent work [31]. In this scheme of quenching, the time interval between two successive passages through the quantum critical points can be small enough for the presence of non-trivial effects of interference in the dynamics of the system. It has been shown that for a single passage through a quantum critical point, the interference effects do not contribute leading to a kink density which goes as $n \sim \sqrt{\omega}$. On the other hand, repeated passages through the quantum critical points result in an oscillatory behavior of the kink density as well as the entropy density.

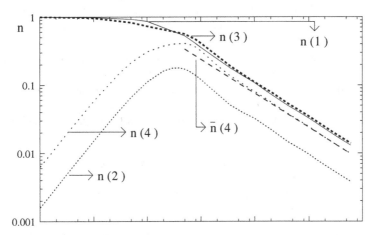

Fig. 3.5 Plot of kink density n after l half cycles as a function of τ, for $J_x - J_y = 1$ and $l = 1, 2, 3, 4$. $\bar{n}(4)$ denotes the defect density as obtained from the analytical expression. In the limit of large τ, the numerical results match perfectly with the analytical results

3.3 Quenching Through a Multicritical Point

We shall now propose a general scaling scheme valid for quenching through a multicritical point as well as a critical point [42] using the LZ non-adiabatic transition probability [48, 53, 54] discussed before. We begin with a generic d-dimensional model Hamiltonian of the form

$$H = \sum_{\mathbf{k}} \psi^{\dagger}(\mathbf{k}) \left[(\lambda(t) + b(\mathbf{k})) \, \sigma^z + \Delta(\mathbf{k}) \, \sigma^+ + \Delta^*(\mathbf{k}) \, \sigma^- \right] \psi(\mathbf{k}), \quad (3.20)$$

where $\sigma^{\pm} = \sigma^x \pm i\sigma^y$, $b(\mathbf{k})$, and $\Delta(\mathbf{k})$ are model-dependent functions and $\psi(\mathbf{k})$ denotes the fermionic operators $(\psi_1(\mathbf{k}), \psi_2(\mathbf{k}))$. The above Hamiltonian represents, for example, a one-dimensional transverse Ising or XY spin chain [50–52] as shown in Eq. (3.8), or an extended Kitaev model in $d = 2$ written in terms of Jordan–Wigner fermions [29, 30, 55]. The excitation spectrum takes the form

$$\varepsilon_k^{\pm} = \pm\sqrt{(\lambda(t) + b(\mathbf{k}))^2 + |\Delta(\mathbf{k})|^2}. \quad (3.21)$$

We assume that the parameter $\lambda(t)$ varies linearly as t/τ and vanishes at $t = 0$. The parameters $b(\mathbf{k})$ and $\Delta(\mathbf{k})$ are assumed to vanish at the quantum critical point in a power-law fashion given by

$$b(\mathbf{k}) \sim |\mathbf{k}|^{z_1} \quad \text{and} \quad \Delta(\mathbf{k}) \sim |\mathbf{k}|^{z_2}, \quad (3.22)$$

where we have taken the critical mode to be at $\mathbf{k}_0 = 0$ without any loss of generality. Equation (3.22) also implies that the system crosses the gapless point at $t = 0$ when $\lambda = 0$, and $b(\mathbf{k})$ and $\Delta(\mathbf{k})$ also vanish for the critical mode $\mathbf{k} = 0$. Many of the models described by Eq. (3.20) exhibit a quantum phase transition with the exponents associated with the quantum critical point being $z_1 > z_2$ and hence $z = z_2 = 1$. We shall, however, explore the more general case encountered at a multicritical point where the dynamical exponent z is not necessarily given by z_2.

The Schrödinger equation describing the time evolution of the system when λ is quenched is given by $i\partial\psi_{\mathbf{k}}/\partial t = H\psi_{\mathbf{k}}$, where we have once again defined $\psi_{\mathbf{k}}$ as $\psi_{\mathbf{k}} = \tilde{C}_{1,\mathbf{k}}|0\rangle + \tilde{C}_{2,\mathbf{k}}|\mathbf{k}, -\mathbf{k}\rangle$. Using the Hamiltonian in Eq. (3.20), we can write

$$i\frac{\partial \tilde{C}_{1,\mathbf{k}}}{\partial t} = \left(\frac{t}{\tau} + b(\mathbf{k})\right) \tilde{C}_{1,\mathbf{k}} + \Delta(\mathbf{k}) \, \tilde{C}_{2,\mathbf{k}},$$

$$i\frac{\partial \tilde{C}_{2,\mathbf{k}}}{\partial t} = -\left(\frac{t}{\tau} + b(\mathbf{k})\right) \tilde{C}_{2,\mathbf{k}} + \Delta^*(\mathbf{k}) \, \tilde{C}_{1,\mathbf{k}}. \quad (3.23)$$

One can now remove $b(\mathbf{k})$ from the above equations by redefining $t/\tau + b(\mathbf{k}) \to t/\tau$; thus the exponent z_1 defined in Eq. (3.22) does not play any role in the following calculations. Defining a new set of variables $C_{1,\mathbf{k}} = \tilde{C}_{1,\mathbf{k}} \exp\left(i \int^t dt' \, t'/\tau\right)$ and $C_{2,\mathbf{k}} = \tilde{C}_{2,\mathbf{k}} \exp\left(-i \int^t dt' \, t'/\tau\right)$, we arrive at a time evolution equation for $C_1(\mathbf{k})$

given by

$$\left(\frac{d^2}{dt^2} - 2i\frac{t}{\tau}\frac{d}{dt} + |\Delta(\mathbf{k})|^2 \right) C_{1,\mathbf{k}} = 0. \tag{3.24}$$

Further rescaling $t \to t\tau^{1/2}$ leads to

$$\left(\frac{d^2}{dt^2} - 2it\frac{d}{dt} + |\Delta(\mathbf{k})|^2\tau \right) C_{1,\mathbf{k}} = 0. \tag{3.25}$$

If the system is prepared in its ground state at the beginning of the quenching, i.e., $C_1(\mathbf{k}) = 1$ at $t = -\infty$, the above equation suggests that the probability of the non-adiabatic transition, $p_{\mathbf{k}} = \lim_{t\to+\infty} |C_{1,\mathbf{k}}|^2$, must have a functional dependence on $|\Delta(\mathbf{k})|^2\tau$ of the form

$$p_{\mathbf{k}} = f(|\Delta(\mathbf{k})|^2\tau). \tag{3.26}$$

The analytical form of the function f is given by the general LZ formula [48, 53, 54]. The defect density in the final state is therefore given by [42]

$$n = \int \frac{d^d k}{(2\pi)^d} f(|\Delta(\mathbf{k})|^2\tau) = \int \frac{d^d k}{(2\pi)^d} f(|\mathbf{k}|^{2z_2}\tau). \tag{3.27}$$

The scaling $|\mathbf{k}| \to |\mathbf{k}|^{2z_2}\tau$ finally leads to a scaling of the defect density given by

$$n \sim 1/\tau^{d/(2z_2)}. \tag{3.28}$$

Let us recall the quenching dynamics of the transverse XY spin chain discussed before [21, 25]. When the system is quenched across the Ising or anisotropic critical line by linearly changing h or J_x as t/τ, $\Delta(\mathbf{k})$ vanishes at the critical point as $\Delta(\mathbf{k}) \sim |\mathbf{k}|$ yielding $z_2 = z = 1$; hence the generalized scaling form given in Eq. (3.28) matches with the Kibble–Zurek prediction with $\nu = z = 1$ as mentioned before. The situation, however, is different when the system is swept across a multicritical point.

Putting $h = 2J_y$ in the Hamiltonian in (3.15), let us analyze the scaling of the diagonal and off-diagonal terms of the Hamiltonian at the multicritical point. When $J_x = J_y$, the diagonal term of the Hamiltonian scales as $-J_y(\pi - k)^2$ whereas the off-diagonal term $|\Delta(k)| \sim |\pi - k|^3$ when expanded around the critical mode at $k = \pi$. The dynamical exponent is obtained from the diagonal term so that $z = z_1 = 2$. As discussed above, the off-diagonal term or more precisely the exponent z_2 determines the scaling of the defect density. The exponent $z_2 = 3$ for the XY multicritical point; hence the defect density scales as $1/\tau^{1/6}$. The density of defects obtained by numerical integration of the Schrödinger equation in (3.23) as shown in Fig. 3.6 supports this scaling behavior. For a non-linear quench across a multicritical point [47], when

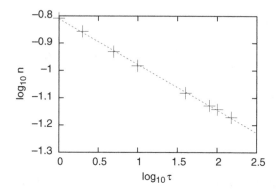

Fig. 3.6 n vs τ obtained by numerically solving Eq. (3.23) at the multicritical point of the XY model in a transverse field, with $h = 10$ and $J_y = 5$. The line has a slope of -0.16

the parameter λ is quenched as $\lambda \sim (t/\tau)^\alpha sgn(t)$, both z_1 and z_2 come into play and the scaling of defect density gets altered to $n \sim \tau^{-d\alpha\nu/[\alpha(z_2\nu+1)+z_1\nu(1-\alpha)]}$; this reduces to the form presented above for $\alpha = 1$ and $z_2\nu = 1$.

3.4 Quenching Along a Gapless Line

We now shift our focus to quenching the XY Hamiltonian in (3.7) along the gapless line $h = J_x + J_y$ by varying the anisotropy parameter $\gamma = J_x - J_y$ as t/τ, keeping $J_x + J_y$ fixed [37]. For convenience, let us set $J_x + J_y = h = 1$. The excitation gap vanishes along this line for the mode $k = \pi$. We rewrite Eq. (3.8) with the present notation in the form

$$H_k = \begin{bmatrix} 1 + \cos k & i\gamma \sin k \\ -i\gamma \sin k & -1 - \cos k \end{bmatrix}. \tag{3.29}$$

We once again encounter a situation in which the off-diagonal terms are time dependent. Noting that the asymptotic form of the Hamiltonian at $t \to \pm\infty$ is given by

$$H = \frac{t}{\tau} \sin k \, \hat{\sigma}^x, \tag{3.30}$$

we make a basis transformation to a representation in which σ^x is diagonal. The Hamiltonian in (3.29) then gets modified to the form

$$\begin{bmatrix} (t/\tau) \sin k & 1 + \cos k \\ 1 + \cos k & -(t/\tau) \sin k \end{bmatrix}, \tag{3.31}$$

where the time dependence has been shifted to the diagonal terms only. Applying the LZ transition probability formula, the probability of excitations is found to be

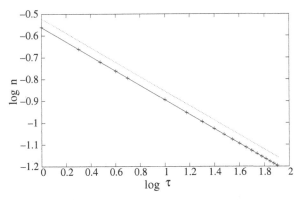

Fig. 3.7 n vs τ obtained by numerically solving Schrödinger equation keeping $h = 2J_y$. Also shown is a line with slope 1/3 for comparison

$$p_k = e^{-\pi\tau|\Delta_k|^2/\sin k}, \tag{3.32}$$

where $|\Delta_k|^2 = |1 + \cos k|^2 = (\pi - k)^4/4$ when expanded about $k = \pi$. Integrating p_k over the Brillouin zone, we find that the density of defects falls off as $n \sim 1/\tau^{1/3}$. Figure 3.7 which shows n vs τ obtained by numerical integration of the Schrödinger equation confirms the $\tau^{-1/3}$ behavior.

It is to be noted that the Kibble–Zurek formalism cannot address defect generation along a gapless line. We will now present another general scaling argument for moving along a gapless line in a d-dimensional system.

Let the excitations on the gapless quantum critical line be of the form $\omega_\mathbf{k} \sim \lambda|\mathbf{k}|^z$, where z is the dynamical exponent, and the parameter $\lambda = t/\tau$ is quenched from $-\infty$ to ∞. Using a perturbative method involving Fermi's Golden rule along with the fact that the system is initially prepared in the ground state, the defect density can be approximated as [28]

$$n \simeq \int \frac{d^d k}{(2\pi)^d} \left| \int_{-\infty}^{\infty} d\lambda \, \langle\mathbf{k}| \frac{\partial}{\partial\lambda} |0\rangle \, e^{i\tau \int^\lambda \delta\omega_\mathbf{k}(\lambda')d\lambda'} \right|^2. \tag{3.33}$$

Let us assume a general scaling form for the instantaneous excitation, $\delta\omega_\mathbf{k}(\lambda') = k^a f(\frac{\lambda k^z}{k^a})$, where $k = |\mathbf{k}|$ and k^a denotes the higher order term in the excitation spectrum on the gapless line. Defining a new variable $\xi = \lambda k^{z-a}$, we obtain the scaling behavior of the defect density as [37]

$$n \sim 1/\tau^{d/(2a-z)}. \tag{3.34}$$

The case $d = 1$, $a = 2$, and $z = 1$ has been discussed in this section. Note that the correlation length exponent ν does not appear in the expression in Eq. (3.34) because our quench dynamics always keeps the system on a critical line. The scaling for quenching along a gapless line is less universal in comparison to the general

Kibble–Zurek prediction. For a non-linear quench [56] $\lambda \sim (t/\tau)^\alpha \, sgn(t)$, this scaling form gets modified to $n \sim \tau^{-d\alpha/[a(\alpha+1)-z]}$; this reduces to $1/\tau^{1/3}$ for $\alpha = d = z = 1$ and $a = 2$.

In passing, let us introduce a variant of the XY Hamiltonian with an alternating field given by [39, 57, 58]

$$
H = -\frac{1}{2} [\sum_j (J_x + J_y)(\sigma_j^x \sigma_{j+1}^x + \sigma_j^y \sigma_{j+1}^y)
$$

$$
+ (J_x - J_y)(\sigma_j^x \sigma_{j+1}^x - \sigma_j^y \sigma_{j+1}^y) + (h - (-1)^j \delta)\sigma_j^z]. \quad (3.35)
$$

The strength of the transverse field alternates between $h + \delta$ and $h - \delta$ on the odd and even sites, respectively. For $\delta = 0$, we recover the conventional XY model in a transverse field. This Hamiltonian can also be solved exactly by a Jordan–Wigner transformation by taking care of the two underlying sublattices, i.e., by defining a_k and b_k as two different Jordan–Wigner fermions. The Hamiltonian in the basis $(a_k^\dagger, a_{-k}, b_k^\dagger, b_{-k})$ can be written as

$$
H_k = \begin{bmatrix} h + J\cos k & i\gamma \sin k & 0 & -\delta \\ -i\gamma \sin k & -h - J\cos k & \delta & 0 \\ 0 & \delta & J\cos k - h & i\gamma \sin k \\ -\delta & 0 & -i\gamma \sin k & -J\cos k + h \end{bmatrix}. \quad (3.36)
$$

The excitation spectrum of the above Hamiltonian is

$$
\Lambda_k^\pm = [\, h^2 + \delta^2 + J^2 \cos^2 k + \gamma^2 \sin^2 k
$$

$$
\pm 2\sqrt{h^2\delta^2 + h^2 J^2 \cos^2 k + \delta^2 \gamma^2 \sin^2 k}\,]^{1/2}, \quad (3.37)
$$

where $J = J_x + J_y$ and $\gamma = J_x - J_y$, with four eigenvalues given by $\pm \Lambda_k^\pm$. In the ground state, $-\Lambda_k^+$ and $-\Lambda_k^-$ are filled. The vanishing of Λ_k^- dictates the quantum critical point, and the critical exponents are obtained by studying the behavior of Λ_k^- in the vicinity of the critical point. The minimum energy gap in the excitation spectrum occurs at $k = 0$ and $k = \pi/2$. The corresponding phase boundaries $h^2 = \delta^2 + J^2$ and $\delta^2 = h^2 + \gamma^2$ signal quantum phase transitions from a paramagnetic to a ferromagnetic phase, respectively, as shown in Fig. 3.8. We extend the study of quenching through a gapless phase in this Hamiltonian by varying γ as before along the phase boundary $h^2 = \delta^2 + J^2$. On this gapless line, the dispersion of the low-energy excitations at $k \to 0$ can be approximated as

$$
\Lambda_k^- = \sqrt{\frac{J^4 k^4}{4(\delta^2 + J^2)} + \frac{\gamma^2 J^2 k^2}{\delta^2 + J^2}}. \quad (3.38)
$$

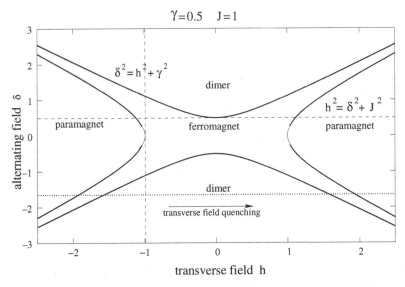

Fig. 3.8 Phase diagram of the *XY* chain in an alternating field with $\gamma = 0.5$ and $J = 1$. Two special points $(h = \pm J, \delta = 0)$ and $(h = 0, \delta = \pm \gamma)$ are shown on the phase boundaries. The spin chain undergoes a quantum phase transition with $\nu = 2$ and $z = 1$ when these points are approached along the *dashed line*. On the other hand, the *dotted line* shows the direction of the quenching of the transverse field. We quench the system along a gapless line parallel to the γ-axis (perpendicular to the plane of the paper) passing through the phase boundary $h^2 = \delta^2 + J^2$

This suggests a truncation of the Hamiltonian in (3.36) to a 2×2 LZ Hamiltonian

$$h_k = \begin{bmatrix} \tilde{\gamma}(t)k & \tilde{J}^2 k^2/2 \\ \tilde{J}^2 k^2/2 & -\tilde{\gamma}(t)k \end{bmatrix}, \tag{3.39}$$

where $\tilde{\gamma}$ and $\tilde{J}$ are renormalized parameters given by $\tilde{\gamma} = \gamma J/\sqrt{\delta^2 + J^2}$ and $\tilde{J}^2 = J^2/\sqrt{\delta^2 + J^2}$. The argument which justifies the truncation of a 4×4 matrix to a 2×2 matrix is presented in [37]. The diagonal terms in Eq. (3.39) describe two time-dependent levels approaching each other linearly in time (since $\gamma = t/\tau$), while the minimum gap is given by the off-diagonal term $\tilde{J}^2 k^2/2$. The probability of excitations p_k from the ground state to the excited state for the kth mode is given by the LZ transition formula [48, 53, 54]

$$p_k = \exp\left[-\frac{2\pi \tilde{J}^4 k^4}{8kd\tilde{\gamma}(t)/dt}\right] = \exp\left[-\frac{\pi J^3 k^3 \tau}{4\sqrt{\delta^2 + J^2}}\right], \tag{3.40}$$

and we get back the $\tau^{-1/3}$ scaling form of the defect density as discussed above.

3.5 Conclusions

The non-equilibrium dynamics of a quantum system driven through a quantum critical point is indeed an exciting and fascinating area of research. The possibility of experimental realizations of quenching dynamics adds to the importance of the theoretical research. In this review we have discussed the scaling relations for the defect density in the final state following a slow quench across a multicritical point and along a gapless line. In both cases, the relations obtained here using the LZ transition formula are not directly derivable using the Kibble–Zurek argument. But in all the cases, the defect density scales as a power law with the rate of quenching. It should be noted that the idea of a dominant critical point has also been invoked to justify the scaling along a gapless line in one-dimensional spin models [39]. A generalization of the studies presented here to higher dimensional quantum systems and to systems with quenched disorder are some of the future directions of this subject.

Acknowledgments The authors acknowledge the organizers of the conference "Quantum Annealing and Quantum Computation." We also thank A. Polkovnikov, S. Sachdev, G. E. Santoro, and K. Sengupta for useful discussions on different occasions. AD also acknowledges R. Moessner and hospitality of MPIPKS where some part of the work discussed in the review was done.

References

1. S. Sachdev, *Quantum Phase Transitions* (Cambridge University Press, Cambridge, 1999).
2. B.K. Chakrabarti, A. Dutta and P. Sen, *Quantum Ising Phases and Transitions in Transverse Ising Models*, **m41** (Springer-Verlag, Berlin, 1996).
3. K. Sengupta, S. Powell and S. Sachdev, Phys. Rev. A **69**, 053616 (2004)
4. A. Das, K. Sengupta, D. Sen and B.K. Chakrabarti, Phys. Rev. B **74**, 144423 (2006).
5. W.H. Zurek, U. Dorner and P. Zoller, Phys. Rev. Lett. **95**, 105701 (2005)
6. B. Damski, Phys. Rev. Lett. **95**, 035701 (2005).
7. A. Polkovnikov, Phys. Rev. B **72**, 161201(R) (2005).
8. L.E. Sadler, J.M. Higbie, S.R. Leslie, M. Vengalattore and D.M. Stamper-Kurn, Nature **443**, 312 (2006).
9. L.-M. Duan, E. Demler and M.D. Lukin, Phys. Rev. Lett. **91**, 090402 (2003)
10. A. Micheli, G.K. Brennen and P. Zoller, Nat. Phys. **2**, 341 (2006)
11. E. Altman, E. Demler and M.D. Lukin, Phys. Rev. A **70**, 013603 (2004).
12. I. Bloch, J. Dalibard and W. Zwerger, Rev. Mod. Phys. **80**, 885 (2008).
13. T. Kadowaki and H. Nishimori, Phys. Rev. E **58**, 5355 (1998).
14. P. Calabrese and J. Cardy, J. Stat. Mech: Theory Expt. P04010 (2005), and Phys. Rev. Lett. **96**, 136801 (2006)
15. R. Schützhold, M. Uhlmann, Y. Xu and U.R. Fischer, Phys. Rev. Lett. **97**, 200601 (2006)
16. C. Kollath, A.M. Läuchli and E. Altman, Phys. Rev. Lett. **98**, 180601 (2007)
17. S.R. Manmana, S. Wessel, R.M. Noack and A. Muramatsu, Phys. Rev. Lett. **98**, 210405 (2007)
18. M. Eckstein and M. Kollar, Phys. Rev. Lett. **100**, 120404 (2008).
19. J. Dziarmaga, Phys. Rev. Lett. **95**, 245701 (2005).
20. B. Damski and W.H. Zurek, Phys. Rev. A **73**, 063405 (2006).
21. R.W. Cherng and L.S. Levitov, Phys. Rev. A **73**, 043614 (2006).
22. J. Dziarmaga, Phys. Rev. B **74**, 064416 (2006)

23. T. Caneva, R. Fazio and G.E. Santoro, Phys. Rev. B **76**, 144427 (2007).
24. F.M. Cucchietti, B. Damski, J. Dziarmaga and W.H. Zurek, Phys. Rev. A **75**, 023603 (2007).
25. V. Mukherjee, U. Divakaran, A. Dutta and D. Sen, Phys. Rev. B **76**, 174303 (2007).
26. U. Divakaran and A. Dutta, J. Stat. Mech: Theory Expt. P11001 (2007).
27. A. Fubini, G. Falci and A. Osterloh, New J. Phys **9**, 134 (2007).
28. A. Polkovnikov and V. Gritsev, Nat. Phys. **4**, 477 (2008).
29. K. Sengupta, D. Sen and S. Mondal, Phys. Rev. Lett. **100**, 077204 (2008)
30. S. Mondal, D. Sen and K. Sengupta, Phys. Rev. B **78**, 045101 (2008).
31. D. Sen, K. Sengupta and S. Mondal, Phys. Rev. Lett. **101**, 016806 (2008)
32. R. Barankov and A. Polkovnikov, Phys. Rev. Lett. **101**, 076801 (2008)
33. C. De Grandi, R.A. Barankov and A. Polkovnikov, Phys. Rev. Lett. **101**, 230402 (2008).
34. F. Pellegrini, S. Montangero, G.E. Santoro and R. Fazio, Phys. Rev. B **77**, 140404(R) (2008).
35. D. Patanè, A. Silva, L. Amico, R. Fazio and G.E. Santoro, Phys. Rev. Lett. **101**, 175701 (2008).
36. V. Mukherjee, A. Dutta and D. Sen, Phys. Rev. B **77**, 214427 (2008).
37. U. Divakaran, A. Dutta and D. Sen, Phys. Rev. B **78**, 144301 (2008).
38. T. Caneva, R. Fazio and G.E. Santoro, Phys. Rev. B **78**, 104426 (2008).
39. S. Deng, G. Ortiz and L. Viola, Europhys. Lett. **84**, 67008 (2008).
40. J. Dziarmaga, J. Meisner and W.H. Zurek, Phys. Rev. Lett. **101**, 115701 (2008).
41. A. Bermudez, D. Patanè, L. Amico and M.A. Martin-Delgado, Phys. Rev. Lett. **102**, 135702 (2009).
42. U. Divakaran, V. Mukherjee, A. Dutta and D. Sen, J. Stat. Mech: Theory Expt. P02007 (2009).
43. V. Mukherjee and A. Dutta, J. Stat. Mech: Theory Expt. P05005 (2009).
44. U. Divakaran and A. Dutta, arXiv:0901.3260, to appear in Phys. Rev. B (2009).
45. T.W.B. Kibble, J. Phys. A **9**, 1387 (1976), and Phys. Rep. **67**, 183 (1980).
46. W.H. Zurek, Nature **317**, 505 (1985), and Phys. Rep. **276**, 177 (1996).
47. S. Mondal, D. Sen and K. Sengupta, *Non-equilibrium Dynamics of Quantum Systems: Order Parameter Evolution, Defect Generation, and Qubit Transfer*. Lect. Notes Phys. **802**. Springer, Heidelberg (2010)
48. S. Suzuki and M. Okada, in *Quantum Annealing and Related Optimization Methods*, A. Das and B.K. Chakrabarti (eds.) (Springer-Verlag, Berlin, 2005), p. 185.
49. J.E. Bunder and R.H. McKenzie, Phys. Rev. B **60**, 344 (1999).
50. E. Lieb, T. Schultz and D. Mattis, Ann. Phys. **16**, 407 (1961)
51. E. Barouch and B.M. McCoy, Phys. Rev. A **3**, 786 (1971)
52. J.B. Kogut, Rev. Mod. Phys. **51**, 659 (1979).
53. C. Zener, Proc. Roy. Soc. London Ser A **137**, 696 (1932)
54. L.D. Landau and E.M. Lifshitz, *Quantum Mechanics: Non-relativistic Theory*, Second Edition (Pergamon Press, Oxford, 1965).
55. A. Kitaev, Ann. Phys. **321**, 2 (2006).
56. S. Mondal, K. Sengupta and D. Sen, Phys. Rev. B **79**, 045128 (2009).
57. J.H.H. Perk, H.W. Capel and M.J. Zuilhof, Physica **81 A**, 319 (1975)
58. K. Okamoto and K. Yasumura, J. Phys. Soc. Jap. **59**, 993 (1990).

Chapter 4
Adiabatic Perturbation Theory: From Landau–Zener Problem to Quenching Through a Quantum Critical Point

C. De Grandi and A. Polkovnikov

4.1 Introduction

Dynamics in closed systems recently attracted a lot of theoretical interest largely following experimental developments in cold atom systems (see e.g., [1] for a review). Several spectacular experiments already explored different aspects of non-equilibrium dynamics in interacting many-particle systems [2–8]. Recent theoretical works in this context focused on various topics, for instance: connection of dynamics and thermodynamics [9–11, M. Rigol, unpublished], dynamics following a sudden quench in low dimensional systems [11–23, L. Mathey and A. Polkovnikov, unpublished; A. Iucci and M.A. Cazalilla, unpublished], adiabatic dynamics near quantum critical points [24–37, D. Chowdhury et al., unpublished; K. Sengupta and D. Sen, unpublished; A.P. Itin and P. Törmä, unpublished; F. Pollmann et al., unpublished] and others. Though there is still very limited understanding of the generic aspects of non-equilibrium quantum dynamics, it has been recognized that such issues as integrability, dimensionality, universality (near critical points) can be explored to understand the non-equilibrium behavior of many-particle systems in various specific situations.

The aim of this chapter is to address some generic aspects of nearly adiabatic dynamics in many-particle systems. In particular, we will discuss in detail the scaling of density of generated quasiparticles n_{ex}, entropy S_d, and heating Q (non-adiabatic part of the energy) with the quenching rate in several situations. It has been already understood that this scaling is universal near quantum critical points [24, 25, 27], and more generally in low-dimensional gapless systems [28]. The universality comes from the fact that if the system is initially prepared in the ground state or in a state with small temperature, then under slow perturbations very few transitions happen and the system effectively explores only the low-energy part of the spectrum. This low-energy part can be described by a small number of

C. De Grandi (✉)
Department of Physics, Boston University, Boston, MA 02215, USA, degrandi@buphy.bu.edu

A. Polkovnikov
Department of Physics, Boston University, Boston, MA 02215, USA, asp@buphy.bu.edu

De Grandi, C., Polkovnikov, A.: *Adiabatic Perturbation Theory: From Landau–Zener Problem to Quenching Through a Quantum Critical Point*. Lect. Notes Phys. **802**, 75–114 (2010)
DOI 10.1007/978-3-642-11470-0_4   © Springer-Verlag Berlin Heidelberg 2010

parameters characterizing some effective low-energy theory (typically field theory). The situation can become different, however, in high-dimensional systems, where the effects of non-adiabaticity are dominated by transitions to the high-energy states and the universality is lost [28, 38]. This happens because of a typically small density of low-energy states, which for, e.g., for free quasiparticles scales as $\rho(\varepsilon) \propto \varepsilon^{d/z-1}$, where d is the dimensionality and z is the dynamical exponent determining scaling of energy ε with momentum k at small k: $\varepsilon(k) \sim k^z$. As a result transitions to high-energy states dominate the dynamics and the universality is lost. Similar situation happens for sudden quenches near critical points [38]. In low dimensions excitations of low-energy quasiparticles determine the (universal) scaling of various thermodynamic quantities. However, in high dimensions transitions to the high-energy states following the quench become predominant and the universality is lost. In this case one can use ordinary perturbation theory or linear response which predicts that n_{ex}, S_d, and Q become analytic (quadratic) functions of the rate for slow quenches and of the quench amplitude for sudden quenches.

In this chapter we will explain in detail how the transition between quadratic and universal regimes can be understood as a result of breakdown of the linear response. More specifically we will illustrate how exactly the crossover between different scaling regimes occurs in the situations where the system can be well described by quasiparticle excitations. We will concentrate on the slow, linear in time quenches, and briefly mention the situation with fast quenches in the end. First we will discuss adiabatic perturbation theory and its implications to many-particle systems. Then using this theory we will analyze a simple driven two-level system (Landau–Zener problem [39, 40]) where the coupling linearly changes in time in the finite range. We will show how quadratic scaling of the transition probability with the rate emerges from this perturbation theory. Then we will consider a more complicated situation where the system consists of free gapless quasiparticle excitations. We will show that in low dimensions, $d \leq 2z$, the scaling of the density of excitations and entropy is universal $n_{ex}, S_d \sim |\delta|^{d/z}$, while in high dimensions the quadratic scaling is restored. The quadratic scaling can be understood as the result of multiple Landau–Zener transitions to high-energy quasiparticle states. We will illustrate our argument with a specific model of coupled harmonic oscillators. Next we will consider a more complicated situation where the system is quenched through a second-order quantum phase transition . We will show how the universal scaling $n_{ex} \sim |\delta|^{d\nu/(z\nu+1)}$ [24, 25] emerges from combining adiabatic perturbation theory and universal scaling form of energies and matrix elements near the quantum critical point. We will also show how this scaling law breaks down and is substituted by a simple quadratic relation $n_{ex} \sim \delta^2$ when the exponent $d\nu/(z\nu+1)$ exceeds two. We will illustrate these results using specific exactly solvable models. We will also discuss the connection between adiabatic and sudden quenches near the quantum critical point. In particular, we will show that in low dimensions, $d\nu < 2$, the density of excited quasiparticles for a slow quench can be understood as a result of a sudden quench, if one correctly identifies the quench amplitude $\lambda^\star$ (for the sudden quench) with the quench rate δ (for the slow quench): $\lambda^\star \sim |\delta|^{1/(z\nu+1)}$. This analogy is very similar to the Kibble–Zurek picture of topological defect formation for quenches through classical phase transitions [41, 42], where one assumes that below certain energy (temperature) scale topological excitations essentially freeze. However at higher dimensions, $dv > 2$, this analogy becomes misleading since for sudden quenches scaling of n_{ex} is no longer determined by low-energy excitations. This work mostly focuses on the situation in which the system is initially in the ground state. In the end of this chapter we will discuss what happens if the system is initially prepared at finite temperature. We will argue that the statistics of low-energy quasiparticles strongly affects the response of the system to fast or slow quenches, enhancing the non-adiabatic effects (compared to the zero-temperature case) in the bosonic case and suppressing them in the fermionic case. We will discuss corrections to the universal scaling laws if the low-energy quasiparticles are described by either bosonic or fermionic statistics.

4.2 Adiabatic Perturbation Theory

We consider a very general setup where the system is described by the Hamiltonian $\mathcal{H}(t) = \mathcal{H}_0 + \lambda(t)V$, where $\mathcal{H}_0$ is the stationary part and $\lambda(t)V$ is the time-dependent part of the Hamiltonian. Our purpose is to characterize the dynamics of the system resulting from the time-dependent perturbation. We consider the situation in which the system is in a pure state. More general situations, where the state is mixed, can be addressed similarly by either solving von Neumann's equation or averaging solutions of the Schrödinger equation with respect to the initial density matrix. We assume that $\lambda(t)$ is a linear function of time:

$$\lambda(t) = \begin{cases} \lambda_i & t < 0, \\ \lambda_i + t\delta(\lambda_f - \lambda_i) & 0 \leq t \leq 1/\delta, \\ \lambda_f & t > 1/\delta. \end{cases} \tag{4.1}$$

Here δ is the rate of change of the parameter $\lambda(t)$: $\delta \rightarrow 0$ corresponds to the adiabatic limit, while $\delta \rightarrow \infty$ corresponds to the sudden quench . In principle the values λ_i and λ_f can be arbitrarily far from each other, therefore we cannot rely on the conventional perturbation theory in the difference between couplings $|\lambda_f - \lambda_i|$.

In the limit of slow parametric changes, we can use δ as a small parameter and find an approximate solution of the Schrödinger equation

$$i\partial_t |\psi\rangle = \mathcal{H}(t)|\psi\rangle, \tag{4.2}$$

where $|\psi\rangle$ is the wave function. In this chapter we will stick to the convention where $\hbar = 1$. This can be always achieved by rescaling either energy or time units. Our analysis will be similar to that of [43] but we will present here the details of the derivation for completeness. It is convenient to rewrite the Schrödinger equation (4.2) in the adiabatic (instantaneous) basis:

$$|\psi(t)\rangle = \sum_n a_n(t)|\phi_n(t)\rangle, \tag{4.3}$$

where $|\phi_n(t)\rangle$ are instantaneous eigenstates of the Hamiltonian $\mathcal{H}(t)$:

$$\mathcal{H}(t)|\phi_n(t)\rangle = E_n(t)|\phi_n(t)\rangle \tag{4.4}$$

corresponding to the instantaneous eigenvalues $E_n(t)$. These eigenstates implicitly depend on time through the coupling $\lambda(t)$. Substituting this expansion into the Schrödinger equation and multiplying it by $\langle\phi_m|$ (to shorten notations we drop the time label t in $|\phi_n\rangle$) we find:

$$i\partial_t a_m(t) + i\sum_n a_n(t)\langle\phi_m|\partial_t|\phi_n\rangle = E_m(t)a_m(t). \tag{4.5}$$

Next we will perform a gauge transformation:

$$a_n(t) = \alpha_n(t)\exp\left[-i\Theta_n(t)\right], \tag{4.6}$$

where

$$\Theta_n(t) = \int_{t_i}^{t} E_n(\tau)d\tau. \tag{4.7}$$

The lower limit of integration in the expression for $\Theta_n(t)$ is arbitrary. We choose it to be equal to t_i for convenience. Then the Schrödinger equation becomes

$$\dot{\alpha}_n(t) = -\sum_m \alpha_m(t)\langle n|\partial_t|m\rangle \exp\left[i(\Theta_n(t) - \Theta_m(t))\right]. \tag{4.8}$$

In turn this equation can be rewritten as an integral equation

$$\alpha_n(t) = -\int_{t_i}^{t} dt' \sum_m \alpha_m(t')\langle n|\partial_{t'}|m\rangle e^{i(\Theta_n(t') - \Theta_m(t'))}. \tag{4.9}$$

If the energy levels $E_n(\tau)$ and $E_m(\tau)$ are not degenerate, the matrix element $\langle n|\partial_t|m\rangle$ can be written as

$$\langle n|\partial_t|m\rangle = -\frac{\langle n|\partial_t\mathcal{H}|m\rangle}{E_n(t) - E_m(t)} = -\dot{\lambda}(t)\frac{\langle n|V|m\rangle}{E_n(t) - E_m(t)}, \tag{4.10}$$

where we emphasize that the eigenstates $|n\rangle$ and eigenenergies $E_n(t)$ are instantaneous. If $\lambda(t)$ is a monotonic function of time then in Eq. (4.9) one can change variables from t to $\lambda(t)$ and derive

$$\alpha_n(\lambda) = -\int_{\lambda_i}^{\lambda} d\lambda' \sum_m \alpha_m(\lambda')\langle n|\partial_{\lambda'}|m\rangle e^{i(\Theta_n(\lambda') - \Theta_m(\lambda'))}, \tag{4.11}$$

where

$$\Theta_n(\lambda) = \int_{\lambda_i}^{\lambda} d\lambda' \frac{E_n(\lambda')}{\dot{\lambda'}}. \tag{4.12}$$

Formally exact Eqs. (4.9) and (4.11) cannot be solved in the general case. However, they allow for the systematic expansion of the solution in the small parameter $\dot{\lambda}$. Indeed, in the limit $\dot{\lambda} \to 0$ all transition probabilities are suppressed because the phase factors are strongly oscillating functions of λ. The only exception to this statement occurs for degenerate energy levels [44], which we do not consider in this work. In the leading order in $\dot{\lambda}$ only the term with $m = n$ should be retained in the sums in Eqs. (4.9) and (4.11). This term results in the emergence of the Berry phase [45]:

$$\Phi_n(t) = -i \int_{t_i}^{t} dt' \langle n|\partial_{t'}|n\rangle = -i \int_{\lambda_i}^{\lambda(t)} d\lambda' \langle n|\partial_{\lambda'}|n\rangle, \tag{4.13}$$

so that

$$a_n(t) \approx a_n(0) \exp[-i\Phi_n(t)]. \tag{4.14}$$

In many situations, when we deal with real Hamiltonians the Berry phase is identically equal to zero. In general, the Berry phase can be incorporated into our formalism by doing a unitary transformation $\alpha_n(t) \to \alpha_n(t) \exp[-i\Phi_n(t)]$ and changing $\Theta_n \to \Theta_n + \Phi_n$ in Eqs. (4.9) and (4.11).

We now compute the first-order correction to the wave function assuming for simplicity that initially the system is in the pure state $n = 0$, so that $\alpha_0(0) = 1$ and $\alpha_n(0) = 0$ for $n \neq 0$. In the leading order in $\dot{\lambda}$ we can keep only one term with $m = 0$ in the sums in Eqs. (4.9) and (4.11) and derive

$$\alpha_n(t) \approx -\int_{t_i}^{t} dt' \langle n|\partial_{t'}|0\rangle e^{i(\Theta_n(t')-\Theta_0(t'))} \tag{4.15}$$

or alternatively

$$\alpha_n(\lambda) \approx -\int_{\lambda_i}^{\lambda} d\lambda' \langle n|\partial_{\lambda'}|0\rangle e^{i(\Theta_n(\lambda')-\Theta_0(\lambda'))}. \tag{4.16}$$

The transition probability from the level $|\phi_0\rangle$ to the level $|\phi_n\rangle$ as a result of the process is determined by $|\alpha_n(\lambda_f)|^2$.

Expression (4.16) can be further simplified in the case where the initial coupling λ_i is large and negative and the final coupling λ_f is large and positive, employing the stationary-phase approximation. The complex roots of the equation $E_n(\lambda^\star) - E_0(\lambda^\star) = 0$ define the stationary point. In turn the dominant contribution to the transition probability is determined by the negative imaginary part of the phase difference $\Theta_n - \Theta_0$ evaluated at these roots [44]:

$$|\alpha_n|^2 \propto \exp[-2\Im(\Theta_n(\lambda^\star) - \Theta_0(\lambda^\star))]. \tag{4.17}$$

In particular, for linearly changing coupling $\lambda(t) = \delta t$, we obtain

$$|\alpha_n|^2 \propto \exp\left(-\frac{2}{\delta}\Im\int^{\lambda^*}[E_n(\lambda') - E_0(\lambda')]d\lambda'\right) \quad (4.18)$$

and the transition probability exponentially vanishes as $\delta \to 0$.

There are many situations, however, when the coupling λ_i or λ_f or both are finite. In this case Eq. (4.17) is no longer valid and the asymptotic values of the integrals in Eqs. (4.15) and (4.16) are determined by the initial and finite times of evolution. Using the standard rules for evaluating the integrals of fast oscillating functions we find:

$$\alpha_n(t_f) \approx \left[i\frac{\langle\phi_n|\partial_t|\phi_0\rangle}{E_n(t) - E_0(t)} - \frac{1}{E_n(t) - E_0(t)}\frac{d}{dt}\frac{\langle\phi_n|\partial_t|\phi_0\rangle}{E_n(t) - E_0(t)} + \ldots\right]e^{i(\Theta_n(t) - \Theta_0(t))}\Big|_{t_i}^{t_f}$$

$$= \left[i\dot\lambda\frac{\langle\phi_n|\partial_\lambda|\phi_0\rangle}{E_n(\lambda) - E_0(\lambda)} - \ddot\lambda\frac{\langle\phi_n|\partial_\lambda|\phi_0\rangle}{(E_n(\lambda) - E_0(\lambda))^2} - \dot\lambda^2\frac{1}{E_n(\lambda) - E_0(\lambda)}\frac{d}{d\lambda}\frac{\langle\phi_n|\partial_\lambda|\phi_0\rangle}{E_n(\lambda) - E_0(\lambda)} + \ldots\right]$$

$$e^{i(\Theta_n(\lambda) - \Theta_0(\lambda))}\Big|_{\lambda_i}^{\lambda_f}. \quad (4.19)$$

In the following analysis we will retain only the first non-vanishing term in $\dot\lambda = \delta$. The terms proportional to higher powers of the expansion parameter, such as $\ddot\lambda$, $(\dot\lambda)^2$ as well as nonanalytic terms similar to Eq. (4.17), will be neglected assuming sufficiently small $\delta \to 0$.

The probability of the transition to the nth level is approximated by

$$|\alpha_n(\lambda_f)|^2 \approx \delta^2\left[\frac{|\langle\phi_n|\partial_{\lambda_i}|\phi_0\rangle|^2}{(E_n(\lambda_i) - E_0(\lambda_i))^2} + \frac{|\langle\phi_n|\partial_{\lambda_f}|\phi_0\rangle|^2}{(E_n(\lambda_f) - E_0(\lambda_f))^2}\right]$$

$$-2\delta^2\frac{\langle\phi_n|\partial_{\lambda_i}|\phi_0\rangle}{E_n(\lambda_i) - E_0(\lambda_i)}\frac{\langle\phi_n|\partial_{\lambda_f}|\phi_0\rangle}{E_n(\lambda_f) - E_0(\lambda_f)}\cos[\Delta\Theta_{n0}], \quad (4.20)$$

where $\Delta\Theta_{n0} = \Theta_n(\lambda_f) - \Theta_0(\lambda_f) - \Theta_n(\lambda_i) + \Theta_0(\lambda_i)$ is the phase difference between the states $|\phi_n\rangle$ and $|\phi_0\rangle$ accumulated during the time evolution. This phase difference is usually very large and thus the last term in Eq. (4.20) is a highly oscillating function, which can be typically dropped because of statistical or time averaging.

4.2.1 Application to the Landau–Zener Problem

We now apply the general formalism above to the notorious Landau–Zener (LZ) problem [39, 40]. The Hamiltonian is given by a 2×2 matrix, which is most conveniently expressed through the Pauli matrices:

$$\mathcal{H} = \lambda \sigma_z + g \sigma_x, \tag{4.21}$$

where

$$\sigma_z = \begin{bmatrix} 1 & 0 \\ 0 & -1 \end{bmatrix}; \ \sigma_x = \begin{bmatrix} 0 & 1 \\ 1 & 0 \end{bmatrix}. \tag{4.22}$$

This system has two eigenstates:

$$|-\rangle = \begin{pmatrix} \sin(\theta/2) \\ -\cos(\theta/2) \end{pmatrix}, \quad |+\rangle = \begin{pmatrix} \cos(\theta/2) \\ \sin(\theta/2) \end{pmatrix}, \tag{4.23}$$

where $\tan \theta = g/\lambda$, with corresponding energies $E_\pm = \pm\sqrt{\lambda^2 + g^2}$.

We assume that the coupling λ linearly changes in time: $\lambda = \delta t$. The system is initially (at $t = t_i$) prepared in the ground state and the process continues until $t = t_f$. In the limit $t_i \to -\infty$ and $t_f \to \infty$ the probability to occupy the excited state $|+\rangle$ [39, 40] is a non-analytic function of δ:

$$|a_+|^2 = \exp\left[-\frac{\pi g^2}{\delta}\right]. \tag{4.24}$$

However, in the general case, where t_i or t_f is finite, the probability contains both non-analytic and analytic contributions in δ.

In principle the LZ problem can be solved exactly for arbitrary t_i and t_f [46, 47] (see also Appendix), however, the general solution is quite cumbersome. Here we illustrate how asymptotical behavior of the transition probability at small δ can be recovered employing the adiabatic perturbation theory . The only non zero matrix element which enters Eq. (4.16) is

$$\langle +|\partial_t|-\rangle = \dot{\theta}/2 = -\frac{1}{2}\frac{\dot{\lambda}g}{\lambda^2 + g^2}. \tag{4.25}$$

We first apply Eq. (4.15) to the case $t_i \to -\infty$ and $t_f \to \infty$ which corresponds to the classic LZ problem. Then Eq. (4.16) gives

$$\alpha_+(\infty) \approx \frac{1}{2}\int\limits_{-\infty}^{\infty} dt \frac{\delta g}{g^2 + (\delta t)^2} \exp\left[2i\int_0^t d\tau\sqrt{(\delta\tau)^2 + g^2}\right]. \tag{4.26}$$

The asymptotic behavior of this expression at small $\delta \ll g^2$ can be derived by studying the analytic properties of the integrand in the complex plane of variable $t = t' + it''$. We notice that the phase factor has a branch-cut singularity in the upper and lower-half planes along the imaginary axis that starts at $t'' = \pm 1$ and go to infinity at $t'' = \pm\infty$. Deforming the contour of integration to include a singularity,

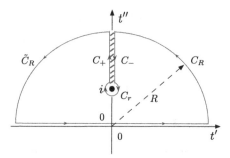

Fig. 4.1 The contour of integration in Eq. (4.26) in the complex t-plane. Integration over the real axis is given by several contributions: the one along C_r around the point $t = i$ that gives the first term in Eq. (4.27), C_+ and C_- from the two sides of the branch represented by second term in Eq. (4.27), and also C_R and $\tilde{C}_R$ that vanish in the limit $R \to \infty$

e.g., in the upper-half plane, we find that the integral has two contributions: one is provided by $t'' = 1$ point, and another one is given by a combination of two paths that run from the complex infinity to $t'' = 1$ and backward with the corresponding phase shift (Fig. 4.1). As a result, we obtain

$$\alpha_+(\infty) \approx \frac{\pi}{2} \exp\left[-\frac{\pi g^2}{2\delta}\right] \tag{4.27}$$
$$\left(1 - \frac{2}{\pi}\Im \int_0^\infty \frac{dx}{\sinh x} \exp\left[i\frac{g^2}{\delta}\left(x - \frac{1}{2}\sinh(2x)\right)\right]\right),$$

where we changed the variables $t\delta/g = i\cosh x$ to simplify the expressions. In the limit $\delta \ll g^2$, the integral in the brackets is a constant equal to $\pi/6$, which leads to the following asymptotic behavior of the transition probability:

$$|\alpha_+(\infty)|^2 \approx \frac{\pi^2}{9} \exp\left[-\frac{\pi g^2}{\delta}\right]. \tag{4.28}$$

This expression correctly reproduces the exponential dependence of the transition probability on the Landau–Zener parameter g^2/δ. However, the exponential prefactor is larger than the exact value of unity [46]. The reason for this discrepancy is that the exponential dependence is a result of very delicate interference of the transition amplitudes in time, which cannot be obtained within perturbative approach. Conversely, as we argue below, in the case of finite t_i or t_f the asymptotical behavior of the transition probability with δ is analytic allowing for the systematic treatment within the adiabatic perturbation theory.

Let us now turn to such situation with finite t_i or t_f or both (positive or negative) and sufficiently small δ. Then using Eq. (4.20) and ignoring the fast oscillating term we find

$$\alpha_+(t_f)|^2 \approx \frac{\delta^2}{16g^4} \left(\frac{g^6}{(g^2 + \lambda_i^2)^3} + \frac{g^6}{(g^2 + \lambda_f^2)^3} \right). \qquad (4.29)$$

In the case of $\lambda_i = -\infty$ and $\lambda_f = 0$, i.e., $t_i = -\infty$ and $t_f = 0$ this gives

$$|\alpha_+(t_f = 0)|^2 \approx \frac{\delta^2}{16g^4} \qquad (4.30)$$

as it can be found solving exactly the LZ problem (see Appendix). We would like to emphasize that in agreement with the general prediction (4.20) in the adiabatic limit the transition probability (4.29) is quadratic in the rate δ. We note that more accurate asymptotic includes additional exponential non-perturbative term (4.24) if λ_i and λ_f have the opposite signs [47]. With this additional term Eq. (4.29) will give the correct asymptotic for $|\alpha_+(t_f)|^2$ even when both $|\lambda_i|$ and $|\lambda_f|$ are large.

4.3 Adiabatic Dynamics in Gapless Systems with Quasiparticle Excitations

In the previous section we applied adiabatic perturbation theory to a two-level system with a time-dependent gap separating the eigenstates of the Hamiltonian. Now we are interested in extending the analysis to the case of a many-particle system. If we consider a situation in which the system (initially prepared in the ground state) is characterized by gapped quasiparticle excitations, then clearly the previous analysis applies to each mode independently. Then under slow quench different quasiparticle states can be excited. Since the transition probability to each quasiparticle state quadratically depends on the quench rate δ, we can expect that the density of created quasiparticles and other thermodynamic quantities will also quadratically depend on δ. The situation becomes more complicated and more interesting if we consider a system with gapless excitations. Then for the low-energy states adiabatic conditions are effectively always violated (unless we consider quench rates which go to zero with the system size) and in principle the adiabatic perturbation theory and the quadratic scaling can break down. Although the adiabatic perturbation theory in this case is no longer quantitatively correct, here we will show that it nevertheless can be very useful in finding the scaling of various quantities with the quench rate and finding the critical dimension above which the scaling becomes quadratic. Let us assume that we are dealing with a homogeneous system of quasiparticles characterized by a dispersion relation:

$$\varepsilon_k = c(\lambda)k^z, \qquad (4.31)$$

where z is the dynamical exponent and $c(\lambda)$ is a prefactor depending on the external parameter λ. We assume that λ changes linearly in time, $\lambda = \delta t$, between the initial and the final values λ_i and λ_f, respectively. The parameterization (4.31) explicitly demonstrates that during the evolution the details of the quasiparticle spectrum do not change, i.e., the system remains gapless and the exponent z stays the same. Another interesting possibility where the system crosses a singularity such as a critical point, which violates this assumption, will be considered in the next section. We also limit ourselves in this chapter to global uniform quenches, where the coupling λ is spatially independent.

In this section we consider the situation where the ground state corresponds to the state with no quasiparticles. Such situation naturally appears in a variety of physical systems, e.g., in bosonic systems with short-range interactions, in fermionic systems without Fermi surfaces such as gapless semiconductors [48], graphene [49] at zero voltage bias, some one-dimensional spin chains that can be mapped to systems of fermions using the Jordan–Wigner transformation [50, 51], system of hard-core bosons in one dimension (or Tonks gas) [51–55]. The situation with Fermi systems with the ground state corresponding to the filled Fermi sea requires special attention and will not be considered here. We do not expect any qualitative differences in the response to slow quenches in these Fermi systems.

Dimensional analysis allows us to estimate the scaling of various thermodynamic quantities with the quench rate. On the one hand, in a gapless system transitions to the low-energy states are unavoidable since any change in the coupling λ looks fast (diabatic) with respect to them. On the other hand, transitions to the high-energy states are suppressed for small δ because of the fast-oscillating phase factor entering the expression for the transition amplitude (4.16), so that the dynamics with respect to these states is adiabatic. It is straightforward to estimate the boundary separating diabatic and adiabatic states. Namely, if throughout the evolution $\dot{\varepsilon}_k(\lambda)$ becomes larger or comparable to $\varepsilon_k^2(\lambda)$, then the corresponding energy level is diabatic and quasiparticles are easily created. On the other hand if $\dot{\varepsilon}_k(\lambda) \ll \varepsilon_k^2(\lambda)$ in the whole interval of $\lambda \in [\lambda_i, \lambda_f]$ then the transitions are suppressed. Of course the implicit assumption here is that there is no kinematic constraint preventing creation of quasiparticles, i.e., that the matrix element for the transition is non-zero. In spatially uniform systems single quasiparticles typically cannot be created because of the momentum conservation, so that quasiparticles can be created only in pairs with opposite momenta. This implies that one should use $2\varepsilon_k(\lambda)$ instead of $\varepsilon_k(\lambda)$ in the argument, however, this extra factor of two is not important for the qualitative discussion. The crossover energy $\tilde{\varepsilon}_k(\lambda)$ is found by equating the two expressions:

$$\tilde{k}^z \frac{\partial c(\lambda)}{\partial \lambda} \delta \sim c(\lambda)^2 \tilde{k}^{2z} \tag{4.32}$$

or alternatively

$$\tilde{\varepsilon}_k(\lambda) \sim \delta \frac{\partial \ln c(\lambda)}{\partial \lambda}; \quad \tilde{k}^z \sim \delta \frac{1}{c(\lambda)} \frac{\partial \ln c(\lambda)}{\partial \lambda}. \tag{4.33}$$

More precisely one needs to find the maximal momentum satisfying the equation above in the whole interval of $\lambda \in [\lambda_i, \lambda_f]$. If there are no singularities during the evolution the derivative $\partial_\lambda \ln c(\lambda)$ remains non-zero and bounded. Thus we see that the crossover energy $\tilde{\varepsilon}_k(\lambda)$ scales linearly with the rate δ. Then we can estimate, for example, the total number of quasiparticles created during the process as

$$n_{\text{ex}}(\delta) \sim \int_0^{\tilde{\varepsilon}} \rho(\varepsilon)d\varepsilon = \int_{k \leq \tilde{k}} \frac{d^d k}{(2\pi)^d} \propto |\delta|^{d/z}. \tag{4.34}$$

We expect this scaling to be valid only when $d/z \leq 2$. Otherwise (as we will argue below) the contribution from high-energy quasiparticles with $\varepsilon > \tilde{\varepsilon}$ will dominate, resulting in the quadratic scaling of n_{ex}.

The result (4.34) can be derived more accurately using the adiabatic perturbation theory . Namely, let us perform the scaling analysis of Eq. (4.16). Because quasiparticles can be typically created only in pairs with opposite momenta one should use twice the quasiparticle energy in the dynamical phase Θ_n in Eq. (4.16). The number of quasiparticles should be also multiplied by two, but at the same time the sum over momenta should only go over half the available states to avoid double counting of pairs with momenta k and $-k$. So this second factor of two can be absorbed extending integration over momenta to the whole spectrum. Thus we get

$$n_{\text{ex}} \approx \frac{1}{L^d} \sum_k |\alpha_k(\lambda_f)|^2 = \int \frac{d^d k}{(2\pi)^d}$$
$$\left| \int_{\lambda_i}^{\lambda_f} d\lambda \, \langle k|\partial_\lambda|0\rangle \exp\left[\frac{2ik^z}{\delta} \int_{\lambda_i}^{\lambda} c(\lambda')d\lambda' \right] \right|^2. \tag{4.35}$$

Rescaling the momentum k as $k = \delta^{1/z}\eta$ we find

$$n_{\text{ex}} \approx |\delta|^{\frac{d}{z}} \int \frac{d^d \eta}{(2\pi)^d} \left| \int_{\lambda_i}^{\lambda_f} d\lambda \, \langle \eta\delta^{1/z}|\partial_\lambda|0\rangle e^{2i\eta^z \int_{\lambda_i}^{\lambda} c(\lambda')d\lambda'} \right|^2. \tag{4.36}$$

If the integral over η converges at large η and the matrix element $\langle k|\partial_\lambda|0\rangle$ goes to a non-zero constant as $k \to 0$, then we are getting the desired scaling. The second condition means that there is no kinematic suppression of the transitions to the low-momentum low-energy states. Absence of this suppression was implicitly assumed in the elementary derivation of Eq. (4.34). The first condition of convergence of the integral in Eq. (4.36) over η implies that only low-energy modes contribute to the total number of generated quasiparticles since $k \sim \eta\delta^{1/z}$. This happens only in the case where $d/z < 2$, otherwise the quadratic scaling $n_{\text{ex}} \sim \delta^2$, which we derived in the previous section, coming from excitations to all energy scales, is restored.

Technically the crossover can be seen from the fact that the transition probability to the state with momentum k at large $k \gg \delta^{1/z}$ corresponding to $\eta \gg 1$ scales as $1/\eta^{2z}$. This follows from combining Eqs. (4.20) and (4.31). Clearly then only if $d \leq 2z$ the integral over η in Eq. (4.36) converges at large η. Therefore after the rescaling $k = \delta^{1/z}\eta$ the upper limit of integration over η can be sent to infinity. Otherwise the quasiparticle excitations with large momenta independent of δ dominate n_{ex} and we obtain the quadratic scaling according to the general result (4.20).

One can check that under the same conditions the excess energy or heat [56] per unit volume in the system scales as

$$
Q \approx \frac{1}{L^d} \sum_k \varepsilon_k(\lambda_f)|\alpha_k(\lambda_f)|^2 = |\delta|^{\frac{d+z}{z}} c(\lambda_f)
$$

$$
\times \int \frac{d^d \eta}{(2\pi)^d} \eta^z \left| \int_{\lambda_i}^{\lambda_f} d\lambda \, \langle \eta \delta^{1/z}|\partial_\lambda|0 \rangle e^{2i\eta^z \int_{\lambda_i}^\lambda c(\lambda')d\lambda'} \right|^2 .
\tag{4.37}
$$

This scaling is now valid provided that $(d + z)/z \leq 2$. Otherwise the energy absorption comes from high-energy states with $k \gg \delta^{1/z}$ and in this case $Q \propto \delta^2$.

Since we assume that quasiparticles are created independently in different channels, i.e., the probability to create a pair of quasiparticles with momentum k is uncorrelated with the probability to create a pair of quasiparticles with momentum $k' \neq k$, we can easily find the scaling of the (diagonal) entropy density of the system [R. Barankov and A. Polkovnikov, unpublished]:

$$
S_d \approx -\frac{1}{2L^d} \sum_k |\alpha_k(\lambda_f)|^2 \ln |\alpha_k(\lambda_f)|^2,
\tag{4.38}
$$

where the factor of $1/2$ comes from the fact that we need to account each state characterized by momenta q and $-q$ once and sum only over half of the momentum modes to avoid over-counting. As before one can instead sum over all momentum modes but multiply the result by $1/2$. In the domain of validity of the adiabatic perturbation theory $|\alpha_k(\lambda_f)|^2 \ll 1$, the expression for the entropy is very similar to the expression for the number of quasiparticles. The extra logarithm clearly does not affect the scaling with δ and we thus expect that $S_d \propto |\delta|^{d/z}$ for $d/z \leq 2$.

Example: harmonic chain. Let us consider a specific example from [28], where one slowly changes mass of particles in a harmonic chain. Specifically we consider the following Hamiltonian:

$$
\mathcal{H} = \sum_k \frac{\rho_s k^2}{2}|x_k|^2 + \frac{\kappa(t)}{2}|p_k|^2,
\tag{4.39}
$$

where x_k and p_k are conjugate coordinate and momentum. We will assume that κ, playing the role of inverse mass, linearly increases in time: $\kappa(t) = \kappa_i + \delta t$. For simplicity we will also assume that $\kappa_f \gg \kappa_i$. In [28] a more complicated situation with κ dependent on k was analyzed. However, this k-dependence is only important if $\kappa_i \to 0$ corresponding to a singularity in the spectrum. In this section we are interested in non-singular situations so we assume that κ_i is finite and the extra possible dependence of κ on k is irrelevant. The problem was analyzed both exactly and perturbatively in [28]. Here, we briefly repeat the analysis and present additional results. Within the adiabatic perturbation theory changing κ in time generates pairs of quasiparticle excitations with opposite momenta. It is straightforward to check that the matrix element $\langle k, -k | \partial_\kappa | 0 \rangle = 1/(4\sqrt{2}\kappa)$ is independent of k satisfying the second requirement (see the sentence after Eq. (4.36)) necessary to get the correct scaling of excitations. Substituting this matrix element together with the dispersion $\varepsilon_k = k\sqrt{\kappa\rho_s}$ into Eq. (4.16) we find that within the perturbation theory the probability to excite a pair of quasiparticles with momenta k and $-k$ is

$$
\begin{aligned}
|\alpha_k|^2 &\approx \frac{1}{32} \left| \int_{\kappa_i}^{\kappa_f} \frac{d\kappa}{\kappa} \exp\left[\frac{4i}{3\delta} k\sqrt{\rho_s}\kappa^{3/2} \right] \right|^2 \\
&\approx \frac{1}{72} \left| \Gamma\left(0, -i\frac{4}{3}\frac{k\kappa_i\sqrt{\rho_s\kappa_i}}{\delta} \right) \right|^2,
\end{aligned}
\tag{4.40}
$$

where $\Gamma(0, x)$ is the incomplete Γ-function. In the equation above we used the fact that $\kappa_f \gg \kappa_i$ so the upper limit in integration over κ can be effectively extended to infinity. The expression depends on the single dimensionless parameter

$$
\xi_k = \frac{k}{k_\delta},
$$

where

$$
k_\delta = \delta / \sqrt{\kappa_i^3 \rho_s}.
$$

When $\xi_k \gg 1$, which corresponds to the high momentum modes, we have

$$
|\alpha_k|^2 \approx \frac{1}{128\xi_k^2} = \frac{1}{128}\frac{k_\delta^2}{k^2}
\tag{4.41}
$$

and in the opposite limit Eq. (4.40) gives

$$
|\alpha_k|^2 \approx \frac{1}{72} |\ln \xi_k|^2.
\tag{4.42}
$$

Hence at high energies the transition probability is proportional to δ^2 as expected from the discussion in the previous section, while at small momenta $k \to 0$ the transition probability diverges. This is of course unphysical and indicates the breakdown of the perturbation theory. Note, however, that because of the small prefactor $1/72$ this divergence occurs at very small values of ξ_k.

This problem can also be solved exactly (see details in [28]). The initial ground state wave function can be written as a product of Gaussians:

$$\Psi_i(\{x_k\}) = \prod_k \frac{1}{(2\pi\sigma_k^0(\kappa_i))^{1/4}} \exp\left[-\frac{|x_k|^2}{4\sigma_k^0(\kappa_i)}\right], \tag{4.43}$$

where

$$\sigma_k^0(\kappa) = \frac{1}{2k}\sqrt{\frac{\kappa}{\rho_s}}.$$

One particularly useful property of Gaussian functions is that for arbitrary time dependence of $\kappa(t)$ (or $\rho_s(t)$) the wave function remains Gaussian with σ_k satisfying the first-order differential equation:

$$i\frac{d\sigma_k(t)}{dt} = 2\rho_s k^2\sigma_k^2(t) - \frac{1}{2}\kappa(t). \tag{4.44}$$

The solution of this equation which satisfies the proper initial condition can be expressed through the Airy functions . It is convenient to introduce the effective width of the wave function at time t_f defining the transition probability

$$\sigma_k^\star = \frac{1}{\Re(\sigma_k^{-1})}.$$

Then it can be shown that [28]

$$\frac{\sigma_k^\star}{\sigma_k^0} = \frac{1 + |f_k|^2}{2\Im f_k}, \tag{4.45}$$

where

$$f_k = -\frac{\sqrt[3]{\xi_k}\,\mathrm{Bi}(-\xi_k^{2/3}) - i\mathrm{Bi}'(-\xi_k^{2/3})}{\sqrt[3]{\xi_k}\,\mathrm{Ai}(-\xi_k^{2/3}) - i\mathrm{Ai}'(-\xi_k^{2/3})}. \tag{4.46}$$

As it occurs in the perturbative treatment, the solution is expressed entirely through the single parameter ξ_k. It is easy to show that the average number of excited particles with momenta k and $-k$ in the Gaussian state characterized by the width $\sigma_k^\star$ is

$$n_k = \frac{1}{2}\left[\frac{\sigma_k^\star}{\sigma_k^0} - 1\right]. \tag{4.47}$$

This expression has the following asymptotics: for $\xi_k \gg 1$

$$n_k \approx \frac{1}{64\xi_k^2} \tag{4.48}$$

and for $\xi_k \ll 1$

$$n_k \approx \frac{\pi}{3^{2/3}\Gamma^2(1/3)}\frac{1}{\sqrt[3]{\xi_k}}. \tag{4.49}$$

For large momenta the exact asymptotic clearly coincides with the approximate one (4.41) (noting that $n_k = 2|\alpha_k|^2$) while at low energies as we anticipated the adiabatic perturbation theory fails. The perturbative result clearly underestimates the number of excitations. This fact is hardly surprising because the adiabatic perturbation theory neglects the tendency of bosonic excitations to bunch leading to the enhancement of the transitions. We can anticipate that the result will be opposite in the fermionic case and in the next section we will see that this is indeed the case.

Density of quasiparticles. The density of quasiparticles created during the process can be obtained by summing n_k over all momenta:

$$n_{\text{ex}} = \frac{1}{2}\int\frac{d^d k}{(2\pi)^d}n_k. \tag{4.50}$$

We remind that the factor of $1/2$ is inserted to avoid double counting of pairs with momenta k and $-k$. In low dimensions $d \leq 2$ this sum is dominated by low momenta $k \sim k_\delta$ and the upper limit can be send to ∞:

$$n_{\text{ex}} \approx \frac{k_\delta^d}{2}\int\frac{d^d\xi}{(2\pi)^d}n_k(\xi). \tag{4.51}$$

This gives the correct scaling $n_{\text{ex}} \sim \delta^{d/z}$ anticipated from the general argument since $z = 1$. On the other hand, in the above two dimensions the integral over k is dominated by high momenta close to the high-energy cutoff and we have

$$n_{\text{ex}} \approx \frac{1}{128}k_\delta^2\int\frac{d^d k}{(2\pi)^d}\frac{1}{k^2}. \tag{4.52}$$

This result again confirms our expectations that the number of excited quasiparticles scales as δ^2 in high dimensions. Note that when $d = 2$ the integral above is still valid but it should be cutoff at small $k \sim k_\delta$ leading to the additional log-dependence of n_{ex} on δ. We point that the perturbative analysis predicts very similar qualitative picture: it correctly predicts the density of excitations in high dimensions $d \geq 2$ and gives the correct scaling in low dimensions $d < 2$. However, in the latter case the perturbative analysis gives a wrong prefactor. We note that there are situations when the adiabatic perturbation theory can fail completely predicting even incorrect scaling [28]. This can happen, for example, if the initial coupling κ_0 is very small. Then the integral (4.50) can become infrared divergent at small momenta and should be cutoff by the inverse system size or another large spatial scale (e.g., mean free path).

The smallest (and the only) physical dimension where the scaling $n_{ex} \propto |\delta|^{d/z}$ is valid in our situation is $d = 1$. Accurate evaluation of the quasiparticle density in this case gives

$$n_{ex} \approx 0.0115 k_\delta. \tag{4.53}$$

For completeness we also quote the perturbative result obtained by integrating Eq. (4.40):

$$n_{ex}^{pert} \approx 0.0104 k_\delta. \tag{4.54}$$

Clearly the difference between the exact and perturbative results is very minor in this case.

In two dimensions both perturbative and exact treatments give the same result

$$n_{ex} \approx \frac{1}{256\pi} k_\delta^2 \ln\left(\frac{\Lambda}{k_\delta}\right), \tag{4.55}$$

where Λ is the short distance cutoff. At higher dimensions, as we mentioned above, results of perturbative and exact treatments are identical and non-universal:

$$n_{ex} \approx C k_\delta^2 \Lambda^{d-2}. \tag{4.56}$$

It is interesting to note that adiabatic conditions are determined now by the high momentum cutoff: $\delta \sim 1/\Lambda^{d/2-1}$. This sensitivity to Λ might be an artifact of sudden change in the rate of change of the parameter κ resulting in the infinite acceleration $\ddot{\lambda}$ at initial (and final) times. However, analyzing this issue in detail is beyond the scope of this work.

Heat (excess energy). Quasiparticle density is not always an easily detectable quantity, especially if the system is interacting and the number of quasiparticles is not conserved. A more physical quantity is the energy change in the system during the process, which is equal to the external work required to change the coupling κ. This energy consists of two parts: adiabatic, which is related to the dependence of the ground state energy on κ, and heat, i.e., the additional energy pumped into the system due to excitations of higher energy levels. The first adiabatic contribution depends only on the initial and final couplings κ_i and κ_f but not on the details of the process. On the contrary, the heat (per unit volume) Q is directly related to the rate δ. The microscopic expression for Q can be obtained by a simple generalization of Eq. (4.50), as shown in Eq. (4.37):

$$Q = \frac{\sqrt{\kappa_f \rho_s}}{2} \int \frac{d^d k}{(2\pi)^d} \, k \, n_k. \tag{4.57}$$

This integral converges at large k only for $d < 1$. Therefore in all physical dimensions it is dominated by the high-energy asymptotic (4.41) so $Q \propto \delta^2$. In one dimension there is an extra logarithmic correction

$$Q_{d=1} \approx \frac{k_\delta^2}{256\pi} \sqrt{\kappa_f \rho_s} \ln \left(\frac{\Lambda}{k_\delta} \right). \tag{4.58}$$

Entropy. Finally let us evaluate the generated diagonal entropy in the system. The latter is formally defined as (see [R. Barankov and A. Polkovnikov, unpublished])

$$S_d = -\frac{1}{L^d} \sum_n \rho_{nn} \log \rho_{nn}, \tag{4.59}$$

where ρ_{nn} are the diagonal elements of the density matrix in the eigenbasis of the (final) Hamiltonian. At finite temperatures it is this entropy which is connected to heat via the standard thermodynamic relation $\Delta Q = T \Delta S_d$. However, at zero initial temperature this relation breaks down and one should analyze $S_d \equiv \Delta S_d$ separately.

Since in our problem different momentum states are decoupled the d-entropy is additive:

$$S_d = \frac{1}{2L^d} \sum_k s_k = \frac{1}{2} \int \frac{d^d k}{(2\pi)^d} s_k, \tag{4.60}$$

where the factor of $1/2$ is again coming in order to avoid double counting of quasiparticle excitations created in pairs. Within the adiabatic perturbation theory only one pair of quasiparticles can be excited. Therefore $s_k^{\text{pert}} \approx -|\alpha_k|^2 \ln |\alpha_k|^2$. Note that at large k the entropy per mode falls off with k essentially in the same manner as $|\alpha_k|^2$ so we conclude that the entropy is dominated by small energies below two dimensions. Then we find

$$S_d^{\text{pert}} \approx -\frac{1}{2} k_\delta^d \int \frac{d^d \xi}{(2\pi)^d} |\alpha(\xi)|^2 \ln |\alpha(\xi)|^2. \tag{4.61}$$

In dimensions higher than two the entropy is dominated by high momenta so

$$S_d^{\text{pert}} \approx -\frac{1}{256}k_\delta^2 \int \frac{d^d k}{(2\pi)^d}\frac{1}{k^2} \ln\left(\frac{k_\delta^2}{128k^2}\right). \tag{4.62}$$

In one dimension Eq. (4.61) gives

$$S_{d=1}^{\text{pert}} \approx 0.02k_\delta. \tag{4.63}$$

To calculate the exact value of the entropy density one needs to project the Gaussian wave function describing each momentum mode to the eigenbasis and calculate the sum (4.59). There is, however, a simple shortcut allowing to instantly write the answer. One can easily check that the probabilities to occupy different eigenstates are identical to those of the equilibrium thermal ensemble. Therefore the entropy per mode can be expressed through the average number of excited quasiparticle pairs $n_k/2$ as

$$s_k = -(n_k/2)\ln(n_k/2) + (1 + n_k/2)\ln(1 + n_k/2). \tag{4.64}$$

Using explicit solution for n_k we find

$$S_{d=1} \approx 0.026k_\delta. \tag{4.65}$$

The result is again quite close to the perturbative.

In two dimensions the entropy density is also readily available from the expressions above:

$$S_{d=2} \approx \frac{k_\delta^2}{512\pi}\left[\ln(128e)\ln\left(\frac{\Lambda}{k_\delta}\right) + \ln^2\left(\frac{\Lambda}{k_\delta}\right)\right]. \tag{4.66}$$

4.4 Adiabatic Dynamics Near a Quantum Critical Point

Let us apply the analysis of the previous section to the case of crossing a quantum critical point. As before we consider the situation in which the system is prepared in the ground state, characterized by some initial coupling λ_i, then this coupling is linearly tuned in time to a finite value λ_f. We assume that the system undergoes a second-order quantum phase transition at $\lambda = 0$. We differentiate the two situations: (i) when λ_i is finite and negative and λ_f is finite and positive and (ii) when either $\lambda_i = 0$ and λ_f is large and positive, or $\lambda_f = 0$ and λ_i is large and negative. As in the previous section we will first give general discussion closely following [24] and then analyze specific examples.

4.4.1 Scaling Analysis

Non-adiabatic effects are especially pronounced in the vicinity of the quantum critical point (QCP), where one can expect universality in the transition rates. Near QCP for $\lambda \neq 0$ the system develops a characteristic energy scale Δ, vanishing at the critical point as $\Delta \sim |\lambda|^{z\nu}$, where z is the dynamical exponent and ν is the critical exponent of the correlation length, $\xi \sim |\lambda|^{-\nu}$ [51]. This energy scale can be either a gap or some crossover scale at which the energy spectrum qualitatively changes.

As before, let us perform the scaling analysis of Eq. (4.16) assuming that quasiparticles are created in pairs of opposite momenta. We first rewrite the dynamical phase factor entering the expression for the transition amplitude (4.16) as

$$\Theta_k(\lambda) - \Theta_0(\lambda) = \frac{1}{\delta} \int_{\lambda_i}^{\lambda} d\lambda' (\varepsilon_k(\lambda') - \varepsilon_0(\lambda')). \tag{4.67}$$

Near the QCP the quasiparticle energy can be rewritten using the scaling function F as

$$\varepsilon_k(\lambda) - \varepsilon_0(\lambda) = \lambda^{z\nu} F(k/\lambda^{\nu}). \tag{4.68}$$

For $x \gg 1$ the function $F(x)$ should have asymptotic $F(x) \propto x^z$, reflecting the fact that at large momenta the energy spectrum should be insensitive to λ. At small x, corresponding to small k, the asymptotical behavior of the scaling function $F(x)$ depends on whether the system away from the singularity is gapped, then $F(x) \rightarrow$ const at $x \rightarrow 0$, or gapless, then $F(x) \propto x^{\alpha}$ with some positive exponent α. The scaling Eq. (4.68), inserted in Eq. (4.67), suggests change of variables

$$\lambda = \xi \delta^{\frac{1}{z\nu+1}}, \quad k = \eta \delta^{\frac{\nu}{z\nu+1}}. \tag{4.69}$$

Furthermore to analyze Eq. (4.16) we introduce the scaling ansatz for the matrix element

$$\langle k|\partial_{\lambda}|0\rangle = -\frac{\langle k|V|0\rangle}{\varepsilon_k(\lambda) - \varepsilon_0(\lambda)} = \frac{1}{\lambda} G(k/\lambda^{\nu}), \tag{4.70}$$

where $G(x)$ is another scaling function. This scaling dimension of the matrix element, which is the same as the engineering dimension $1/\lambda$, follows from the fact that the ratio of the two energies $\langle k|\lambda V|0\rangle$ and $\varepsilon_k(\lambda) - \varepsilon_0(\lambda)$ should be a dimensionless quantity. At large momenta $k \gg \lambda^{\nu}$ this matrix element should be independent of λ so $G(x) \propto x^{-1/\nu}$ at $x \gg 1$. This statement must be true as long as the matrix element $\langle k|V|0\rangle$ is non-zero at the critical point, which is typically the case. In the opposite limit $x \ll 1$ we expect that $G(x) \propto x^{\beta}$, where β is some non-negative number.

These two scaling assumptions allow one to make general conclusions on behavior of the density of quasiparticles and other thermodynamic quantities with the quench rate for the adiabatic passage through a quantum critical point . Thus

$$n_{\text{ex}} \sim \int \frac{d^d k}{(2\pi)^d} |\alpha_k|^2 = |\delta|^{\frac{dv}{zv+1}} \int \frac{d^d \eta}{(2\pi)^d} |\alpha(\eta)|^2, \tag{4.71}$$

where, after the rescaling (4.69),

$$\alpha(\eta) = \int_{\xi_i}^{\xi_f} d\xi \frac{1}{\xi} G\left(\frac{\eta}{\xi^v}\right) \exp\left[i \int_{\xi_i}^{\xi} d\xi_1 \xi_1^{zv} F(\eta/\xi_1^v)\right]. \tag{4.72}$$

Note that if $\lambda_i < 0$ and $\lambda_f > 0$ then ξ_i is large and negative and ξ_f is large and positive. If we start (end) exactly at the critical point then $\xi_i = 0$ ($\xi_f = 0$). Note that the integral over ξ is always convergent because: at large ξ we are dealing with a fast oscillating function, and at $\xi \sim 0$ there are no singularities because of the scaling properties of $G(x)$. The only issue which has to be checked carefully is the convergence of the integral over η at large η. If this integral converges then Eq. (4.71) gives the correct scaling relation for the density of quasiparticles with the rate δ found in earlier works [24, 25]. If the integral does not converge then the density of created defects is dominated by high energies and from general arguments of Sect. 4.2 we expect that $n_{\text{ex}} \propto \delta^2$. As it is evident from Eq. (4.71), the crossover between these two regimes happens at the critical dimension $d_c = 2z + 2/v$.[1] To see how this critical dimension emerges from Eq. (4.71) we analyze the asymptotical behavior of $\alpha(\eta)$ at $\eta \gg \xi^v$. In this limit the integral over ξ can be evaluated in accord with the discussion given in Sect. 4.2 because the exponent in Eq. (4.72) is a rapidly oscillating function of ξ. Using explicit asymptotics of scaling functions $F(x)$ and $G(x)$ at large x we find

$$\alpha(\eta) \propto \frac{1}{\eta^{z+1/v}}. \tag{4.73}$$

Then the integral over η in Eq. (4.71) converges at large η precisely when $d \leq d_c = 2z + 2/v$. This indicates that the scaling $n_{\text{ex}} \propto |\delta|^{dv/(zv+1)}$ breaks down for $d > d_c$ and instead we get $n_{\text{ex}} \propto \delta^2$. As in the previous section we can expect logarithmic corrections at $d = d_c$. Let us note that especially in low dimensions the exponent v itself can depend on the dimensionality, so d_c can be a function of d. The statement about critical dimension should be rather understood in the sense that: if $d > d_c(d)$ then we expect quadratic scaling of n_{ex}, while in the opposite case we expect the scaling given by Eq. (4.71) to be valid.

One can similarly analyze dependence of heat Q and diagonal entropy S_d on the rate δ. The entropy has the same scaling as the number of quasiparticles . We note that in general the heat is universal only if the process ends at the critical point $\lambda_f = 0$, where $\epsilon_k \propto k^z$. Then it is easy to see that

---

[1] We note that there is a small error in Ref. [24], which gives a different expression for d_c.

$$Q \propto |\delta|^{\frac{(d+z)\nu}{z\nu+1}}. \tag{4.74}$$

This scaling is valid for $d \leq \tilde{d}_c = z + 2/\nu$, which is different from d_c just introduced for n_{ex}. Above $\tilde{d}_c$ the scaling of Q with δ becomes quadratic. If the final coupling λ_f is away from the critical point then the δ-dependence of Q becomes sensitive to the behavior of the spectrum. Thus if there is a gap in the spectrum, Q has the same scaling as n_{ex} and S_d, since each excitation roughly carries the same energy equal to the gap. If the spectrum at λ_f is gapless then the scaling (4.74) remains valid.

4.4.2 Examples

Here we will consider several specific models illustrating the general predictions above. In particular we will analyze the transverse field Ising model, which serves as a canonical example of quantum phase transitions [51] and which was used as an original example where the general scaling (4.71) was tested [24–26]. The transverse field Ising model also maps to the problem of loading one-dimensional hard-core bosons or non-interacting fermions into a commensurate optical lattice potential [32] and describes the so-called Toulouse point in the sine-Gordon model, where this model can be mapped to free spinless fermions [50].

4.4.2.1 Transverse Field Ising and Related Models

The transverse field Ising model is described by the following Hamiltonian:

$$\mathcal{H}_I = -\sum_j \left[g(t)\sigma_j^x + \sigma_j^z \sigma_{j+1}^z \right]. \tag{4.75}$$

For simplicity we will focus only on the domain of non-negative values of the transverse field g. This model undergoes a quantum phase transition at $g = 1$ [51] with $g \geq 1$ corresponding to the transversely magnetized phase and $g \leq 1$ corresponding to the phase with longitudinal magnetization. It is thus convenient to use $\lambda(t) = g(t) - 1$ as the tuning parameter. Under the Jordan–Wigner transformation:

$$\sigma_j^z = -(c_j + c_j^\dagger) \prod_{i<j}(1 - 2c_i^\dagger c_i), \quad \sigma_j^x = 1 - 2c_j^\dagger c_j, \tag{4.76}$$

the Hamiltonian assumes the quadratic form:

$$\mathcal{H}_I = -\sum_j c_j^\dagger c_{j+1} + c_{j+1}^\dagger c_j + c_j^\dagger c_{j+1}^\dagger + c_{j+1}c_j - 2g(t)c_j^\dagger c_j \tag{4.77}$$

and can be diagonalized using the Bogoliubov transformation in the momentum space:

$$c_k = \gamma_k \cos(\theta_k/2) + i \sin(\theta_k/2)\gamma^\dagger_{-k},$$ (4.78)

where

$$\tan \theta_k = \frac{\sin(k)}{\cos(k) - g(t)}.$$ (4.79)

After this transformation the Hamiltonian becomes

$$\mathcal{H}_I = \sum_k \varepsilon_k \gamma^\dagger_k \gamma_k,$$ (4.80)

where

$$\varepsilon_k = 2\sqrt{1 + g^2 - 2g\cos(k)}.$$ (4.81)

The ground state of this Hamiltonian factorizes into the product:

$$|\Omega_0\rangle = \prod_k \left(\cos(\theta_k/2) + i\sin(\theta_k/2)c^\dagger_k c^\dagger_{-k}\right)|0\rangle.$$ (4.82)

The excited states can be obtained by applying various combinations of operators $\gamma^\dagger_k$ to the ground state above. However, because of the momentum conservation, only the excited states obtained by acting on the ground state by the products $\gamma^\dagger_k \gamma^\dagger_{-k}$ are relevant. Because excitations to different momentum states are independent, the problem effectively splits into a sum of independent Landau–Zener problems [26] and can be exactly solved. In the case when g_i and g_f lie on the opposite sides of the quantum critical point the transition probability in the slow limit is approximately given by [26]

$$p_k \approx \exp\left[-\frac{2\pi k^2}{\delta}\right].$$ (4.83)

The density of the excited quasiparticles is then

$$n_{ex} = \frac{1}{2\pi} \int_{-\infty}^{\infty} p_k dk \approx \frac{\sqrt{\delta}}{2\pi\sqrt{2}} \approx 0.11\sqrt{\delta}.$$ (4.84)

Similarly one can find the entropy density generated during the process

$$S_d \approx -\frac{1}{4\pi} \int_{-\infty}^{\infty} dk \left[p_k \ln p_k + (1 - p_k)\ln(1 - p_k)\right]$$
$$\approx 0.052\sqrt{\delta}.$$ (4.85)

Both expressions for n_{ex} and S_d agree with the general scaling law (4.71) with $d = z = \nu = 1$. Since in the final state all excitations are gapped the heat in this case is approximately equal to the number of excited quasiparticles multiplied by the gap in the final state and thus has the same scaling as n_{ex}.

The same problem can be also solved using the adiabatic perturbation theory. Let us note that as it follows from Eq. (4.82) the transition matrix element reads [24]

$$\langle k, -k|\partial_\lambda|0\rangle = \langle k, -k|\partial_g|0\rangle = \frac{i}{2}\frac{\sin k}{1 + g^2 - 2g\cos(k)}. \tag{4.86}$$

In the limit of small δ only the transitions happening in the vicinity of the critical point contribute to n_{ex}. In this case we have

$$\langle k, -k|\partial_\lambda|0\rangle \approx \frac{i}{2}\frac{k}{\lambda^2 + k^2}, \tag{4.87}$$

which clearly satisfies the scaling (4.70). Under the same approximation we can use that $\varepsilon_k \approx 2\sqrt{k^2 + \lambda^2}$, which in turn satisfies the energy scaling (4.68). Substituting these expansions into Eq. (4.16) and extending the limits of integration over λ to $(-\infty, \infty)$ we find that

$$\alpha_k \approx -\frac{i}{2}\int_{-\infty}^{\infty} d\lambda \frac{k}{\lambda^2 + k^2} \exp\left[\frac{4i}{\delta}\int_0^\lambda d\lambda'\sqrt{k^2 + \lambda'^2}\right], \tag{4.88}$$

where the additional factor of two in the exponent comes from the fact that two quasiparticles are created during each transition. It is now straightforward to evaluate the perturbative expression for n_{ex}^{pert}:

$$n_{ex}^{\text{pert}} \approx \int_{-\infty}^{\infty} \frac{dk}{2\pi}|\alpha_k|^2 \approx 0.21\sqrt{\delta}. \tag{4.89}$$

Note that unlike the bosonic case discussed in the previous section, the adiabatic perturbation theory now overestimates the number of created quasiparticles. This happens because this perturbation theory does not take into account Pauli blocking which prevents more than one pair of quasiparticles with momenta $k, -k$ to be excited. Technically the perturbative transition probability $|\alpha_k|^2$ can exceed unity. This prevents one from computing the generated entropy using Eq. (4.38) because there will be spurious negative contributions. To get a sensible expression one needs to integrate only over the momenta satisfying $|\alpha_k|^2 \leq 1$. Then one finds

$$S_d^{\text{pert}} \approx -\frac{1}{2\pi}\int_{k_{min}}^{\infty} dk|\alpha_k|^2 \ln|\alpha_k|^2 \approx 0.022\sqrt{\delta}. \tag{4.90}$$

One can somewhat improve the perturbative argument explicitly using the fact that the quasiparticles are fermions and adding an additional contribution coming from $(1 - |\alpha_k|^2)\ln(1 - |\alpha_k|^2)$. In this case $S_d^{\text{pert}} \approx 0.038\sqrt{\delta}$, which is closer to the exact result.

In a similar spirit we can consider the situation when either $\lambda_i = 0$ or $\lambda_f = 0$, i.e., when the initial or final state exactly corresponds to the quantum critical point.

This situation requires fine tuning from the point of view of crossing a quantum phase transition . However, it naturally appears in other contexts. For example, the problem of loading hard-core bosons into commensurate periodic potential exactly describes this situation [32]. Since we consider $\lambda \in (0, \infty)$ then Eq. (4.88) becomes

$$\alpha_k \approx -\frac{i}{2} \int_0^\infty d\lambda \frac{k}{\lambda^2 + k^2} \exp\left[\frac{4i}{\delta} \int_0^\lambda d\lambda' \sqrt{k^2 + \lambda'^2}\right], \qquad (4.91)$$

then it is easy to see that $n_{ex}^{pert}(0, \infty) = \frac{1}{4} n_{ex}^{pert}(-\infty, \infty)$, therefore the scaling of the density of quasiparticles (and entropy) remains the same $n_{ex} \propto \sqrt{\delta}$. This problem can be also solved exactly and the results remain very close to the perturbative case. Because as we mentioned the setup where we start (end) at the critical point more naturally appears in the context of turning on commensurate periodic potential we will postpone a careful analysis of this problem until the next section.

4.4.2.2 sine-Gordon Model: Toulouse Point and the Limit of Free Massive Bosons

A convenient playground to test the general scaling laws presented in Sect. 4.4.1 is the sine-Gordon model (SG) – a one-dimensional model described by the Hamiltonian density:

$$\mathcal{H} = \frac{1}{2} \int dx \left[\Pi(x)^2 + (\partial_x \phi)^2 - 4\lambda \cos(\beta \phi)\right]. \qquad (4.92)$$

Here $\Pi(x)$ and $\phi(x)$ are conjugate fields, λ is the tuning parameter, and $\beta = 2\sqrt{\pi K_{SG}}$ is a constant. From the renormalization group (RG) analysis it is known that the cosine term is a relevant perturbation to the quadratic model only if $0 \leq K_{SG} < 2$, and therefore the system is gapped at any finite λ [50], while for $K_{SG} > 2$ the system remains gapless. Turning on interaction in the regime $K_{SG} < 2$ is akin starting at the critical point and driving the system into the new gapped phase. So the scaling of the density of excitations and other quantities should be described by the critical exponents in accord with Eq. (4.71) as it was indeed shown in [32]. The spectrum of SG model consists of solitons for $1 \leq K_{SG} \leq 2$ and solitons and breathers for $K_{SG} < 1$. The point $K_{SG} = 1$ is the Toulouse point described by free fermions and in the limit $K_{SG} \to 0$ the SG model effectively describes a system of free massive bosons with B_1 breathers being the only surviving excitations (see, e.g., [57]). The energy spectrum for each momentum k is of the form:

$$\varepsilon_k = \sqrt{k^2 + m_s^2}, \qquad (4.93)$$

where m_s is the soliton (or breather) mass that scales with the external parameter as $m_s \sim \lambda^{1/(2-K_{SG})}$. Therefore it is evident how this model fulfills the assumption (4.68) with the critical exponents $z = 1$ and $\nu = 1/(2 - K_{SG})$.

The SG model gives the correct low-energy description of (i) interacting bosons in a commensurate periodic potential and (ii) two one-dimensional condensates (Luttinger liquids) coupled by a tunneling term [32]. In the former case increasing λ in time corresponds to loading bosons into an optical lattice and in the latter case increasing λ describes turning on tunneling. We note that ϕ describes density modulation in the first situation and the relative phase between the two superfluids in the second. The SG model also naturally appears in many other one-dimensional systems [50]. To simplify the analysis we will consider only the situation where either $\lambda_i = 0$ and λ_f large or vice versa. For a generic value of K_{SG} the problem can only be solved perturbatively [32] and the corresponding discussion is beyond the scope of this work. Here we will discuss only two specific solvable limits $K_{SG} \ll 1$ and $K_{SG} = 1$.

$K_{SG} \ll 1$: free bosons.

This limit can be naturally realized in the situation of merging two weakly interacting one-dimensional condensates. Then $K_{SG} = 1/(2K)$, where $K \gg 1$ is the Luttinger liquid parameter describing the individual condensates [32]. Since K_{SG} and hence β in the Hamiltonian (4.92) are small we can expand the cosine term and get a quadratic Hamiltonian that in the Fourier space has the form:

$$\mathcal{H} = \frac{1}{2} \sum_k |\Pi_k|^2 + \kappa_k(t)|\phi_k|^2, \tag{4.94}$$

with $\kappa_k(t) = k^2 + 2\lambda(t)\beta^2$, where as before we assume that $\lambda(t) = \delta t$. We see that in this case the problem is equivalent to the harmonic chain (4.39) already considered in Sect. 4.3. The only difference is in the excitations spectrum $\varepsilon_k = \sqrt{\kappa_k(t)}$ which is now gapped:

$$\varepsilon_k = \sqrt{k^2 + c\lambda}, \tag{4.95}$$

with $c = 2\beta^2$. The perturbative and exact solutions follow the similar steps showed in Sect. 4.3. In the perturbative case we find

$$|\alpha_k|^2 \approx \frac{1}{32} \left| \int_{\kappa_i}^{\kappa_f} \frac{d\kappa}{\kappa} \exp\left[\frac{4i}{3c\delta}\kappa^{3/2} \right] \right|^2$$

$$\approx \frac{1}{72} \left| \Gamma\left(0, -i\frac{4}{3}\frac{k^3}{c\delta}\right) \right|^2, \tag{4.96}$$

where we used the fact that $\kappa_i = k^2$. It is convenient to introduce the rescaled momentum

$$\zeta_k = \frac{k}{k_\delta}, \tag{4.97}$$

where

$$k_\delta = (c\delta)^{1/3}. \qquad (4.98)$$

Then all the results can be found from those of the harmonic chain (cf. Eq. (4.40)), performing the mapping

$$\xi_k \longleftrightarrow \zeta_k^3.$$

In the limit $\zeta_k \gg 1$, corresponding to the high momentum modes, the transition probability becomes

$$|\alpha_k|^2 \approx \frac{1}{128\zeta_k^6} = \frac{1}{128}\frac{(c\delta)^2}{k^6} \qquad (4.99)$$

and in the opposite limit

$$|\alpha_k|^2 \approx \frac{1}{72}\left|\ln\zeta_k^3\right|^2. \qquad (4.100)$$

For the exact solution of the problem we again follow the Gaussian functions ansatz as in Eq. (4.43), with initial value

$$\sigma_k^0(\kappa) = \frac{\sqrt{\kappa_k(0)}}{2}$$

and time dependence satisfying the equation:

$$i\frac{d\sigma_k(t)}{dt} = 2\sigma_k^2(t) - \frac{1}{2}\kappa(t). \qquad (4.101)$$

The solution of this equation is analogous to that of Eq. (4.44) with the only difference that the function f_k (cf. Eq. (4.46)) now becomes

$$f_k = -\frac{\zeta_k \,\text{Bi}(-\zeta_k^2) - i\text{Bi}'(-\zeta_k^2)}{\zeta_k \,\text{Ai}(-\zeta_k^2) - i\text{Ai}'(-\zeta_k^2)}, \qquad (4.102)$$

which gives the following asymptotics for the average number of excited particle pairs with momenta k and $-k$:

$$n_k \approx \frac{1}{64\zeta_k^6} = \frac{1}{64}\frac{(c\delta)^2}{k^6}, \qquad (4.103)$$

for $\zeta_k \gg 1$ and

$$n_k \approx \frac{\pi}{3^{2/3}\Gamma^2(1/3)}\frac{1}{\zeta_k} = \frac{\pi}{3^{2/3}\Gamma^2(1/3)}\frac{(c\delta)^{1/3}}{k} \qquad (4.104)$$

in the opposite limit.

Density of quasiparticles. We note that the exact result gives weak logarithmic divergence of the density of quasiparticles with the system size coming from the $1/k$ dependence of n_k at small k (see Eq. (4.104)):

$$n_{\text{ex}}^{\text{exact}} \approx \frac{(c\delta)^{1/3} \log(k_\delta L)}{2\sqrt[3]{3^2}\Gamma^2(1/3)} \approx 0.033(c\delta)^{1/3} \log(k_\delta L), \tag{4.105}$$

while the perturbative solution gives

$$n_{\text{ex}}^{\text{pert}} \approx 0.068(c\delta)^{1/3}. \tag{4.106}$$

As previously found, the perturbative result underestimates the number of excitated quasiparticles, since it does not take into account bosonic enhancement of transitions. Emergence of the length dependence in the expression for n_{ex} indicates the breakdown of the adiabatic perturbation approach and corresponds to a different (non-adiabatic) response of the system according to the classification of [28]. Physically this divergence comes from the effect of bunching of bosons at small momenta and overpopulation of low-momentum modes, which cannot be captured in the lowest order of the adiabatic perturbation theory. Similar analysis can be also performed for the heat and the entropy. In the situation in which the system starts at the critical point ($\lambda_i = 0$) and λ_f is finite, both the entropy and the heat have a very similar behavior as n_{ex}, i.e., proportional to $\delta^{1/3}$ and showing weak logarithmic divergence with the system size. In the opposite case where λ_i is finite but $\lambda_f = 0$ the expressions for the density of excitations and entropy do not change, while the expression for the heat becomes different because each mode now carries energy $\varepsilon_k \sim k$ proportional to the momentum. This removes the logarithmic divergence and both perturbative and exact results give $Q \sim k_\delta^2 \sim (c\delta)^{2/3}$

$K_{SG} = 1$: Tonks–Girardeau gas.

In the limit of $K_{SG} = 1$ the repulsive interaction between bosons is infinitely strong, therefore the particles behave as impenetrable spheres (hard-core bosons). It is well known that in this limit the system known as the Tonks–Girardeau gas [52] can be mapped into an equivalent system of free spinless fermions (the corresponding limit of the SG model describes the so-called Toulouse point). Therefore the dynamical problem described in the scenario **(i)** of loading hard-core bosons into a commensurate optical lattice can instead be approached with the much simpler analysis of free fermions in a periodic potential.

To understand the dynamics in this case we need to solve the Schrödinger equation of free fermions in a periodic potential with time-dependent amplitude $V(x, t) = V(t) \cos(2k_f x)$, where $k_f = \pi/a$ is the Fermi momentum and a is the lattice spacing. The potential $V(t)$ is related to the coupling $\lambda(t)$ in the SG Hamiltonian (4.92) according to $V(t)n = 4\lambda(t)$, where $n = k_f/\pi$ is the electron density [32, 58]. We assume $V(t) = \delta_V t$, therefore the rate δ_V is related to the rate δ by simple rescaling: $\delta_V = 4\delta/n$. We restrict the analysis to the two lowest bands of the Brillouin

Fig. 4.2 TG-gas: the loading
into a commensurate lattice
problem can be mapped into
a two-level system composed
of lower filled band and
upper empty band, the
excitations are the particles
hopping on the upper band
(figure taken from [32])

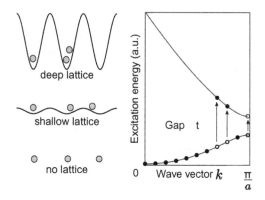

zone (Fig. 4.2) and we linearize the spectrum close to the Fermi momentum. These
two approximations, justified for analyzing the low-energy excitations that we are
interested in, make the fermion problem equivalent to the SG problem. It follows
that for each momentum k the problem is described by a Landau–Zener Hamiltonian

$$\mathcal{H}_k = \begin{bmatrix} V(t)/2 & \Delta_k \\ \Delta_k & -V(t)/2 \end{bmatrix}, \tag{4.107}$$

where $\Delta_k = (\varepsilon_k - \varepsilon_{k-2k_f})/2$ is half the energy difference between the first and
the second band. In the linearized approximation (under which the mapping to SG
model is valid) we have $\varepsilon_k \approx v_f(k - k_f)$, where v_f is the Fermi velocity. Choosing
the units where $v_f = 1$ makes the free fermion problem identical to the SG problem
with the Hamiltonian (4.92).

The problem of turning on the potential amplitude from zero maps then to a direct
sum of *half* LZ problem ($t \in [0, +\infty)$) (see Appendix) and can be solved exactly
(see also [59]). The probability of exciting a particle from the lower to the upper
band during the loading process is

$$p_{ex}(\tau_k) = 1 - 2\frac{e^{-\pi\tau_k/8}}{\pi\tau_k}\sinh\left(\frac{\pi\tau_k}{4}\right)$$

$$\times \left| \Gamma\left(1 + \frac{i\tau_k}{8}\right) + \sqrt{\frac{\tau_k}{8}}\Gamma\left(\frac{1}{2} + \frac{i\tau_k}{8}\right)e^{i\pi/4} \right|^2 \tag{4.108}$$

with

$$\tau_k = \frac{4\Delta_k^2}{\delta_V}. \tag{4.109}$$

To find the total density of excited particles we have to sum the probability $p_{ex}(k)$
over all the momenta in the first Brillouin zone:

$$n_{\text{ex}} = \frac{1}{L} \sum_{k \in [-\pi/a, \pi/a]} p_{\text{ex}}(k). \tag{4.110}$$

It is convenient to introduce a shifted momentum $q = (k - k_f)$. Then we have $\tau_k = (q/k_\delta)^2$ with $k_\delta = \frac{\sqrt{\delta_V}}{2}$. In the limit of small δ_V the upper limit of the integral over q can be sent to infinity, therefore

$$n_{\text{ex}} \approx 2 \int_0^\infty \frac{dq}{2\pi} p_{\text{ex}}(q^2/k_\delta^2) \approx 0.12 k_\delta. \tag{4.111}$$

We see that the exact solution confirms the general scaling $n_{\text{ex}} \propto \sqrt{\delta}$ (see Eq. (4.71)) provided that $d = z = \nu = 1$.

This problem can be also solved perturbatively. In fact the loading problem becomes equivalent to the transverse field Ising model. For example, from the mapping to the LZ problem it is easy to see that the matrix element appearing in Eq. (4.16) is

$$\langle +_k | \partial_\lambda | -_k \rangle = \frac{1}{2} \frac{2\Delta_k.}{(2\Delta_k)^2 + V^2}, \tag{4.112}$$

which is similar to Eq. (4.87). The situation is also analogous for the energy spectrum. The net result of the calculation is

$$n_{\text{ex}}^{\text{pert}} \approx 0.14 k_\delta. \tag{4.113}$$

The results for the entropy and heat for the loading problem have identical scaling with the loading rate.

4.5 Sudden Quenches Near Quantum Critical Points

So far we focused exclusively on the adiabatic regime of small δ. For simplicity we restrict the discussion in this section only to the situation where we start at the critical point and increase λ in time: $\lambda(t) = \delta t$. We argued that in low dimensions, when $d\nu/(z\nu + 1) < 2$, the dynamics is dominated by low-energy excitations with energies $\varepsilon \lesssim \varepsilon^\star = |\delta|^{z\nu/(z\nu+1)}$. The energy scale $\varepsilon^\star$ corresponds to the value of the tuning parameter $\lambda^\star = |\delta|^{1/(z\nu+1)}$ (because of the general scaling relation $\varepsilon \sim \lambda^{z\nu}$). So equivalently we can say that in low dimensions non-adiabatic effects are dominated by the transitions occurring in the vicinity of the critical point $|\lambda| \lesssim \lambda^\star$. A simple qualitative way to understand the adiabatic dynamics in this case is to split the time evolution into two domains: $t < t^\star = \lambda^\star/\delta \sim 1/|\delta|^{z\nu/(z\nu+1)}$ and $t > t^\star$. Then in the first domain the dynamics can be thought as approximately fast (sudden) and in the second domain the dynamics is adiabatic (see also discussion in [32]). Then we can think about the slow quench as a sudden quench with the amplitude of the quench being equal to $\lambda^\star$; i.e., one can expect that the scaling of the density of excitations and other thermodynamic quantities can be approximately obtained by projecting the initial ground state corresponding to the critical point ($\lambda = 0$) to the eigenstates of the new quenched Hamiltonian with $\lambda = \lambda^\star$. Such approach was indeed successfully applied to the problem of quenching the system through the BCS–BEC crossover [60]. This argument implies that for sudden quenches we should have that $n_{ex} \sim |\lambda^\star|^{dv}$ [38]. For $dv < 2$ this is a non-analytic function of the quench rate, which cannot be obtained within the ordinary perturbation theory. As we will show below this scaling, however, immediately follows from the adiabatic perturbation theory applied to sudden quenches. Also we will show that when $dv > 2$ this scaling fails and one gets back to the perturbative analytic result $n_{ex} \sim (\lambda^\star)^2$. As in the case of slow quenches the analytic quadratic dependence comes from dominance of high-energy excitations [38].

Let us return to the discussion of Sect. 4.2, where we introduced the adiabatic perturbation theory . We emphasize that the word adiabatic only means that we are working in the instantaneous (adiabatic) basis. The small parameter in this theory is the probability to excite higher energy levels. For slow quenches the excitation probability is small because the rate δ is small, while for fast quenches this probability is small because the quench amplitude $\lambda^\star$ is small. Thus for sudden quenches of small amplitude we can still use Eq. (4.16), with further simplification that the phase factor $\Theta_n - \Theta_0 \to 0$ since it is inversely proportional to the rate $\delta \to \infty$. Thus instead of Eq. (4.16) we can write [38]

$$\alpha_n(\lambda^\star) \approx - \int_0^{\lambda^\star} d\lambda' \langle n|\partial_{\lambda'}|0\rangle = \int_0^{\lambda^\star} d\lambda' \frac{\langle n|V|0\rangle}{E_n(\lambda') - E_0(\lambda')}. \tag{4.114}$$

Note that in the case where there is a finite gap in the spectrum, the difference $E_n(\lambda') - E_0(\lambda')$ remains large for all values of $\lambda' \in [0, \lambda^\star]$. If the same is true for the matrix element then this expression reduces to the one from the ordinary perturbation theory:

$$\alpha_n(\lambda^\star) \approx \lambda^\star \frac{\langle n|V|0\rangle_{\lambda=0}}{E_n(0) - E_0(0)}. \tag{4.115}$$

The advantage of using the expression (4.114) for α_n over the standard perturbative result (4.115) is that Eq. (4.114) does not give explicit preference to the initial state $\lambda = 0$ over the final state $\lambda = \lambda^\star$ and it does not assume the analytic behavior of α_n with $\lambda^\star$. As we will see both these points are necessary for studying quench dynamics near QCP.

Now let us apply Eq. (4.114) to the problem of quenching starting from the critical point. Using the scaling ansatz for the matrix element (4.70), we find that the density of excitations is approximately given by

$$n_{\text{ex}} \approx \int \frac{d^d k}{(2\pi)^d} \left| \int_0^{-\lambda^*} \frac{d\lambda}{\lambda} G\left(\frac{k}{\lambda^\nu}\right) \right|^2 . \tag{4.116}$$

To analyze this expression it is convenient to change variables as $\lambda = \lambda^* \xi$, $k = (\lambda^*)^\nu \eta$. Then we obtain

$$n_{\text{ex}} \approx |\lambda^*|^{d\nu} \int \frac{d^d \eta}{(2\pi)^d} \left| \int_0^1 \frac{d\xi}{\xi} G\left(\frac{\eta}{\xi^\nu}\right) \right|^2 . \tag{4.117}$$

This expression gives the desired scaling $n_{\text{ex}} \sim (|\lambda^*|)^{d\nu}$ provided that the integral over η converges at large η. From the asymptotics of the scaling function $G(x) \propto 1/x^{1/\nu}$ at large x, discussed earlier, we see that the necessary condition for convergence of the integral is $d < 2/\nu$ or $d\nu < 2$. In the opposite limit $d\nu > 2$ high-energy quasiparticles dominate the total energy and one can use the ordinary perturbation theory (linear response) which gives

$$n_{\text{ex}}^{lr} \approx |\lambda^*|^2 \int \frac{d^d k}{(2\pi)^d} \frac{|\langle k|V|0\rangle_0|^2}{|\varepsilon_k^0 - \varepsilon_0^0|^2} , \tag{4.118}$$

where all the quantities are evaluated at the critical point $\lambda = 0$. We note that the quantity multiplying $|\lambda^*|^2$ is called the fidelity susceptibility [61]. Thus the regime of validity of universal scaling (4.117) corresponds to divergent fidelity susceptibility. One can make similar analysis of the scaling of the heat and the entropy:

$$Q \propto |\lambda^*|^{(d+z)\nu}, \quad S_d \propto |\lambda^*|^{d\nu}. \tag{4.119}$$

As in the case of n_{ex} these scaling laws are valid as long as the corresponding exponents are less than two. One can verify that these scalings are indeed reproduced for the models we analyzed in the previous section. Both adiabatic perturbation theory and the exact calculation give the same scaling, however, the perturbative approach gives a mistake in the prefactor.

From the above analysis we see that in low dimensions $d\nu < 2$ there is indeed a direct analogy between slow and sudden quenches . One gets the same scaling if one correctly identifies the quench rate δ in the former case and the quench amplitude λ^* in the latter: $\lambda^* \sim |\delta|^{1/(z\nu+1)}$. This analogy has direct similarity to the original arguments by Kibble and Zurek who predicted the scaling (4.71) in the context of topological defect formation while crossing classical phase transition with the parameter δ playing the role of temperature quenching rate [41, 42]. The main argument in the Kibble–Zurek (KZ) mechanism is that there is a divergent relaxation time scale near the critical point and a corresponding divergent length scale. The topological excitations with distances larger than this length scale do not have a chance to thermalize and remain in the system for a long time even if they are thermodynamically forbidden. A simple estimate gives that this length scale behaves precisely as $\xi \sim 1/|\delta|^{\nu/(z\nu+1)}$ [42], resulting in a density of topological defects $n_{\text{ex}} \sim 1/\xi^d$, which is the same as in Eq. (4.71). One can thus think about the KZ mechanism as adiabatic quenching of the temperature up to the scale $T^\star$, corresponding to the correlation length ξ, followed by sudden quench of the temperature to zero, so that remaining topological excitations essentially freeze. Essentially the same argument we used here in the quantum case and it indeed works qualitatively right for $d\nu < 2$.

The situation becomes quite different in the regime $d\nu/(z\nu+1) < 2 < d\nu$. In this case for slow quenches the universal scaling is still applicable, i.e., only low-energy excitations created at $\lambda \lesssim \lambda^\star$ are important for the scaling of n_{ex} while for sudden quenches this is no longer the case. Namely, the scaling of n_{ex} with the quench amplitude becomes quadratic and non-universal, i.e., sensitive to the high-energy cutoff. So in this regime the analogy between slow and sudden quenches becomes misleading. This makes an important difference with the Kibble–Zurek mechanism where such issues do not arise. If $d\nu/(z\nu+1) > 2$ then the scaling becomes quadratic in both cases (with the rate δ for slow quenches and with the amplitude $\lambda^\star$ for sudden quenches), with the main contribution to excitations coming from high-energy quasiparticles.

4.6 Effect of the Quasiparticle Statistics in the Finite Temperature Quenches

One can try to extend the analysis we carried through (both for sudden and slow quenches) to the finite temperature situation. We note that because we consider isolated systems with no external bath, the temperature only describes the initial density matrix. To realize this situation one can imagine that the system weakly couples to a thermal bath and reaches some thermal equilibrium. Then, on the timescales of the dynamical processes we are interested in, this coupling has a negligible effect and the dynamics of the system is essentially Hamiltonian. One does not even have to assume coupling to the external bath if we are dealing with ergodic systems, since they are believed to reach thermal equilibrium states by themselves. By now this problem remains unsolved in the most general case. As we will argue below at finite temperatures statistics of low-energy quasiparticles plays the key role. In general one can have critical points where low-energy excitations have fractional statistics or do not have well-defined statistics at all.

In this section we will consider only a relatively simple situation in which statistics of quasiparticles is either bosonic or fermionic. In particular, the two limits of the sine-Gordon model corresponding to $K_{SG} \ll 1$, where the system maps to a set of independent harmonic oscillators, and $K_{SG} = 1$, where the system is equivalent to non-interacting fermions, are examples of critical systems with bosonic and fermionic statistics, respectively. The other two examples in this chapter (harmonic chain and transverse field Ising model) obviously also fall into the category of a system with well-defined statistics of quasiparticles (bosonic and fermionic, respectively).

For bosonic excitations it is straightforward to show that the Gaussian ansatz (4.43) still holds at finite temperatures for the Wigner function [28, A. Polkovnikov, unpublished]. The width of the Wigner function satisfies a similar equation as the width of the wave function (4.44), (4.101). The only difference with the initial ground state is that this width gets "dressed" as $\sigma_k \coth(\varepsilon_k^0/2T)$ at initial time, where σ_k^0 is the ground state width defined earlier and ε_k^0 is the initial energy of the particle with momentum k. Using this fact it is straightforward to show [28] that

$$\frac{1}{2}\left[\frac{\sigma_k^{\mathrm{eff}}}{\sigma_k^{\mathrm{eq}}} - 1\right] \longrightarrow \frac{1}{2}\left[\frac{\sigma_k^{\mathrm{eff}}}{\sigma_k^{\mathrm{eq}}}\coth\left(\frac{\varepsilon_k^0}{2T}\right) - 1\right].$$

Since we are interested in the excitations created by the dynamical process, we need to subtract from this quantity the initial number of quasiparticle excitations, which were present due to initial thermal fluctuations:

$$n_{\mathrm{ex}}^T(k) = \frac{1}{2}\left[\frac{\sigma_k^{\mathrm{eff}}}{\sigma_k^{\mathrm{eq}}}\coth\left(\frac{\varepsilon_k^0}{2T}\right) - 1\right]$$
$$-\frac{1}{2}\left[\coth\left(\frac{\varepsilon_k^0}{2T}\right) - 1\right] = n_{\mathrm{ex}}^0(k)\coth\left(\frac{\varepsilon_k^0}{2T}\right), \qquad (4.120)$$

where $n_{\mathrm{ex}}^0(k)$ is the number of created quasiparticles with momentum k at zero temperature. Note that this result does not depend on the details of the quench process, i.e., whether it is fast or slow. At low temperatures $T \ll \varepsilon_k^0$ the expression (4.120) obviously reduces to the zero-temperature result. However, at high temperatures $T \gg \varepsilon_k^0$ we have the bosonic enhancement of the transitions

$$n_{\mathrm{ex}}^T(k) \approx n_{\mathrm{ex}}^0(k)\frac{2T}{\varepsilon_k^0}. \qquad (4.121)$$

When we sum $n_{\mathrm{ex}}^T(k)$ over all possible momenta k starting from a gapless system, this extra T/ε_k^0 factor makes the result more infrared divergent. In the regime of validity of the adiabatic perturbation theory it changes the scaling, for example, for the density of excitations to

$$n_{\mathrm{ex}} \propto T|\delta|^{(d-z)\nu/(z\nu+1)}. \qquad (4.122)$$

Also the critical dimension, where the quadratic scaling is restored, becomes higher than in the zero-temperature case and can be determined from the equation $(d_c - z)\nu/(z\nu + 1) = 2$. We note, however, that in small dimensions the response of the system can become non-adiabatic [28] and the scaling above can break down. For the sine-Gordon model in the bosonic limit using Eq. (4.104) it is indeed straightforward to see that

$$n_{\text{ex}} \sim T(c\delta)^{1/3}L, \qquad (4.123)$$

where L is the system size. If we consider massive bosonic theories in higher dimensions we would find that the scaling (4.122) would be restored above two dimensions. Even though in one dimension the adiabatic perturbation theory fails to predict the correct scaling, it unambiguously shows that finite temperatures make the response of the system less adiabatic. For sudden quenches effect of initial temperature on harmonic system was recently considered in [S. Sotiriaolis et al., unpublished] and the authors obtained results consistent with the statements above.

The scenario becomes quite opposite in the fermionic case, $K_{SG} = 1$, where we are dealing with a sum of independent two-level systems. At finite temperature each level is occupied according to the Fermi distribution

$$f_k^{\pm} = \left(\exp\left[\pm \frac{\varepsilon_k^0}{T} \right] + 1 \right)^{-1}.$$

The probability of excitation thus gets corrected as

$$n_{\text{ex}}^T(k) = n_{\text{ex}}^0(k)(f_k^- - f_k^+) = n_{\text{ex}}^0(k) \tanh\left(\frac{\varepsilon_k^0}{2T} \right). \qquad (4.124)$$

The fact that we got tanh factor for fermions (and coth for bosons) is hardly surprising. Similar factors appear in conventional fluctuation–dissipation relations [62]. However, this mere change of coth to tanh factor has an important implication. As in the case of bosons, in the small temperature limit this additional factor reduces to unity and the zero-temperature result is recovered. At high temperatures $T \gg \varepsilon_k^0$ we find

$$n_{\text{ex}}^T(k) \approx n_{\text{ex}}^0(k) \frac{\varepsilon_k^0}{2T}. \qquad (4.125)$$

Therefore the number of created quasiparticles is much smaller than in the zero temperature case. This fact reflects fermionic anti-bunching. In other words preexisting thermal quasiparticles are blocking the transition to the already occupied excited states and the dynamical process becomes more adiabatic. Thus for the density of excitations we expect now the scaling

$$n_{\text{ex}} \propto |\delta|^{(d+z)\nu/(z\nu+1)}/T. \qquad (4.126)$$

As a consequence the critical dimension d_c, where the quadratic scaling is restored, is now lower than in the zero-temperature case. As before d_c is found by equating the exponent in the power of δ in Eq. (4.126) to two.

This analysis is not sensitive to the details of the process, hence the situation remains the same for sudden quenches . Thus instead of the scaling $n_{ex} \sim |\lambda^\star|^{d\nu}$ we will get $n_{ex} \sim |\lambda^\star|^{(d\mp z)\nu} T^{\pm 1}$ where the upper "+" or "−" sign corresponds to bosons and the lower sign corresponds to fermions. The crossover to the quadratic linear response scaling happens when the corresponding exponent becomes two. As in the case of slow quenches one should be careful with the validity of the perturbative scaling in low dimensions for bosonic excitations, where the system size can affect the scaling and change the exponent.

The main conclusion of this section is that at finite initial temperatures the statistic of quasiparticles qualitatively changes the response of the quantum critical system to quenches (independently if they are slow or fast). At zero temperature the quasiparticle statistics does not seem to play an important role, since it does not enter the general scaling exponents and the value of the critical dimension. This suggests that the non-adiabatic response will be extremely interesting in the systems with fractional statistics of excitations. This sensitivity and potential universality of the response of the system to sudden or slow perturbations at finite temperatures might allow one to use non-adiabatic transitions as an experimental probe of the quasiparticle statistics.

4.7 Conclusions

In this work we focused on the analysis of the response of a translationally invariant system, initially prepared in the ground state, to linear quenches, where an external parameter globally coupled to the whole system linearly changes in time. Using the adiabatic perturbation theory we showed how to obtain scaling of various quantities like the density of quasiparticles n_{ex}, heating (excess non-adiabatic energy) Q, and entropy S_d with the quench rate δ for small δ. We started from a simple two-level system, where we showed that the transition probability scales quadratically with δ, at $\delta \to 0$, if the external parameter changes in the finite range. We then showed that this quadratic scaling can be violated in various low-dimensional systems, especially quenched through singularities like quantum critical points. This violation comes because excitations of low-energy levels, for which dynamics is adiabatic, dominate the scaling of n_{ex} and other quantities, which then acquire universal non-analytic dependence on the rate δ. For example, we argued that in generic low-dimensional gapless systems $n_{ex} \propto |\delta|^{d/z}$ as long as $d/z < 2$. In the opposite limit $d/z > 2$ we expect that the high-energy quasiparticles dominate n_{ex} and the quadratic scaling is restored: $n_{ex} \propto \delta^2$. Similar story is true for heat and entropy. We note that the quadratic scaling of the transition probability to highly excited states is specific to linear quenches, where the time derivative of the tuning parameter has a discontinuity at initial and final times. For example we analyzed the situation where $\dot\lambda = \delta$ for $t > t_i$ and $\dot\lambda = 0$ for $t < t_i$. If the tuning parameter is turned on and off smoothly in time then this quadratic scaling is no longer valid. One can argue, however, that at sufficiently high energies there are always dephasing mechanisms for quasiparticles which reset their phase and effectively reset the value of t_i. This resetting of the phase will likely restore the quadratic scaling. However, this issue needs to be investigated separately and is beyond the scope of this work.

Using the same adiabatic perturbation theory and general scaling arguments we showed how the universal behavior of n_{ex}, S_d, and Q emerges for quenching through quantum critical points. In particular, we showed that the density of generated quasiparticles (and entropy) scale as $n_{ex}, S_d \sim |\delta|^{d\nu/(z\nu+1)}$, where ν is the critical exponent characterizing divergence of the correlation length near phase transition agreement with earlier works [24, 25]. Based on the scaling analysis we showed that for linear quenches this scaling is valid as long as $d\nu/(z\nu + 1) < 2$, otherwise the quadratic scaling is restored. We also discussed the connection between adiabatic and sudden quenches near a quantum critical point. We argued that if $d\nu < 2$ the two are qualitatively similar provided that one correctly associates quench amplitude $\lambda^\star$ and the quench rate δ: $\lambda^\star \sim |\delta|^{1/(z\nu+1)}$. Using the adiabatic perturbation theory adopted to sudden quenches of small amplitude we showed that the density of excitations scales with the quench amplitude as $n_{ex} \propto |\lambda^\star|^{d\nu}$. When $d\nu > 2$ then the quadratic scaling, which follows from the conventional perturbation theory (linear response) is restored: $n_{ex} \sim (\lambda^\star)^2$. As in the case of slow quenches the quadratic scaling comes from dominant contribution of high-energy quasiparticles to n_{ex}.

We also discussed the situation where a gapless system is initially prepared at finite temperature. In this case we argued that the statistics of quasiparticles becomes crucial. In particular, for bosonic quasiparticles finite temperature enhances the effects of non-adiabaticity due to bunching effect, while conversely for Fermions the response becomes more adiabatic than at zero temperature due to the Pauli blocking. In the regime of validity of the adiabatic perturbation theory we argued that in the scaling laws presented above (both for sudden and slow quenches) one should change $d \to d - z$ in the bosonic case and $d \to d + z$ in the fermionic case. We note, however, that in the bosonic case the adiabatic perturbation theory can break down due to overpopulation of low-energy bosonic modes and the system can enter a non-adiabatic regime [28]. We illustrated all our general statements with explicit results for various solvable models.

Acknowledgments We would like to acknowledge R. Barankov and V. Gritsev for useful discussions and comments. We also would like to acknowledge R. Barankov for providing Fig. 4.1 and help in derivation of Eq. (4.27). This work was supported by AFOSR YIP and Sloan Foundation.

Appendix

Half Landau–Zener Problem

Here we briefly describe the derivation of the transition probability in the LZ problem, when the system starts at the symmetric point with the smallest gap. Let us consider the Hamiltonian:

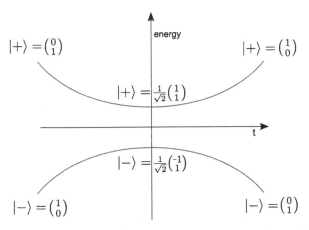

Fig. 4.3 Landau–Zener problem: the eigenvectors of the system are exchanged during the time evolution

$$H = \begin{bmatrix} \lambda(t) & g \\ g & -\lambda(t) \end{bmatrix}, \tag{4.127}$$

where g is a constant and $\lambda(t) = \delta t$. This Hamiltonian has normalized eigenvectors given by Eq. (4.23):

$$|-\rangle = \begin{pmatrix} \sin(\theta/2) \\ -\cos(\theta/2) \end{pmatrix}, \quad |+\rangle = \begin{pmatrix} \cos(\theta/2) \\ \sin(\theta/2) \end{pmatrix}, \tag{4.128}$$

where $\tan\theta(t) = g/\lambda(t)$, corresponding to the eigenergies

$$E_\pm(t) = \pm\sqrt{(\lambda(t))^2 + g^2}. \tag{4.129}$$

The vectors representing the ground state $|-\rangle_t$ and the excited state $|+\rangle_t$ significantly change as the time t is varied, as it is sketched in Fig. 4.3.

Unlike in Sect. 4.2.1 for the exact solution it is convenient to work in the fixed basis:

$$|\psi(t)\rangle = \phi_1(t)\begin{pmatrix} 1 \\ 0 \end{pmatrix} + \phi_2(t)\begin{pmatrix} 0 \\ 1 \end{pmatrix}. \tag{4.130}$$

Then the equations of motion for coefficients $\phi_1(t)$ and $\phi_2(t)$ become

$$i\dot{\phi}_1 = \delta t\phi_1 + g\phi_2, \tag{4.131}$$
$$i\dot{\phi}_2 = g\phi_1 - \delta t\phi_2. \tag{4.132}$$

This system of equations has the following generic solutions for ϕ_1 and ϕ_2:

$$\phi_1(t) = c_1 e^{-i\frac{t^2}{4}} M\left(\frac{ig^2}{4\delta}, \frac{1}{2}, \frac{it^2}{2}\right) + c_2 t e^{-i\frac{t^2}{4}} M\left(\frac{1}{2} + \frac{ig^2}{4\delta}, \frac{3}{2}, \frac{it^2}{2}\right), \quad (4.133)$$

$$\phi_2(t) = c_3 e^{-i\frac{t^2}{4}} M\left(\frac{1}{2} + \frac{ig^2}{4\delta}, \frac{1}{2}, \frac{it^2}{2}\right) + c_4 t e^{-i\frac{t^2}{4}} M\left(1 + \frac{ig^2}{4\delta}, \frac{3}{2}, \frac{it^2}{2}\right), \quad (4.134)$$

where $M(a, b, z)$ is the confluent hypergeometric function otherwise called $_1F_1$. The coefficients c_1, c_2, c_3, c_4 are determined by the initial conditions on the wave function at the initial time t_{in}

$$|\psi(t_{in})\rangle = \phi_1(t_{in}) \begin{pmatrix} 1 \\ 0 \end{pmatrix} + \phi_2(t_{in}) \begin{pmatrix} 0 \\ 1 \end{pmatrix} \quad (4.135)$$

plus two auxiliary conditions, e.g.,

$$i\dot{\phi}_1|_{t=0} = \frac{g}{\sqrt{2\delta}} \phi_2|_{t=0}. \quad (4.136)$$

$$i\dot{\phi}_2|_{t=0} = \frac{g}{\sqrt{2\delta}} \phi_1|_{t=0}. \quad (4.137)$$

The conventional LZ problem [39, 40] considers the case of starting in the ground state at $t = -\infty$ and asks what is the excitation probability after evolving the system to a final time $t = +\infty$, which is

$$p_{ex} = \exp\left[-\frac{\pi g^2}{\delta}\right]. \quad (4.138)$$

This result can be recovered imposing the initial conditions:

$$t_{in} = -\infty \quad |\psi(-\infty)\rangle = |-\rangle = \begin{pmatrix} 1 \\ 0 \end{pmatrix} \quad \begin{array}{l} \phi_1(-\infty) = 1, \\ \phi_2(-\infty) = 0. \end{array} \quad (4.139)$$

and looking at the asymptotic behavior of $|\phi_1(\infty)|^2$.

Another situation that can be straightforwardly considered corresponds to starting from $t = 0$ (instead of $t = -\infty$) and ending at $t = +\infty$, essentially this is *half* of the usual LZ problem. As we have explained in Sect. 4.4.2.2 this situation naturally arises in, e.g., describing the problem of loading hard-core bosons into an optical lattice. Here we need to impose the initial conditions

$$t_{in} = 0 \quad |\psi(0)\rangle = |-\rangle = \frac{1}{\sqrt{2}} \begin{pmatrix} -1 \\ 1 \end{pmatrix} \quad \begin{array}{l} \phi_1(0) = -\frac{1}{\sqrt{2}} \\ \phi_2(0) = \frac{1}{\sqrt{2}} \end{array}. \quad (4.140)$$

The probability of excitation $p_{ex} = \lim_{t \to \infty} |\phi_1(\infty)|^2$ turns out to be (see also [47, 59])

$$p_{ex}(\tau) = 1 - 2\frac{e^{-\pi\tau/8}}{\pi\tau}\sinh\left(\frac{\pi\tau}{4}\right) \cdot \tag{4.141}$$

$$\cdot\left|\Gamma\left(1 + \frac{i\tau}{8}\right) + \sqrt{\frac{\tau}{8}}\Gamma\left(\frac{1}{2} + \frac{i\tau}{8}\right)e^{i\pi/4}\right|^2.$$

where we introduced $\tau = 2g^2/\delta$. This function has the following asymptotic behavior: for $\tau \to 0$

$$p_{ex} \approx \frac{1}{2} - \frac{\sqrt{\pi\tau}}{4}$$

and for $\tau \to \infty$

$$p_{ex} \approx \frac{1}{4\tau^2} = \frac{\delta^2}{16g^4}.$$

As expected the latter slow asymptotic agrees with the result of the adiabatic perturbation theory (4.30).

References

1. I. Bloch, J. Dalibard, and W. Zwerger, Rev. Mod. Phys, **80**, 885 (2008).
2. M. Greiner, O. Mandel, T.W. Hansch and I. Bloch, Nature, **419**, 51 (2002).
3. C.D. Fertig et al., Phys. Rev. Lett. **94**, 120403 (2005).
4. A.K. Tuchman, C. Orzel, A. Polkovnikov and M. Kasevich, Phys. Rev. A, **74**, 051601 (2006).
5. T. Kinoshita, T. Wenger and D.S. Weiss, Nature, **440**, 900 (2006).
6. L.E. Sadler, J.M. Higbie, S.R. Leslie, M. Vengalattore and D.M. Stamper-Kurn, Nature, **443**, 312 (2006).
7. S. Hofferberth, I. Lesanovsky, B. Fischer, T. Schumm and J. Schmiedmayer, Nature, **449**, 324 (2007).
8. S. Trotzky et al., Science, **319**, 295 (2008).
9. M. Rigol, V. Dunjko, and M. Olshanii, Nature, **452**, 854 (2008).
10. P. Reimann, Phys. Rev. Lett. **101**, 190403 (2008).
11. A. Silva, Phys. Rev. Lett. **101**, 236803 (2008).
12. E. Altman and A. Auerbach, *Phys. Rev. Lett.* **89**, 250404 (2002).
13. A. Polkovnikov, S. Sachdev and S.M. Girvin, Phys. Rev. A, **66**, 053607 (2002).
14. R.A. Barankov, L.S. Levitov and B.Z. Spivak, Phys. Rev. Lett. **93**, 160401 (2004).
15. P. Calabrese and J. Cardy, Phys. Rev. Lett., **96**, 136801 (2006).
16. P. Calabrese and J. Cardy, J. Stat. Mech: Theory Exp., **P06008**, (2007).
17. K. Sengupta, S. Powell and S. Sachdev, Phys. Rev. A, **69**, 053616 (2004).
18. C. Kollath, A.M. Läuchli and E. Altman, Phys. Rev. Lett. **98**, 180601 (2007).
19. E.A. Yuzbashyan, B.L. Altshuler, V.B. Kuznetsov, and V.Z. Enolskii, Phys. Rev. B, **72**, 220503(R) (2005).
20. G. Roux., Phys. Rev. A, **79**, 021608 (2009).
21. V. Gritsev, E. Demler, E. Lukin and A. Polkovnikov, Phys. Rev. Lett. **99**, 200404 (2007).
22. S.R. Manmana, S. Wessel, R.M. Noack and A. Muramatsu, Phys. Rev. Lett. **98**, 210405 (2007).

23. P. Barmettler, M. Punk, V. Gritsev, E. Demler and E. Altman, Phys. Rev. Lett. **102**, 130603 (2009).
24. A. Polkovnikov, Phys. Rev. B, **72**, R161201 (2005).
25. W.H. Zurek, U. Dorner and P. Zoller, Phys. Rev. Lett. **95**, 105701 (2005).
26. J. Dziarmaga, Phys. Rev. Lett., **95**, 245701 (2005).
27. R.W. Cherng and L.S. Levitov, Phys. Rev. A, **73**, 043614 (2006).
28. A. Polkovnikov and V. Gritsev, Nature Phys. **4**, 477 (2008).
29. A. Altland and V. Gurarie, Phys. Rev. Lett., **100**, 063602 (2008).
30. R. Barankov and A. Polkovnikov, Phys. Rev. Lett. **101**, 076801 (2008).
31. R. Bistritzer and E. Altman, PNAS, **104**, 9955 (2007).
32. C. De Grandi, R.A. Barankov and A. Polkovnikov, Phys. Rev. Lett. **101**, 230402 (2008).
33. K. Sengupta, D. Sen and S. Mondal, Phys. Rev. Lett. **100**, 077204 (2008).
34. D. Sen, K. Sengupta and S. Mondal, Phys. Rev. Lett. **101**, 016806 (2008).
35. A. Altland, V. Gurarie, T. Kriecherbauer and A. Polkovnikov, Phys. Rev. A, **79**, 042703 (2009).
36. U. Divakaran, V. Mukherjee, A. Dutta and D. Sen., J. Stat. Mech., **P02007** (2009).
37. D. Rossini, A. Silva, G. Mussardo, and G. E. Santoro, Phys. Rev. Lett. **102**, 127204 (2009).
38. C.De. Grandi, R.A. Barankov, A. Polkovnikov and V. Gritsev, In preparation.
39. L. Landau, Phys. Z. Sowj. **2**, 46 (1932).
40. C. Zener, Proc. R. Soc. **137**, 696 (1932).
41. T.W.B Kibble, J. Phys. A, **9**, 1387 (1976).
42. W.H. Zurek, Phys. Rep., **276**, 177 (1996).
43. G. Ortiz, G. Rigolin and V.H. Ponce, Phys. Rev. A, **78**, 052508 (2008).
44. L.D. Landau and E.M. Lifshitz, *Quantum Mechanics: Non-Relativistic Theory, Vol. 3* (3rd ed.). (Pergamon Press, 1977).
45. R. Shankar, *Principles of Quantum Mechanics* (Springer, New York, 1994).
46. N.V. Vitanov and B.M. Garraway, Phys. Rev. A, **53**, 4288 (1996).
47. N.V. Vitanov, Phys. Rev. A, **59**, 988 (1999).
48. B.L. Gelmont, V.I. Ivanovomskii and I.M. Tsidilkovskii, Uspekhi Fizicheskikh Nauk, **120**, 337 (1976).
49. A.H. Castro Neto, F. Guinea, N.M.R. Peres, K.S. Novoselov and A.K. Geim, Rev. Mod. Phys., **81**, 109 (2009).
50. T. Giamarchi, *Quantum Physics in One Dimension* (Clarendon Press, Oxford, 2004).
51. S. Sachdev, *Quantum Phase Transitions* (Cambridge University Press, Cambridge, 1999).
52. M. Girardeau, J. Math .Phys. **1**, 516 (1960).
53. C.N. Yang and Y.P. Yang, J. Math. Phys. **10**, 1115 (1969).
54. L. Tonks, Phys. Rev. **50**, 955 (1936).
55. B. Paredes, A. Widera, V. Murg, O. Mandel, S. Fölling, I. Cirac, G.V. Shlyapnikov, T.W. Hansch and I. Bloch, Nature, **429**, 277 (2004).
56. A. Polkovnikov, Phys. Rev. Lett., **101**, 220402 (2008).
57. V. Gritsev, A. Polkovnikov and E. Demler, Phys. Rev. B, **75**, 174511 (2007).
58. H.P. Büchler, G. Blatter and W. Zwerger, Phys. Rev. Lett., **90**, 130401 (2003).
59. B. Damski and W.H. Zurek, Phys. Rev. A, **73(6)**, 063405 (2006).
60. E. Altman and A. Vishwanath, Phys. Rev. Lett. **95**, 110404 (2005).
61. S.-J. Gu and H.-Q. Lin, Europhys. Lett. **87**, 10003 (2009).
62. A.A. Abrikosov, L.P. Gor'kov and I.E. Dzyaloshinski. *Methods of quantum field theory in statistical physics.* R. A. Silverman (ed.) (Courier Dover Publications, 1975).

Chapter 5
Quench Dynamics of Quantum and Classical Ising Chains: From the Viewpoint of the Kibble–Zurek Mechanism

S. Suzuki

5.1 Introduction

In most macroscopic systems, the state in the equilibrium undergoes a phase transition when the temperature is changed. In the ordinary phase transition, the correlation length and the relaxation time diverge at the critical point. Hence it takes infinitely long time for the state to evolve from a disordered state to an ordered state by lowering the temperature. Such a quasi-static motion of the state is ideal. However, a motion of a state by lowering the temperature with a finite speed is more familiar to us. For instance, a natural magnet is made from lava through a thermal quench. If the temperature is cooled with a finite speed across the critical point, the phase transition ends incompletely. Consequently the symmetry breaking occurs not globally but locally.

A phase transition also takes place at zero temperature. When a parameter which controls a quantum fluctuation in a system is changed across a critical point, the static ground state of the system changes accompanied by a symmetry breaking. For this quantum phase transition, one may consider changing a parameter across the critical point with a finite speed so as to reduce a quantum fluctuation. Then the adiabaticity breaks in the vicinity of the critical point. As a result, an incomplete symmetry breaking occurs, and the correlation length does not extend throughout the whole system even though the parameter lies in an ordered phase after the change.

The Kibble–Zurek mechanism is known as a generic scenario about an evolution of a system across a second-order phase transition. Originally, Kibble proposed this scenario for the evolution of the universe [28, 29]. Zurek discussed the same scenario in condensed matter systems [49]. Let us imagine an evolution from a disordered state to an ordered state with quenching the temperature with a finite rate. We assume that the system is in the equilibrium state initially. With decreasing the temperature, the relaxation time grows. When the temperature comes close to the critical point, the temperature decreases further before the system relaxes into the

S. Suzuki (✉)
Aoyama Gakuin University, Fuchinobe Sagamihara 229-8558 Japan, sei@phys.aoyama.ac.jp

Suzuki, S.: *Quench Dynamics of Quantum and Classical Ising Chains: From the Viewpoint of the Kibble–Zurek Mechanism.* Lect. Notes Phys. **802**, 115–143 (2010)
DOI 10.1007/978-3-642-11470-0_5   © Springer-Verlag Berlin Heidelberg 2010

equilibrium. Therefore the growth of the spatial correlation stops before the temperature passes the critical point. Here we roughly consider that the system maintains the equilibrium state until a time $\hat{t}$ and after that the correlation length remains at the value of the equilibrium state at $T = T(\hat{t})$. The time $\hat{t}$ is estimated by making an equality between the relaxation time and the remaining time until the critical temperature is reached. This is because, once the relaxation time τ_r becomes larger than the remaining time, the system cannot relax into the equilibrium state during that time period. This equality is arranged into a closed equation of the correlation length ξ, if the temperature dependence of ξ and the relation between τ_r and ξ are available. The solution of the equation gives the correlation length when the temperature is lowered below the critical temperature.

The experimental progress in atomic physics enables one to observe the time evolution across a phase transition. A system of Bose atoms trapped in an optical lattice attracts much attentions in particular. This system is effectively explained by the Bose–Hubbard model. When the repulsive interaction energy between atoms at the same site is dominant in the Hamiltonian, the ground state should be a Mott insulator. On the other hand, when the tunneling energy is sufficiently large, a superfruid should be stabler than the Mott insulator. A quantum phase transition separates two distinct ground states. Greiner et al. [23] have demonstrated a time evolution of this system kept in an extremely low temperature from the Mott insulator phase to the superfruid phase by changing the potential height of an optical lattice with time. They observed that the transition from the Mott insulator to the superfruid is more complete for smaller change rate of the potential height. Another important system is a Bose gas composed of atoms with spin degree of freedom. In a sufficiently low temperature, the system forms a Bose–Einstein condensate. Moreover, if parameters of the system are appropriately tuned, a ferromagnetic Bose–Einstein condensate is realized as the ground state. Weiler et al. [48] performed a quench of the temperature across a thermal phase transition point in the system of ^{87}Rb atoms. They observed formation of defects in the spatial distribution of the spin orientation. Sadler et al. [39] performed a quench of the magnetic field in this system and observed imperfection of a quantum phase transition from a paramagnetic ground state to a ferromagnetic ground state. The imperfection was captured by spatial inhomogeneity of the spin orientation. The spontaneous formation of the spatial inhomogeneity in these experiments is interpreted as a consequence of the Kibble–Zurek mechanism. These highly controlled experiments have raised theoretical interests in quench dynamics in condensed matter systems [3–7, 13–16, 35–37, 42–44, 50].

Apart from the condensed matter physics, quench dynamics have attracted much attentions in the computer science. In the computer science, finding the global minimum of a complex high-dimensional cost function is an important problem [21]. The traveling salesman problem is a famous example. Such an optimization problem is regarded as the energy minimization, or identifying the ground state in other words, of a disordered system.

Simulated annealing is known as a method to obtain good candidates for the solution of an optimization problem [8, 29, 30]. To apply simulated annealing, we regard the cost function as a classical Hamiltonian. We then introduce a fictitious temperature and consider the Maxwell–Boltzmann distribution. The distribution at zero temperature yields the ground state, which corresponds to the solution of the problem, with the probability one as far as the solution is unique. If the distribution at a low temperature possesses many local minima separated by a high-potential wall, it takes long time to obtain the equilibrium state. In order to avoid being stuck at local minima, we reduce the temperature from a sufficiently high temperature according to an appropriate schedule. By deforming the distribution gradually from a simple one to a complex one, one can expect to obtain the global minima with high probability when the temperature reaches the absolute zero.

Quantum annealing is an analogous method with simulated annealing [1, 17, 18, 26]. Instead of a fictitious temperature, we introduce a kinetic energy in quantum annealing [46]. The Hamiltonian of the kinetic energy is supposed not to commute with the original Hamiltonian of the problem. When the kinetic Hamiltonian dominates the system, the ground state is the state that minimizes the kinetic energy. Such a state is easily obtained. Suppose that the kinetic term is dominant and the system is in the ground state initially. We reduce the magnitude of the kinetic term gradually with time. If the change rate of the Hamiltonian is sufficiently slow, the state of the system evolves adiabatically. One can expect that the ground state of the original Hamiltonian is obtained with high probability when the kinetic term vanishes [10, 11, 33, 41].

The question whether quantum annealing performs better than simulated annealing or not has been discussed since quantum annealing was proposed [10, 41]. Several numerical studies have reported better performances of quantum annealing [2, 27, 31, 32, 40, 45]. However, there exists no proof that shows the advantage of quantum annealing.

Obviously, simulated annealing and quantum annealing make use of quench dynamics. The former is performed through a thermal quench, while the latter is done through a quantum quench . Hence the study of quench dynamics in quantum and classical systems leads us to a better understanding of quantum annealing and simulated annealing.

In this chapter, we study the quantum and classical quenches in the pure and random Ising models in one dimension. Our interest lies in a difference between the quantum quench and the classical quench in a fixed model. In order to discuss the quantum quench, we consider the Ising chains in the transverse field at zero temperature. As for the classical quench, we introduce the finite temperature without the transverse field. The one-dimensional Ising models enable us to study them analytically, whether they are in the transverse field or not. To study the quench dynamics, we focus on the Kibble–Zurek mechanism. By making use of a phenomenology of the Kibble–Zurek mechanism, we can contrast the quantum quench and the classical quench in the same theoretical framework.

The results we will show in this chapter are summarized as follows. In the pure Ising chain, there is no significant difference between the quantum quench and the classical quench as far as the density of kinks after the quench is concerned. However, a significant difference arises in the quantum quench and the classical quench of the random Ising chain . The density of kinks as a function of the inverse quench rate decays faster in the quantum quench than in the classical quench. These results mean that, though the random Ising chain serves as a trivial optimization problem, the advantage of quantum annealing over simulated annealing is significant [47].

The rest of this chapter is organized as follows. In the next section, we describe the Kibble–Zurek mechanism. Although we focus on the quench dynamics across the standard second-order phase transition in the next section, the argument can be extended to a non-standard phase transition. Studies in subsequent sections rely on the argument of this section. In Sect. 5.3, we discuss the quench dynamics of the pure Ising chain. This section is divided into two parts. The first part is devoted to the study of the classical quench. We calculate several physical quantities at a fixed temperature, and then derive the density of kinks after the classical quench. The second part is for the quantum quench. We derive the energy gap and the correlation length at a fixed transverse field by means of the Jordan–Wigner transformation. Using the argument on the Kibble–Zurek mechanism, we estimate the density of kinks after the quantum quench. Section 5.4 is assigned for the study of the quench dynamics of the random Ising chain. We discuss the classical quench in the first half part of this section. We give an expression for some physical quantities at a fixed temperature, and then we derive the density of kinks and the residual energy after the quench. In the last half of this section, we study the quantum quench. We derive the density of kinks after the quench on the basis of quantities at a fixed transverse field, which we do not derive in detail, and the Kibble–Zurek mechanism. Section 5.5 is the conclusion of this chapter. We compare the results for the classical quench and the quantum quench in this section and mention future problems.

5.2 Kibble–Zurek Mechanism

The Kibble–Zurek mechanism was originally proposed by Kibble [28] for the evolution of the universe. The universe begins to expand after the Bigbang. The temperature of the universe is initially so high that the universe is in a disordered homogeneous state. As the expansion goes on, the temperature of the universe is lowered. When the temperature becomes a critical value, a thermal phase transition occurs. However, since the change rate of the temperature is finite, the phase transition takes place not globally but locally. The universe is decomposed into domains and the symmetry breaking occurs in each domain locally and independently. The order parameter of a different domain takes a different phase. Spatial defects such as domain walls, vortices, and poles form between neighboring domains. This is a scenario of the Kibble–Zurek mechanism for spontaneous formation of spatial defects in the universe.

Zurek proposed a method to verify the above scenario in condensed matter systems [49]. Here we consider the case of a ferromagnet. A ferromagnet is in a paramagnetic state in the equilibrium when the temperature is higher than the Curie temperature. We consider quenching temperature with time with a finite quench speed. The temperature changes from a high temperature to a low temperature across the critical point. We assume that the state is in the equilibrium initially.

When the temperature is sufficiently high, the state maintains the equilibrium at each temperature since the relaxation time is short compared to the inverse of the quench speed. However, when the temperature becomes close to the critical point, a long relaxation time prevents the state from relaxing into the equilibrium state. The temperature is lowered further before the state attains the equilibrium. As a result, the growth in ferromagnetic correlation almost stops before the temperature arrives at the critical point. The state after the quench has no long-range order and consists of ferromagnetic domains.

The following argument brings us an estimation of the quench-rate dependence of the correlation length after the quench. According to the scaling hypothesis of the second-order phase transition, the correlation length ξ is scaled by the dimensionless temperature $\varepsilon = |T - T_C|/T_C$ as

$$\xi(\varepsilon) \sim \xi_0 \varepsilon^{-\nu}, \tag{5.1}$$

where T_C stands for the critical temperature. The relaxation time τ_r is scaled by the correlation length as

$$\tau_r(\varepsilon) \sim \tau_0 \left(\frac{\xi(\varepsilon)}{\xi_0} \right)^z \sim \tau_0 \varepsilon^{-z\nu}, \tag{5.2}$$

where ν and z are the scaling exponents, and ξ_0 and τ_0 are non-universal factors. We assume that the temperature is quenched according to the schedule:

$$T(t) = T_C \left(1 - \frac{t}{\tau} \right),$$

namely,

$$\varepsilon(t) = \frac{|t|}{\tau}, \tag{5.3}$$

where t is time and τ is a parameter, which corresponds to the inverse of the quench rate. We assume that time t moves from $t = -\infty$ to τ. The critical point corresponds to $t = 0$. When the temperature is high, the relaxation time is short and hence the state of the system maintains the equilibrium. With decreasing temperature, the relaxation time grows. Before the temperature reaches T_C, the relaxation time becomes longer than the remaining time $|t|$ to the critical point. This means that the state cannot relax into the equilibrium state by the time when the temperature reaches T_C. As a result, the growth of the correlation length stops before the temperature becomes T_C. The state with a finite correlation length has ferromagnetic domains. Once domains are formed, they should not relax into a completely ordered state even when the temperature is lowered below T_C. We now assume that the evolution of the state maintains the equilibrium before a time $\hat{t}$ and the growth of the correlation length stops at $\hat{t}$. The time $\hat{t}$ is determined by the equality between the relaxation time τ_r and the remaining time to the critical point $|t|$, namely,

$$\tau_r(\varepsilon(\hat{t})) = |\hat{t}|. \tag{5.4}$$

Using Eqs. (5.2) and (5.3), Eq. (5.4) yields

$$\tau_0 \hat{\varepsilon}^{-z\nu} \sim \tau \hat{\varepsilon},$$

where we define $\hat{\varepsilon} \equiv \varepsilon(\hat{t})$. The solution of this equation is written as

$$\hat{\varepsilon} \sim \left(\frac{\tau}{\tau_0}\right)^{-1/(1+z\nu)}. \tag{5.5}$$

Taking Eqs. (5.1) and (5.3) into account, one obtains

$$\hat{\xi} \equiv \xi(\hat{\varepsilon}) \sim \xi_0 \left(\frac{\tau}{\tau_0}\right)^{\nu/(1+z\nu)}. \tag{5.6}$$

Therefore one estimates that the correlation length after the quench is scaled by the quench rate as $\tau^{\nu/(1+z\nu)}$.

In this section we considered the case of a second-order thermal phase transition. However, the above argument can be applied to the case of a quantum phase transition [9, 15, 50]. The same scaling relation in the quantum phase transition has been obtained in [37] by evaluating the number of excitation in the vicinity of the critical point. We focus in the next section on the Kibble–Zurek mechanism in a quench across a quantum phase transition as well as a thermal phase transition.

5.3 Quench in the Pure Ising Chain

In this section, we consider the pure Ising model in one dimension and discuss a classical quench [47] and a quantum quench [50]. The pure Ising chain is represented by

$$H = -J \sum_j \sigma_j \sigma_{j+1}, \tag{5.7}$$

where σ_j is the Ising spin at site j. We suppose $J = 1$ throughout this section. When we discuss finite systems, we assume the periodic boundary condition.

The present model has the ferromagnetic ground states with a twofold degeneracy. The difference between a state and the ground state can be measured by the number of kinks among the neighboring spins. We define the density of kinks by

$$n = \frac{1}{2L} \sum_j \left(1 - \langle \sigma_j \sigma_{j+1} \rangle\right), \tag{5.8}$$

where L is the number of spins in the system. We assume that L is an even number. $\langle \cdots \rangle$ indicates the expectation value with respect to the state under consideration. We rely on the density of kinks to estimate how the state after a quench is close to the ordered ground state of Eq. (5.7).

5.3.1 Classical Quench

5.3.1.1 Properties at Fixed Temperature

The pure Ising chain is a soluble model. The partition function at a temperature T and the inverse temperature $\beta = 1/T$ is written as

$$Z = \sum_{\{\sigma_j = \pm 1\}} e^{\beta \sum_j \sigma_j \sigma_{j+1}} = 2^L (\cosh \beta)^L \left(1 + (\tanh \beta)^L \right). \qquad (5.9)$$

We suppose $k_B = 1$ throughout this chapter. The expectation value of σ_j,

$$\langle \sigma \rangle = \sum_{\{\sigma_i\}} \sigma_j e^{\beta \sum_j \sigma_j \sigma_{j+1}},$$

is zero at any finite T, while $|\langle \sigma_j \rangle| = 1$ at $T = 0$, namely, the ground state is the fully polarized state. One may say that the present system undergoes a discontinuous phase transition at $T = 0$.

Let us consider the correlation length and the energy per spin of the present system. At first, the correlation function is written as

$$\langle \sigma_j \sigma_{j+r} \rangle = \sum_{\{\sigma_j\}} \sigma_j \sigma_{j+r} e^{\beta \sum_j \sigma_j \sigma_{j+1}}$$

$$= \frac{(\tanh \beta)^r + (\tanh \beta)^{L-r}}{1 + (\tanh \beta)^L}$$

In the thermodynamic limit ($L \to \infty$), it is reduced to

$$\langle \sigma_j \sigma_{j+r} \rangle \to (\tanh \beta)^r = \exp \left[-r \log \left(\frac{1}{\tanh \beta} \right) \right]. \qquad (5.10)$$

Hence the correlation length is given by

$$\xi(T) = -r / \ln \langle \sigma_j \sigma_{j+r} \rangle = \left[\log \left(\frac{1}{\tanh \beta} \right) \right]^{-1}. \qquad (5.11)$$

Using Eq. (5.10), the energy per spin is easily obtained:

$$\frac{\langle H \rangle}{L} = -\frac{1}{L} \sum_{j=1}^{L} \langle \sigma_j \sigma_{j+1} \rangle = -\tanh \beta. \tag{5.12}$$

In order to obtain the relaxation time, one has to discuss the dynamics. The dynamics of the Ising chain in a finite temperature is expressed by the Glauber's equation [22]:

$$\frac{d}{dt} \langle \sigma_j \rangle(t) = -\langle \sigma_j \rangle(t) + \frac{1}{2} \gamma \left(\langle \sigma_{j-1} \rangle(t) + \langle \sigma_{j+1} \rangle(t) \right), \tag{5.13}$$

where $\langle \sigma_j \rangle(t)$ is the expectation value of the Ising spin at site j and we define

$$\gamma = \tanh \frac{2}{T}. \tag{5.14}$$

The Glauber's equation is derived as follows [22]. We first define the probability, $p(\sigma_1, \sigma_2 \cdots, \sigma_L; t)$, that the Ising spins are in the state $(\sigma_1, \sigma_2, \cdots, \sigma_L)$ at time t. $p(\sigma_1, \sigma_2 \cdots, \sigma_L; t)$ obeys the following master equation:

$$\frac{d}{dt} p(\sigma_1, \cdots, \sigma_L; t) = -\left(\sum_j w_j(\sigma_j) \right) p(\sigma_1, \cdots, \sigma_L; t) + \sum_j w_j(-\sigma_j) p(\sigma_1, \cdots, \sigma_L; t)$$

$$= -\sum_j \sigma_j \sum_{\sigma_j' = \pm \sigma_j} \sigma_j' w_j(\sigma_j') p(\sigma_1, \cdots, \sigma_j', \cdots, \sigma_L; t), \tag{5.15}$$

where $w_j(\sigma_j)$ is the probability that the spin at site j flips from σ_j to $-\sigma_j$ with the other spins unchanged. We define $p_{eq}(\sigma_1, \cdots, \sigma_L)$ as the probability of the spin state $(\sigma_1, \cdots, \sigma_L)$ in the equilibrium. From the detailed balance principle, we have the following relation between w_j and p_{eq}:

$$\frac{w_j(\sigma_j)}{w_j(-\sigma_j)} = \frac{p_{eq}(\sigma_1, \cdots, -\sigma_j, \cdots, \sigma_L)}{p_{eq}(\sigma_1, \cdots, \sigma_j, \cdots, \sigma_L)} = \frac{\exp[-\beta \sigma_j(\sigma_{j-1} + \sigma_{j+1})]}{\exp[\beta \sigma_j(\sigma_{j-1} + \sigma_{j+1})]}$$

$$= \frac{\cosh[\beta(\sigma_{j-1} + \sigma_{j+1})] - \sigma_j \sinh[\beta(\sigma_{j-1} + \sigma_{j+1})]}{\cosh[\beta(\sigma_{j-1} + \sigma_{j+1})] + \sigma_j \sinh[\beta(\sigma_{j-1} + \sigma_{j+1})]}$$

$$= \frac{1 - \sigma_j \tanh[\beta(\sigma_{j-1} + \sigma_{j+1})]}{1 + \sigma_j \tanh[\beta(\sigma_{j-1} + \sigma_{j+1})]} = \frac{1 - \frac{1}{2} \gamma \sigma_j(\sigma_{j-1} + \sigma_{j+1})}{1 + \frac{1}{2} \gamma \sigma_j(\sigma_{j-1} + \sigma_{j+1})}.$$

Therefore one obtains

$$w_j(\sigma_j) = \frac{\alpha}{2} \left(1 - \frac{1}{2} \gamma \sigma_j(\sigma_{j-1} + \sigma_{j+1}) \right), \tag{5.16}$$

where α is an undetermined factor. Now we define $\langle \sigma_j \rangle(t)$ by

$$\langle\sigma_j\rangle(t) = \sum_{\{\sigma_i\}} \sigma_j p(\sigma_1, \cdots, \sigma_L; t). \tag{5.17}$$

By differentiating this equation by t and using Eqs. (5.15) and (5.16), one obtains

$$\frac{d}{dt}\langle\sigma_j\rangle(t) = -\sum_{\{\sigma_i\}} \sigma_j \sum_k \sigma_k \sum_{\sigma'_k = \pm\sigma_k} \sigma'_k w_k(\sigma'_k) p(\sigma_1, \cdots, \sigma'_k, \cdots, \sigma_L; t)$$

$$= -\sum_{\{\sigma_i\}} \sum_{\sigma'_j = \pm\sigma_j} \sigma'_j w_j(\sigma'_j) p(\sigma_1, \cdots, \sigma'_j, \cdots, \sigma_L; t)$$

$$= -2\sum_{\{\sigma_i\}} \sigma_j \frac{\alpha}{2} \left(1 - \frac{1}{2}\gamma\alpha_j(\sigma_{j-1} + \sigma_{j+1})\right) p(\sigma_1, \cdots, \sigma_L; t)$$

$$= -\alpha\langle\sigma_j\rangle(t) + \frac{\alpha}{2}\gamma\left(\langle\sigma_{i-1}\rangle(t) + \langle\sigma_{j+1}\rangle(t)\right).$$

We here arrange the definition of time so as to replace αt by t. As a result, one obtains the Glauber's equation (5.13).

To solve the Glauber's equation, we expand $\langle\sigma_j\rangle(t)$ into the Fourier series:

$$\langle\sigma_j\rangle(t) = \frac{1}{La} \sum_k e^{ikja} f_k(t), \tag{5.18}$$

where a stands for the lattice constant. The Glauber's equation (5.13) leads to the equation of $f_k(t)$:

$$\frac{d}{dt}f_k(t) = -f_k(t) + \gamma \cos ka f_k(t) = -(1 - \gamma \cos ka)f_k(t). \tag{5.19}$$

The solution of this equation is

$$f_k(t) = f_k(0) \exp\left[-(1 - \gamma \cos ka)t\right]$$

$$= f_k(0)e^{-t} \sum_{l=-\infty}^{\infty} e^{ikla} K_l(\gamma t), \tag{5.20}$$

where $K_l(x)$ is the modified Bessel function. We used an identity: $e^{x(a+a^{-1})/2} = \sum_{l=-\infty}^{\infty} a^l K_l(x)$. From Eqs.(5.18) and (5.20), one obtains

$$\langle\sigma_j\rangle(t) = e^{-t} \sum_{l=-\infty}^{\infty} K_{l-j}(\gamma t)\langle\sigma_l\rangle(0). \tag{5.21}$$

The asymptotic form of the modified Bessel function $K_n(x)$ with a fixed n for $x \to \infty$ is written as $K_n(x) \approx e^x/\sqrt{2\pi x}$. Since $K_{l-j}(\gamma t)$ with large $l-j$ does not contribute to the summation in r.h.s. of Eq. (5.21), the asymptotic form of $\langle\sigma_j\rangle(t)$ for $t \to \infty$ is

$$\langle \sigma_j \rangle (t) \approx \frac{1}{\sqrt{2\pi\gamma t}} e^{-(1-\gamma)t}. \tag{5.22}$$

Thus the relaxation time τ_r is given by

$$\tau_r(T) = \frac{1}{1-\gamma} = \frac{1}{1-\tanh 2\beta}. \tag{5.23}$$

5.3.1.2 Quench Dynamics

Now we consider the quench dynamics of the present system [47]. We assume the following quench schedule:

$$T(t) = -t/\tau, \tag{5.24}$$

where we assume the that time t moves from $-\infty$ to 0. τ stands for the inverse of the quench rate. We introduce the dimensionless temperature measured from the critical point, ε. In the present case, ε is identical to the temperature itself. Hence ε is scheduled as

$$\varepsilon(t) \equiv T(t) = -t/\tau. \tag{5.25}$$

The correlation length and the relaxation time in the present system do not obey the power law. Indeed, from Eqs. (5.11) and (5.23), they are expressed at a low temperature as

$$\xi(\varepsilon) \approx \frac{1}{2}e^{2\beta} = \frac{1}{2}e^{2/\varepsilon}, \tag{5.26}$$

$$\tau_r(\varepsilon) \approx \frac{1}{2}e^{4\beta} = 2\xi^2(\varepsilon). \tag{5.27}$$

Therefore one cannot apply the result of the Kibble–Zurek argument for a second-order phase transition to the present case. We then go back to the equality between the relaxation time τ_r and the remaining time $|t|$ to the critical point (Eq. (5.4)),

$$\tau_r(\hat{\varepsilon}) = |\hat{t}|, \tag{5.28}$$

where $\hat{\varepsilon} \equiv \varepsilon(\hat{t})$. From Eqs. (5.25) and (5.26), r.h.s. of Eq. (5.28) can be written as $|\hat{t}| \approx 2\tau/\ln(2\hat{\xi})$. The l.h.s. is replaced by $2\xi^2$ from Eq. (5.27). Hence one obtains

$$2\hat{\xi}^2 \approx \frac{2\tau}{\ln(\hat{\xi})}. \tag{5.29}$$

The approximation is correct for large τ, because Eqs. (5.26) and (5.27) are correct for small ε, namely, large ξ and τ_r. Equation (5.29) cannot be solved analytically. However, it leads us to

$$\hat{\xi} \approx \frac{\sqrt{\tau}}{\sqrt{\ln\left(2\hat{\xi}\right)}}. \tag{5.30}$$

For a large $\hat{\xi}$, the change of $\ln(2\hat{\xi})$ is negligible compared to the change of $\hat{\xi}$. Hence, ignoring the logarithmic correction, $\hat{\xi}$ scales as $\hat{\xi} \approx \sqrt{\tau}$. The density of kinks is estimated by the inverse of the correlation length. Therefore the density of kinks after the quench is obtained as [47]

$$n \approx \frac{1}{\hat{\xi}} \approx \frac{1}{\sqrt{\tau}}, \tag{5.31}$$

as far as the logarithmic correction is ignored.

5.3.2 Quantum Quench

Let us consider the pure Ising chain in the transverse field. The Hamiltonian is given by

$$H = -\sum_j \sigma_j^x \sigma_{j+1}^x - \Gamma \sum_j \sigma_j^z, \tag{5.32}$$

where σ_i^x and σ_i^z are the x and z components of the Pauli matrices at site j, respectively. Γ is the strength of the transverse field. Remark that we choose the x-axis of the spin as the Ising-spin axis, namely, σ_j^x corresponds to σ in Eq. (5.7). The system represented by this Hamiltonian exhibits a quantum-phase transition at $\Gamma = 1$ [38].

5.3.2.1 Energy Gap

The Hamiltonian (5.32) is mapped to a free fermion model through the Jordan–Wigner transformation . The Jordan–Wigner transformation from the Pauli spin at site j, σ_j^α ($\alpha = x, y, z$), to a spinless fermion represented by C_j and $C_j^\dagger$ is given by

$$\sigma_j^+ \equiv \frac{1}{2}(\sigma_j^x + i\sigma_j^y) = e^{i\pi \sum_{k=1}^{j-1} n_k} C_j^\dagger,$$

$$\sigma_j^- \equiv \frac{1}{2}(\sigma_j^x - i\sigma_j^y) = e^{i\pi \sum_{k=1}^{j-1} n_k} C_j,$$

where $n_j = C_j^\dagger C_j$ is the number operator of the fermion at site j. The operator C_j and $C_j^\dagger$ satisfy the fermionic anti-commutation relation: $\{C_j, C_k^\dagger\} = \delta_{jk}$ and $\{C_j, C_k\} = \{C_j^\dagger, C_k^\dagger\} = 0$. Using this fermion representation, σ_j^z is written as

$$\sigma_j^z = [\sigma_j^+, \sigma_j^-] = [C_j^\dagger, C_j] = 2n_j - 1.$$

The Hamiltonian is written as

$$H = - \sum_{j=1}^{L-1} (C_j^\dagger C_{j+1} + C_{j+1}^\dagger C_j + C_j^\dagger C_{j+1}^\dagger + C_{j+1} C_j + \Gamma(2n_j - 1))$$

$$+ (-1)^{N_F} (C_L^\dagger C_1 + C_1^\dagger C_L + C_1^\dagger C_L^\dagger + C_1 C_L), \tag{5.33}$$

where $N_F = \sum_{j=1}^{L} n_j$ is the total number of the fermion. Since $[N_F, H] = 0$, the Hamiltonians with different N_F are independent. We denote H with an even N_F as H_{even} and that with an odd N_F as H_{odd}. For an even N_F, the anti-periodic boundary condition for C_j, namely, $C_{L+1} = -C_1$ makes H_{even} translationally invariant. For an odd N_F, the periodic boundary condition of C_j assures the translational invariance of H_{odd}.

We expand C_j into the Fourier series:

$$C_j = \frac{1}{L} \sum_k e^{ikja} C_k. \tag{5.34}$$

The wave number k takes different values, depending on whether N_F is even or odd. When N_F is even, possible values of k is shown by

$$k = \frac{\pi}{La}(2n - 1), \quad n = -\frac{L}{2} + 1, -\frac{L}{2} + 2, \cdots, \frac{L}{2}.$$

Remark that L is an even number. The Hamiltonian is expressed in terms of C_k and $C_k^\dagger$ as

$$H_{\text{even}} = - \sum_k \{2 \cos ka C_k^\dagger C_k + (e^{ika} C_k^\dagger C_{-k}^\dagger + e^{-ika} C_{-k} C_k) + 2\Gamma C_k^\dagger C_k - \Gamma\}$$

$$= - \sum_{0 < k < \frac{\pi}{a}} \{2 \cos ka (C_k^\dagger C_k - C_{-k} C_{-k}^\dagger) + 2i \sin ka (C_k^\dagger C_{-k}^\dagger - C_{-k} C_k)$$

$$+ 2\Gamma (C_k^\dagger C_k - C_{-k} C_{-k}^\dagger)\},$$

where we transformed terms with a negative k into those with a positive k by the transformation of $k \to -k$ in the second equality. The Hamiltonian is simplified as follows:

$$H_{\text{even}} = - \sum_{0 < k < \frac{\pi}{a}} \begin{bmatrix} C_k^\dagger & C_{-k} \end{bmatrix} M_k \begin{bmatrix} C_k \\ C_{-k}^\dagger \end{bmatrix}, \tag{5.35}$$

where M_k is defined by

$$M_k \equiv \begin{bmatrix} -2\cos ka - 2\Gamma & -2i\sin ka \\ 2i\sin ka & 2\cos ka + 2\Gamma \end{bmatrix}. \tag{5.36}$$

For an odd N_F, the wave number takes following values:

$$k = \frac{2\pi}{La}n, \quad n = -\frac{L}{2}+1, -\frac{L}{2}+2, \cdots, \frac{L}{2}.$$

The Hamiltonian is written as

$$H_{\text{odd}} = -\sum_{0<k<\frac{\pi}{a}} \begin{bmatrix} C_k^\dagger & C_{-k} \end{bmatrix} M_k \begin{bmatrix} C_k \\ C_{-k}^\dagger \end{bmatrix} - 2(1+\Gamma)C_0^\dagger C_0 + 2(1-\Gamma)C_{\pi/a}^\dagger C_{\pi/a} + 2\Gamma. \tag{5.37}$$

In order to diagonalize the Hamiltonian, one needs to know the eigenvalues and eigenvectors of M_k. They are given as follows:

$$M_k \begin{bmatrix} u_k \\ v_k \end{bmatrix} = \varepsilon_k \begin{bmatrix} u_k \\ v_k \end{bmatrix}, \quad M_k \begin{bmatrix} -v_k^* \\ u_k^* \end{bmatrix} = -\varepsilon_k \begin{bmatrix} -v_k^* \\ u_k^* \end{bmatrix},$$

with

$$u_k = \frac{\varepsilon_k + a_k}{\sqrt{2\varepsilon_k(\varepsilon_k + a_k)}}, \quad v_k = \frac{ib_k}{\sqrt{2\varepsilon_k(\varepsilon_k + a_k)}}, \tag{5.38}$$

where ε_k, a_k, and b_k are defined by

$$\varepsilon_k = \sqrt{a_k^2 + b_k^2} \tag{5.39}$$

$$a_k = -2\cos ka - 2\Gamma, \quad b_k = 2\sin ka. \tag{5.40}$$

Using them, we introduce the Bogoliubov transformation:

$$\gamma_k = u_k^* C_k + v_k^* C_{-k}^\dagger \tag{5.41}$$

$$\gamma_{-k}^\dagger = -v_k C_k + u_k C_{-k}^\dagger. \tag{5.42}$$

Using Eqs. (5.38)–(5.40), one can easily show that the operators γ_k and $\gamma_k^\dagger$ satisfy usual fermionic anti-commutation relation. The Hamiltonian is diagonalized in terms of γ_k:

$$H_{\text{even}} = \sum_{0<k<\frac{\pi}{a}} \varepsilon_k(\gamma_k^\dagger \gamma_k + \gamma_{-k}^\dagger \gamma_{-k} - 1), \tag{5.43}$$

$$H_{\text{odd}} = \sum_{0<k<\frac{\pi}{a}} \varepsilon_k(\gamma_k^\dagger \gamma_k + \gamma_{-k}^\dagger \gamma_{-k} - 1) - 2(1+\Gamma)C_0^\dagger C_0 + 2(1-\Gamma)C_{\pi/a}^\dagger C_{\pi/a} + 2\Gamma. \tag{5.44}$$

The ground energy with an even fermion number is

$$E_{\text{even}}^{\text{g.s.}} = -\sum_{0<k<\frac{\pi}{a}} \varepsilon_k.$$

The ground state with an odd fermion number must contain one quasi-particle. Since $-2(1+\Gamma) < 2(1-\Gamma) < \varepsilon_k$, one should have $C_0^\dagger C_0 = 1$ with $C_{\pi/a}^\dagger C_{\pi/a} = \gamma_k^\dagger \gamma_k = 0$. Hence the ground energy with an odd fermion number is

$$E_{\text{odd}}^{\text{g.s.}} = -\sum_{0<k<\frac{\pi}{a}} \varepsilon_k - 2.$$

At last one can evaluate the energy gap from the true ground state to the first excited state. While the true ground state corresponds to the ground state of H_{even}, the first excited state comes from the ground state of H_{odd}. The energy gap is evaluated as follows:

$$
\begin{aligned}
\Delta &= E_{\text{odd}}^{\text{g.s.}} - E_{\text{even}}^{\text{g.s.}} \\
&\approx -2 - La \int_{2\pi/La}^{\pi/a} \frac{dk}{2\pi} \varepsilon_k + La \int_{\pi/La}^{(1+1/L)\pi/a} \frac{dk}{2\pi} \varepsilon_k \\
&= -2 + \frac{1}{2}\varepsilon_{\pi/La} + \frac{1}{2}\varepsilon_{\pi/a} \\
&\xrightarrow{L\to\infty} -2 + \frac{1}{2}\varepsilon_0 + \frac{1}{2}\varepsilon_{\pi/a} \\
&= \begin{cases} 2(\Gamma - 1) \ (\Gamma \geq 1) \\ 0 \qquad\quad (\Gamma < 1) \end{cases}
\end{aligned}
\tag{5.45}
$$

We define the dimensionless transverse field ε measured from the critical point by

$$\varepsilon = \Gamma - 1.$$

Using this parameter, the energy gap is scaled as

$$\Delta = 2\varepsilon \tag{5.46}$$

for $\varepsilon \geq 0$.

5.3.2.2 Correlation Length

We next investigate the correlation function, $\langle \sigma_i^x \sigma_{i+r}^x \rangle$, at zero temperature. The detailed analysis is available in [38]. For long-distance properties of the correlation function, the lattice structure is not significant. Hence we apply a continuous approximation.

We define the fermion field operator by

$$\psi^\dagger(aj) = \frac{(-1)^{j-1}}{\sqrt{a}} C_j, \quad \psi(aj) = \frac{(-1)^{j-1}}{\sqrt{a}} C_j^\dagger.$$

The Hamiltonian, Eq. (5.33), is arranged into

$$H = -\sum_j \left\{ C_j^\dagger(C_{j+1} + C_j) - C_j^\dagger C_j + (C_{j+1}^\dagger + C_j^\dagger)C_j - C_j^\dagger C_j \right.$$

$$+ C_j^\dagger(C_{j+1}^\dagger + C_j^\dagger) - C_j(C_{j+1} + C_j) + 2\Gamma C_j^\dagger C_j - \Gamma \right\}$$

$$= -a\sum_j \left\{ \left(\psi^\dagger(aj+a) - \psi(aj)\right)\psi(aj) + \psi^\dagger(aj)\left(\psi(aj+a) - \psi(aj)\right) \right.$$

$$- \psi(aj)\left(\psi(aj+a) - \psi(aj)\right) + \psi^\dagger(aj)\left(\psi^\dagger(aj+a) - \psi^\dagger(aj)\right)$$

$$+ 2(1-\Gamma)\psi^\dagger(aj)\psi(aj) - 2 + \Gamma \right\}, \tag{5.47}$$

where we neglect the boundary terms. Taking the continuous limit and performing the partial integral, the Hamiltonian is reduced to

$$H \xrightarrow{a\to 0} -\int dx \left\{ a\left(\frac{d\psi^\dagger(x)}{dx}\psi(x) + \psi^\dagger(x)\frac{d\psi(x)}{dx} + \psi^\dagger(x)\frac{d\psi^\dagger(x)}{dx} - \psi(x)\frac{d\psi(x)}{dx}\right) \right.$$

$$\left. + 2(1-\Gamma)\psi^\dagger(x)\psi(x) \right\}$$

$$= -\int dx \left\{ a\left(\psi^\dagger(x)\frac{d\psi^\dagger(x)}{dx} - \psi(x)\frac{d\psi(x)}{dx}\right) + 2(1-\Gamma)\psi^\dagger(x)\psi(x) \right\}, \tag{5.48}$$

where we omitted the constant terms. We introduce the Fourier transformation of the field operator

$$\psi(x) = \int \frac{dk}{2\pi} e^{ikx}\psi_k, \quad \psi^\dagger(x) = \int \frac{dk}{2\pi} e^{-ikx}\psi_k^\dagger.$$

The Hamiltonian is arranged into the wave-number representation as follows:

$$H = -\int \frac{dk}{2\pi} \left\{ aik\left(\psi_k^\dagger\psi_{-k}^\dagger + \psi_k\psi_{-k}\right) + 2(1-\Gamma)\psi_k^\dagger\psi_k \right\}$$

$$= -\int_0^\infty \frac{dk}{2\pi} \left\{ aik(\psi_k^\dagger\psi_{-k}^\dagger - \psi_{-k}^\dagger\psi_k^\dagger + \psi_k\psi_{-k} - \psi_{-k}\psi_k) \right.$$

$$\left. + 2(1-\Gamma)(\psi_k^\dagger\psi_k + \psi_{-k}^\dagger\psi_{-k}) \right\}.$$

The Bogoliubov transformation from ψ_k to $\tilde{\gamma}_k$ diagonalizes the above Hamiltonian:

$$H = \int_0^\infty \frac{dk}{2\pi} \tilde{\varepsilon}_k \left(\tilde{\gamma}_k^\dagger \tilde{\gamma}_k - \tilde{\gamma}_{-k} \tilde{\gamma}_{-k}^\dagger \right), \tag{5.49}$$

where the energy eigenvalue is given by

$$\tilde{\varepsilon}_k = \sqrt{(2ak)^2 + (2 - 2\Gamma)^2}. \tag{5.50}$$

The Bogoliubov transformation is defined by

$$\tilde{\gamma}_k = \tilde{u}_k^* \psi_k + \tilde{v}_k^* \psi_{-k}^\dagger, \quad \tilde{\gamma}_{-k}^\dagger = -\tilde{v}_k \psi_k + \tilde{u}_k \psi_{-k}^\dagger$$

with

$$\tilde{u}_k = \frac{\tilde{\varepsilon}_k - 2(1 - \Gamma)}{\sqrt{2\tilde{\varepsilon}_k(\tilde{\varepsilon}_k - 2(1 - \Gamma))}}, \quad \tilde{v}_k = \frac{2aik}{\sqrt{2\tilde{\varepsilon}_k(\tilde{\varepsilon}_k - 2(1 - \Gamma))}}.$$

The dispersion relation of Eq. (5.50) is reminiscent of a relativistic particle with mass $|2 - 2\Gamma|$. Indeed, the system represented by Eq. (5.49) possesses the Lorentz invariance. To see the Lorentz invariance of Eq. (5.49), let us consider the Lagrangian [25]:

$$L = i\psi^* \frac{\partial}{\partial t} \psi + a \left(\psi^* \frac{\partial}{\partial x} \psi^* - \psi \frac{\partial}{\partial x} \psi \right) - 2(1 - \Gamma)\psi^* \psi, \tag{5.51}$$

where ψ and ψ^* are the complex Grassmann number. We introduce two real Grassmann numbers, ψ_1 and ψ_2, to express the complex Grassmann number:

$$\psi = \psi_1 + i\psi_2.$$

The Lagrangian is written in terms of ψ_1 and ψ_2 as

$$L = i(\psi_1 \frac{\partial}{\partial t} \psi_1 + \psi_2 \frac{\partial}{\partial t} \psi_2) + 2ia(\psi_1 \frac{\partial}{\partial x} \psi_2 + \psi_2 \frac{\partial}{\partial x} \psi_1) + 4i(1 - \Gamma)\psi_1 \psi_2$$
$$= \begin{pmatrix} \psi_1 & \psi_2 \end{pmatrix} \left(i\frac{\partial}{\partial t} I - 2ia\frac{\partial}{\partial x}\sigma^x + 4(1 - \Gamma)\sigma^y \right) \begin{pmatrix} \psi_1 \\ \psi_2 \end{pmatrix},$$

where I is the unit matrix, and σ^α ($\alpha = x, y$, and z) are the Pauli matrices. We define a matrix $\varsigma \equiv \frac{1}{\sqrt{2}}(\sigma^x + \sigma^z)$. This matrix transforms σ^x to σ^z and σ^y to $-\sigma^y$, namely,

$$\varsigma\sigma^x\varsigma = \sigma^z, \quad \varsigma\sigma^y\varsigma = -\sigma^y.$$

We introduce a transformation by ς as

$$\begin{pmatrix} \psi_1' \\ \psi_2' \end{pmatrix} = \varsigma \begin{pmatrix} \psi_1 \\ \psi_2 \end{pmatrix} = \frac{1}{\sqrt{2}} \begin{pmatrix} \psi_1 + \psi_2 \\ \psi_1 - \psi_2 \end{pmatrix}.$$

The Lagrangian is arranged into

$$L = \begin{pmatrix} \psi_1' & \psi_2' \end{pmatrix} \left(i\frac{\partial}{\partial t}I - 2ia\frac{\partial}{\partial x}\sigma^z - 4(1 - \Gamma)\sigma^y \right) \begin{pmatrix} \psi_1' \\ \psi_2' \end{pmatrix}.$$

This Lagrangian yields the equations of motion:

$$\left(i\frac{\partial}{\partial t} - 2ia\frac{\partial}{\partial x} \right) \psi_1' + 4i(1 - \Gamma)\psi_2' = 0,$$

$$\left(i\frac{\partial}{\partial t} + 2ia\frac{\partial}{\partial x} \right) \psi_2' - 4i(1 - \Gamma)\psi_1' = 0,$$

or equivalently,

$$\left(-\frac{\partial^2}{\partial t^2} + 4a^2\frac{\partial^2}{\partial x^2} - (4(1 - \Gamma))^2 \right) \psi_i' = 0, \qquad (i = 1, 2).$$

This equation is Lorentz invariant. Therefore it turns out that the Hamiltonian represented by Eq. (5.49) is relativistically invariant.

Now we consider the dynamical structure factor:

$$S(k, \omega) = \int dx \int dt \langle \sigma_j^x(t)\sigma_{j+r}^x(0) \rangle e^{-ikr} e^{i\omega t}$$

$$= \frac{2\pi}{a} \sum_\nu \int \frac{dk'}{2\pi} \langle g|\sigma_{k'}^x|\nu \rangle \langle \nu|\sigma_{-k}^x|g \rangle e^{i(k'-k)ja} \delta(\omega - E_\nu + E_g), \quad (5.52)$$

where $|g\rangle$ and $|\nu\rangle$ are the ground and excited eigenstates. E_g and E_ν are eigenenergies of $|g\rangle$ and $|\nu\rangle$, respectively. σ_k^x is the Fourier transform of σ_j^x. The energy eigenstates are represented by the Slater determinant of momentum eigenstates of the quasi-particle $\tilde{\gamma}_k$. A simple perturbative argument shows that σ_k^x does not mix quasi-particle states with different momentum when $\Gamma \gg 1$. Note that the long-distance, i.e., small wave-number property of the system with $\Gamma > 1$ is not different from that with $\Gamma \gg 1$. Hence Eq. (5.52) is reduced to

$$S(k, \omega) = \frac{2\pi}{a^2} \sum_\nu |_k\langle g|\sigma_k|\nu \rangle_k|^2 \delta(\omega - \nu\tilde{\varepsilon}_k),$$

where $|g\rangle_k$ and $|\nu\rangle_k$ stand for the ground and excited states of the quasi-particle state with momentum k, respectively. We assume that the first excited state has the dominant contribution to $S(k, \omega)$. Then we have

$$S(k, \omega) \approx \frac{2\pi}{a^2} |_k\langle g|\sigma_k^x|1 \rangle_k|^2 \delta(\omega - \tilde{\varepsilon}_k). \qquad (5.53)$$

This assumption is correct for $\Gamma \gg 1$. Now the dynamical structure factor must have the relativistic invariance. Hence $S(k, \omega)$ should be written in the form of

$$S(k, \omega) \approx A\delta(\omega^2 - \tilde{\varepsilon}_k^2), \quad (\omega > 0), \tag{5.54}$$

where A is a factor independent of k and ω. This dynamical structure factor yields the same-time correlation function as follows:

$$\langle \sigma_j^x(0)\sigma_{j+r}^x(0)\rangle \approx \int \frac{\omega}{2\pi} \frac{dk}{2\pi} A\delta(\omega^2 - \tilde{\varepsilon}_k^2)e^{ikra}$$

$$= \int_{-\infty}^{\infty} \frac{dk}{2\pi} \frac{A}{2\tilde{\varepsilon}_k} e^{ikra} = \frac{A}{4\pi a} K_0(r|1 - \Gamma|)$$

$$\approx \frac{A}{4\pi a}\sqrt{\frac{\pi}{2r|1 - \Gamma|}} e^{-r|1-\Gamma|},$$

where $K_0(x)$ is the modified Bessel function. From this expression of the correlation function, one obtains the correlation length for $\Gamma > 1$ as

$$\xi \approx 1/(\Gamma - 1).$$

Using the dimensionless transverse field, ε, it turns out

$$\xi \approx \varepsilon^{-\nu} \quad \text{with} \quad \nu = 1. \tag{5.55}$$

Since the present system is a dissipationless system, the relaxation by the contact with the thermal bath does not occur. Instead, the coherence time τ_{coh} defined by the energy gap should play the role of the relaxation time τ_r discussed in the previous section. From Eqs. (5.46) and (5.55), one obtains the dynamical critical exponent as follows:

$$\tau_{\text{coh}} \approx \frac{1}{2}\xi^z \approx \frac{1}{2}\varepsilon^{-z\nu} \quad \text{with} \quad z = 1. \tag{5.56}$$

5.3.2.3 Quench Dynamics

From now we discuss the Kibble–Zurek mechanism of the present system [50]. We assume that the transverse field is quenched with the schedule:

$$\Gamma(t) = 1 - t/\tau, \tag{5.57}$$

where τ stands for the inverse of the quench rate. The time t is assumed to move from $-\infty$ to τ. Using the dimensionless transverse field, the quench schedule is given by

$$\varepsilon(t) = -t/\tau. \tag{5.58}$$

Note that the critical point, $\Gamma = 1$ or $\varepsilon = 0$, corresponds to $t = 0$.

The argument in Sect. 5.2 with Eqs. (5.55) and (5.56) immediately leads us to the scaling relation between the correlation length after the quench and the inverse quench rate τ:

$$\hat{\xi} \approx \sqrt{2\tau} \tag{5.59}$$

One can estimate the density of kinks by the inverse of the correlation length. Equation (5.59) yields the scaling relation of the density of kinks after the quench as

$$n \approx \frac{1}{\sqrt{2\tau}}. \tag{5.60}$$

Dziarmaga has analytically solved the Schrödinger equation [15]:

$$i\frac{d}{dt}|\Psi(t)\rangle = H(t)|\Psi(t)\rangle,$$

with

$$H(T) = -\sum_i \sigma_i^x \sigma_{i+1}^x - \Gamma(t)\sum_i \sigma_i^z,$$

$$\Gamma(t) = -t/\tau.$$

According to the solution that is asymptotically exact for $\tau \to \infty$, the scaling relation of the density of kinks is given by

$$n = \frac{1}{\sqrt{2\pi\tau}}. \tag{5.61}$$

Comparing Eqs. (5.60) and (5.61), these two relations are identical except for the difference of a factor $1/\sqrt{\pi}$. Therefore it turns out that the Kibble–Zurek argument yields correct scaling relation between n and τ.

5.4 Quench in the Random Ising Chain

The random Ising chain we discuss in the present section is represented by the Hamiltonian:

$$H = -\sum_j J_j \sigma_j \sigma_{j+1}. \tag{5.62}$$

The coupling constant J_j is random and assumed to obey the uniform distribution between 0 and 1, namely,

$$P(J_i) = \begin{cases} 1 & \text{for } J_i \in [0, 1] \\ 0 & \text{otherwise} \end{cases}. \tag{5.63}$$

Same as the pure Ising chain, the ground states of the present model are the fully spin-polarized states with the twofold degeneracy. We measure the difference of the state after a quench from the ground states by the density of kinks and the residual energy per spin. The density of kinks is defined by Eq. (5.8). The residual energy is defined by

$$\varepsilon_{\text{res}} = \frac{[\langle H \rangle_{\text{qch}}]_{\text{av}}}{L} - \varepsilon_g, \tag{5.64}$$

where $\langle \cdots \rangle_{\text{qch}}$ stands for the expectation value with respect to the state after the quench, and $[\cdots]_{\text{av}}$ for the average over the randomness in the Hamiltonian. ε_g is the ground-state energy per spin. Note that the density of kinks and the residual energy per spin are not identical in the random Ising chain, though they are in the pure Ising chain.

5.4.1 Classical Quench

5.4.1.1 Properties at Fixed Temperature

We first derive the correlation length and the energy of the present model at a fixed temperature. The partition function with fixed $\{J_j\}$ is given by

$$Z = \sum_{\{\sigma_j\}} e^{\beta \sum_j J_j \sigma_j \sigma_{j+1}} = 2^L \prod_{j=1}^{L} (\cosh \beta J_j) \left(1 + \prod_j \tanh \beta J_j \right). \tag{5.65}$$

The correlation function with fixed $\{J_j\}$ is written as

$$\langle \sigma_j \sigma_{j+r} \rangle = \frac{\prod_{k=j}^{j+r-1} \tanh \beta J_k + \prod_{k=1}^{j-1} \tanh \beta J_k \prod_{k=j+r}^{L} \tanh \beta J_k}{1 + \prod_{k=1}^{L} \tanh \beta J_k}.$$

In the thermodynamics limit, this expression is reduced to

$$\langle \sigma_j \sigma_{j+r} \rangle \rightarrow \prod_{k=j}^{j+r-1} \tanh \beta J_k. \tag{5.66}$$

The averaged correlation function is written as follows:

$$[\langle \sigma_j \sigma_{j+r} \rangle]_{\text{av}} = \int \left[\prod_j dJ_j P(J_j) \right] \langle \sigma_j \sigma_{j+r} \rangle$$

$$= \left[\int dJ P(J) \tanh \beta J \right]^r = \left[\frac{1}{\beta} \ln \cosh \beta \right]^r. \tag{5.67}$$

From this, one obtains the correlation length

$$\xi = -r / \ln [\langle \sigma_j \sigma_{j+r} \rangle]_{\text{av}} = -\frac{1}{\ln \left[\frac{1}{\beta} \ln \cosh \beta \right]}. \tag{5.68}$$

In the low-temperature limit ($T \ll 1$), it is reduced to

$$\xi(T) \approx \frac{1}{T \ln 2}. \tag{5.69}$$

The energy with fixed $\{J_j\}$ is given by

$$\langle H \rangle = -\sum_{j=1}^{L} J_j \langle \sigma_j \sigma_{j+1} \rangle \xrightarrow{L \to \infty} -\sum_j J_j \tanh \beta J_j,$$

where we used Eq. (5.66) with $r = 1$. We take the average over $P(J_j)$.

$$\begin{aligned}
\varepsilon_{eq} \equiv \frac{[\langle H \rangle]_{av}}{L} &= -\frac{1}{L} \sum_{j=1}^{L} \int \left[\prod_{j=1}^{L} dJ_j P(J_j) \right] J_j \langle \sigma_j \sigma_{j+1} \rangle \\
&\xrightarrow{L \to \infty} -\int dJ P(J) J \tanh \beta J \\
&= -\frac{1}{2} - \frac{1}{\beta} \ln(1 + e^{-2\beta}) + \frac{\pi^2}{24} \frac{1}{\beta^2} - \frac{1}{2\beta^2} \sum_{n=1}^{\infty} \frac{(-1)^{n-1}}{n^2} e^{-2n\beta}.
\end{aligned}$$

$$\tag{5.70}$$

In the low-temperature limit, it is reduced to

$$\varepsilon_{eq} \approx -\frac{1}{2} + \frac{\pi^2}{24} \frac{1}{\beta^2}. \tag{5.71}$$

The ground state of the present system is the fully polarized state, while the paramagnetic state is realized in the equilibrium at any finite temperature. Hence the present system exhibits a discontinuous phase transition at zero temperature.

Now we consider the dynamics of the present system. Same as the pure case, we denote the transition probability of the Ising spin at site j from σ to $-\sigma$ as $w_j(\sigma)$. We also denote the probability distribution of the spin state $(\sigma_1, \cdots, \sigma_L)$ at the equilibrium as $p_{eq}(\sigma_1, \cdots, \sigma_L)$. From the principle of the detailed balance, one has a relation between w_j and p_{eq} as

$$\begin{aligned}
\frac{w_j(\sigma_j)}{w_j(-\sigma_j)} &= \frac{p_{eq}(\sigma_1, \cdots, -\sigma_j, \cdots, \sigma_L)}{p_{eq}(\sigma_1, \cdots, \sigma_j, \cdots, \sigma_L)} = \frac{\exp[-\beta \sigma_j (J_{j-1}\sigma_{j-1} + J_j \sigma_{j+1})]}{\exp[\beta \sigma_j (J_{j-1}\sigma_{j-1} + J_j \sigma_{j+1})]} \\
&= \frac{1 - \sigma_j (C_j^- \sigma_{j-1} + C_j^+ \sigma_{j+1})}{1 + \sigma_j (C_j^- \sigma_{j-1} + C_j^+ \sigma_{j+1})},
\end{aligned}$$

where we defined

$$C_j^{\pm} = \frac{1}{2} \left(\tanh\{\beta(J_{j-1} + J_j)\} \mp \tanh\{\beta(J_{j-1} - J_j)\} \right).$$

Therefore $w_j(\sigma_j)$ is written as

$$w_j(\sigma_j) = \frac{\alpha}{2}\left\{1 - \sigma_j(C_j^-\sigma_{j-1} + C_j^+\sigma_{j+1})\right\},$$

where α is an undetermined factor. We denote the probability of the spin state $(\sigma_1, \cdots, \sigma_L)$ at time t as $p(\sigma_1, \cdots, \sigma_L;t)$ and the expectation value of the Ising spin at a site j as $S_j(t)$, namely,

$$S_j(t) \equiv \langle\sigma_j\rangle(t) = \sum_{\{\sigma_j\}} \sigma_j P(\sigma_1, \cdots, \sigma_L;t).$$

From the Master equation for $p(\sigma_1, \cdots, \sigma_L;t)$, Eq. (5.15), one obtains the equation of motion of $S_j(t)$:

$$
\begin{aligned}
\frac{d}{dt}S_j(t) &= \sum_{\{\sigma_i\}} \sigma_j\left\{-\left(\sum_k w_k(\sigma_k)\right)p(\sigma_1, \cdots, \sigma_L;t)\right. \\
&\quad \left. + \sum_k w_k(-\sigma_k)p(\sigma_1, \cdots, -\sigma_k, \cdots, \sigma_L;t)\right\} \\
&= \sum_{\{\sigma_i\}} \sigma_j \sum_k \sigma_k \sum_{\sigma_k'=\pm\sigma_k} (-\sigma_k')w_k(\sigma_k')p(\sigma_1, \cdots, \sigma_k', \cdots, \sigma_L;t) \\
&= -2\sum_{\{\sigma_i\}} \sigma_j w_j(\sigma_j)p(\sigma_1, \cdots, \sigma_L;t) \\
&= -\alpha\left(S_j(t) - C_j^- S_{j-1}(t) - C_j^+ S_{j+1}(t)\right).
\end{aligned}
$$

We can remove α by arranging the definition of time. As a result, we obtain the Glauber's equation for the random Ising chain:

$$\frac{d}{dt}S_j(t) = -\left(S_j(t) - C_j^- S_{j-1}(t) - C_j^+ S_{j+1}(t)\right).$$

The motion of $S_j(t)$ is not trivial. Dhar and Barma [12] studied the relaxation of $S_j(t)$, assuming the initial condition that every spin has $\sigma_j = 1$. We do not explain the detail here, but provide the result. The relaxation time is the same as that of the pure system up to the leading order:

$$\tau_r(T) = \frac{1}{1 - \tanh 2/T}. \tag{5.72}$$

When the temperature is sufficiently low, it is reduced to

$$\tau_r(T) \approx \frac{1}{2}e^{4/T} \approx \frac{1}{2}e^{(4\ln 2)\xi}, \tag{5.73}$$

where we used Eq. (5.69) in the last formula.

5.4.1.2 Quench Dynamics

Now we consider a quench of the temperature with time [47]. We assume the schedule:

$$T(t) = -t/\tau, \qquad (5.74)$$

where t moves from $-\infty$ to 0 and τ stands for the inverse quench rate. Denoting the dimensionless temperature measured from the critical point by ε, ε is identical to T and Eq. (5.74) is followed by

$$\varepsilon(t) = -t/\tau. \qquad (5.75)$$

Now we discuss the Kibble–Zurek mechanism on the basis of Eq. (5.4), namely,

$$\tau_{\mathrm{r}}(\varepsilon(\hat{t})) = |\hat{t}|. \qquad (5.76)$$

Using Eqs. (5.69), (5.73), and (5.75), Eq. (5.76) leads us to an equation of ξ:

$$\frac{1}{2} e^{4\hat{\xi}} \approx \frac{\tau}{\hat{\xi} \ln 2}, \qquad (5.77)$$

where we defined $\hat{\xi} \equiv \xi(\varepsilon(\hat{t}))$. This equation cannot be solved analytically. However, taking the logarithm, one obtains from Eq. (5.77)

$$\hat{\xi} \approx \frac{1}{4} \ln \tau - \ln \frac{\hat{\xi} \ln 2}{2}. \qquad (5.78)$$

Since $\ln \hat{\xi}$ is negligible compared to $\hat{\xi}$ for $\hat{\xi} \to \infty$, one has for $\tau \to \infty$

$$\hat{\xi} \approx \frac{1}{4} \ln \tau. \qquad (5.79)$$

The density of kinks after the quench is estimated as

$$n \approx \frac{1}{\hat{\xi}} \approx \frac{1}{\frac{1}{4} \ln \tau - \ln \frac{\ln 2}{2n}} \approx \frac{4}{\ln \tau}. \qquad (5.80)$$

The residual energy after the quench is estimated from making an equation of ε. Using Eqs. (5.73), (5.75), and (5.76), an equation of $\hat{\varepsilon} \equiv \varepsilon(\hat{t})$ is given as

$$\frac{1}{2} e^{4/\hat{\varepsilon}} \approx \tau \hat{\varepsilon}. \qquad (5.81)$$

This equation is arranged as follows:

$$\hat{\varepsilon} \approx \frac{4}{\ln \tau + \ln 2\hat{\varepsilon}}. \tag{5.82}$$

The second term in the denominator of r.h.s. gives a double-logarithmic correction to $\ln \tau$. Since $\beta = 1/\varepsilon$, the residual energy is given by

$$\varepsilon_{\text{res}} \approx \frac{\pi^2}{24}\hat{\varepsilon}^2 \approx \frac{2\pi^2}{3}\frac{1}{(\ln \tau)^2}, \tag{5.83}$$

as far as the double-logarithmic correction is neglected. The scaling relation of the density of kinks and the residual energy obtained above have been confirmed by Monte Carlo simulation [47].

5.4.2 Quantum Quench

5.4.2.1 Properties at Fixed Transverse Field

We consider the random Ising chain in the random transverse field:

$$H = -\sum_j J_j \sigma_j^x \sigma_{j+1}^x - \Gamma \sum_j h_j \sigma_j^z. \tag{5.84}$$

We assume that h_j is a random constant and obeys the same distribution as J_j, namely,

$$P(h_i) = P(J_j) = \begin{cases} 1 \text{ for } J_i, h_i \in [0, 1] \\ 0 \text{ otherwise} \end{cases}. \tag{5.85}$$

The present system undergoes a quantum phase transition at $\Gamma = 1$. We define the dimensionless transverse field measured from the critical point as

$$\varepsilon = \Gamma - 1. \tag{5.86}$$

According to Fisher [19], the typical value of a physical quantity is different from the averaged value in the present system at the critical point. Hence the critical properties of the present system differ from those of the pure model. We do not trace the derivation here but provide only results for the critical exponent ν of the correlation length and the dynamical exponent z.

The correlation length ξ that appears in the averaged correlation function at zero temperature, $[\langle \sigma_j^x \sigma_{j+r}^x \rangle]_{\text{av}}$, is scaled as

$$\xi(\varepsilon) \approx \varepsilon^{-\nu}, \quad \nu = 2, \quad (0 < \varepsilon \ll 1). \tag{5.87}$$

The coherence time τ and the correlation length are related through the dynamical exponent z as

$$\tau_{\text{coh}}(\varepsilon) \approx \xi(\varepsilon)^z, \quad z = \frac{1}{\ln \Gamma} \approx \frac{1}{\varepsilon}, \quad (0 < \varepsilon \ll 1). \tag{5.88}$$

The dynamical exponent z diverges at the critical point, $\varepsilon = 0$. This extraordinary critical behavior reflects that the critical point is the infinite-randomness fixed point [20, 34].

5.4.2.2 Quench Dynamics

Now we consider a quench of the transverse field according to the following schedule:

$$\Gamma(t) = 1 - t/\tau, \tag{5.89}$$

namely,

$$\varepsilon(t) = -t/\tau, \tag{5.90}$$

where t is assumed to move from $-\infty$ to τ. τ stands for the inverse of the quench rate. We are going to discuss the Kibble–Zurek mechanism, according to [16]. Equation (5.4) determines the time $\hat{t}$ at which the coherence time τ_{coh} becomes identical to the remaining time $|t|$ to the critical point;

$$\tau_{\text{coh}}(\varepsilon(\hat{t})) = |\hat{t}|.$$

Expressing the both sides of this equation in terms of $\hat{\xi} = \xi(\varepsilon(\hat{t}))$ with the use of Eqs. (5.87), (5.88), and (5.90), one obtains an equation of ξ:

$$\hat{\xi}^{\hat{\xi}^{1/2}} \approx \tau/\hat{\xi}^{1/2}. \tag{5.91}$$

This equation cannot be solved analytically, but it is reduced to [16]

$$\hat{\xi} \approx \left(\frac{\ln \tau}{\ln \hat{\xi}} \right)^2 \tag{5.92}$$

for $\tau \to \infty$, where we employed an approximation: $\hat{\xi}^{1/2} + \frac{1}{2} \approx \hat{\xi}^{1/2}$ for large $\hat{\xi}$. The variation of $\ln \hat{\xi}$ in the denominator of r.h.s. is moderate compared to $\hat{\xi}$ when $\hat{\xi}$ is large. Hence, as far as the logarithmic correction is ignored, the correlation length $\hat{\xi}$ after the quench behaves $\hat{\xi} \approx (\ln \tau)^2$. We estimate that the density of kinks after the quench is scaled as

$$n \approx \frac{1}{\hat{\xi}} \approx \frac{1}{(\ln \tau)^2}. \tag{5.93}$$

The same result has been obtained in [5] by an analysis using the Landau–Zener formula and the distribution of energy gaps at the critical point. This logarithmic scaling of the density of kinks has been numerically confirmed in [16, 5].

5.5 Conclusion

5.5.1 Summary of the Results

We discussed the classical quench and the quantum quench in the pure and random Ising chains in the previous two sections. We summarize the results here. Throughout the present section, τ stands for the inverse quench rate.

Pure Ising chain: the classical quench leaves the density of kinks, which is scaled as

$$n_C \sim 1/\sqrt{\tau}$$

as far as the logarithmic correction is ignored. As for the quantum quench, the density of kinks after the quench is scaled as

$$n_Q \sim 1/\sqrt{\tau}.$$

There is no significant difference between scaling of n_C and n_Q.

Random Ising chain: the classical quench yields a logarithmic dependence of the density of kinks on τ:

$$n_C \sim 1/(\ln \tau).$$

The residual energy after the classical quench is scaled as

$$\varepsilon_{res,C} \sim 1/(\ln \tau)^2.$$

The quantum quench also yields logarithmic scaling, but the power is different:

$$n_Q \sim 1/(\ln \tau)^2.$$

We neglected a double-logarithmic correction in all cases. The power of $1/\ln \tau$ in n_Q is twice larger than that in n_C. The difference between n_C and n_Q is significant. The analytic result on the residual energy after the quantum quench has not been available. The numerical result given by Caneva et al. is $\varepsilon_{res,Q} \sim 1/(\ln \tau)^\zeta$ with $\zeta \approx 3.4$ [5]. This suggests a significant difference in $\varepsilon_{res,C}$ and $\varepsilon_{res,Q}$.

Comparing the scaling relations of the density of kinks in the random case, the decay rate of the density of kinks is larger in the quantum quench. Regarding the random Ising chain as a cost function of an optimization problem, this fact implies that quantum annealing performs better than simulated annealing significantly. Although

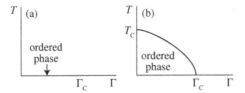

Fig. 5.1 Two types of phase diagrams which involve a quantum phase transition. The *vertical axis* indicates the temperature, while the *horizontal axis* corresponds to the parameter which controls the quantum fluctuation. The Ising chains in the transverse field studied in this chapter provide the left diagram

this problem has a trivial solution, the advantage of quantum annealing over simulated annealing in the sense of a faster decay of a residual error was shown analytically by the present study.

5.5.2 Discussion and Future Problems

The physics of the quench dynamics across a phase transition is different in the quantum quench and the classical quench. The former relates with the quantum adiabatic time evolution at zero temperature. The latter does with the quasi-static evolution with changing the temperature. However, the Kibble–Zurek mechanism tells that the difference between them comes from the critical properties of the correlation length and the relaxation (coherence) time. According to the theory of critical phenomena, the critical properties are classified into universality classes. The different scaling relations in the quantum quench and the classical quench reflect the fact that the universality class is different in the involved quantum phase transition and thermal phase transition.

We here like to ask whether the results presented above can be generalized to other systems. The models which exhibit a quantum phase transition are classified into two types. The first type involves a quantum phase transition but no thermal phase transition at any finite temperature (Fig. 5.1a). The second type involves both a quantum phase transition and a thermal phase transition at a finite temperature (Fig. 5.1b). The Ising chains studied in this chapter are prototypes of the first type. Numerical studies have shown that for the two-dimensional spin-glass model quantum annealing has a faster decay rate of the residual energy than simulated annealing [31, 40]. The spin-glass model studied in [40, 31] is considered to belong to the first type [24]. From the results for these one- and two-dimensional models, we anticipate that there is a significant difference between the quantum quench and the classical quench in the first type of random system. It is a future work to verify this conjecture.

The system in a dimension higher than the lower critical dimension is classified into the second type. The pure Ising ferromagnet in two dimension, the Heisenberg ferromagnet in three dimension, an atomic Bose–Einstein condensate in three dimension, etc., are included in the second type. Systems which represent optimization problems usually involve long-range interactions. The effective dimension of these systems is much higher than the lower critical dimension. Hence they also belong to the second type. For such systems, no analytic result on quench dynamics has been available so far. To reveal the dynamics across a quantum and classical critical point of a system in high dimension is a challenging problem in future.

Acknowledgments The author acknowledges T. Caneva, G.E. Santoro, and H. Nishimori for fruitful discussions and comments. The present work is partially supported by CREST, JST, and by Grant-in-Aid for Scientific Research (No. 20740225) of MEXT, Japan.

References

1. B. Apolloni, C. Carvalho and D. de Falco, Stochastic Process. Appl. **33**, 233 (1989).
2. D.A. Battaglia, G.E. Santoro and E. Tosatti, Phys. Rev. E **71**, 066707 (2005).
3. P. Calabrese and J. Cardy, J. Stat. Mech., P04010 (2005).
4. P. Calabrese and J. Cardy, Phys. Rev. Lett. **96**, 136801 (2006).
5. T. Caneva, R. Fazio and G. E. Santoro, Phys. Rev. B **76**, 144427 (2007).
6. E. Canovi, D. Rossini, R. Fazio and G. E. Santoro, J. Stat. Mech. P03038 (2009).
7. M.A. Cazalilla, Phys. Rev. Lett. **97**, 156403 (2006).
8. V. Černý, J. Optimization Theory Appl. **45**, 41 (1985).
9. B. Damski and W.H. Zurek, Phys. Rev. A **73**, 063405 (2006).
10. A. Das and B.K. Chakrabarti, (eds.), *Quantum Annealing and Related Optimization Methods*, Lect. Notes Phy. (Springer-Verlag, Berlin, 2005).
11. A. Das and B.K. Chakrabarti, Rev. Mod. Phys. **80**, 1061 (2008).
12. D. Dhar and M. Barma, J. Stat. Phys. **22**, 259 (1980).
13. U. Divakaran, A. Dutta and D. Sen, Phys. Rev. B **78**, 144301 (2008).
14. U. Divakaran, V. Mukherjee, A. Dutta and D. Sen, J. Stat. Mech. P02007 (2009).
15. J. Dziarmaga, Phys. Rev. Lett. **95**, 245701 (2005).
16. J. Dziarmaga, Phys. Rev. B **74**, 064416 (2006).
17. E. Farhi, J. Goldstone, S. Gutmann, J. Lapan, A. Lundgren and D. Preda, Science **292**, 472 (2001).
18. A.B. Finnila, M.A. Gomez, C. Sebenik, C. Stenson and J.D. Doll, Chem. Phys. Lett. **219**, 343 (1994).
19. D.S. Fisher, Phys. Rev. B **51**, 6411 (1995).
20. D.S. Fisher, Physica A **263**, 222 (1999).
21. M.R. Garey and D.S. Johnson, *Computers and Intractability: A Guide to the Theory of NP-Completeness* (Freeman, San Francisco, 1979).
22. R.J. Glauber, J. Math. Phys. **4**, 294 (1963).
23. M. Greiner, O. Mandel, T. Esslinger, T.W. Hänsch and I. Bloch, Nature **415**, 39 (2002).
24. A.K. Hartmann and A.P. Young, Phys. Rev. B **64**, 180404(R) (2001).
25. C. Itzykson and J.-M. Drouffe, *Statistical Field Theory* (Cambridge University Press, Cambridge, 1989).
26. T. Kadowaki and H. Nishimori, Phys. Rev. E **58**, 5355 (1998).
27. T. Kadowaki, *Thesis, Tokyo Institute of Technology*, arXiv:quant-ph/0205020.
28. T.W.B. Kibble, J. Phys. A **9**, 1387 (1976).
29. T.W.B. Kibble, Phys. Rep. **67**, 183 (1980).
30. S. Kirkpatrick, C.D. Gelett and M.P. Vecchi, Science **220**, 671 (1983).
31. R. Martoňák, G.E. Santoro and E. Tosatti, Phys. Rev. B **66**, 094203 (2002).
32. R. Martoňák, G.E. Santoro and E. Tosatti, Phys. Rev. E **70**, 057701 (2004).

33. S. Morita and H. Nishimori, J. Math. Phys. **49**, 125210 (2008).
34. O. Motrunich, S.-C. Mau, D.A. Huse and D.S. Fisher, Phys. Rev. B **61**, 1160 (2000).
35. V. Mukherjee, U. Divakaran, A. Dutta and D. Sen, Phys. Rev. B **76**, 174303 (2007).
36. F. Pellegrini, S. Montangero, G.E. Santoro and R. Fazio, Phys. Rev. B **77**, 140404 (2008).
37. A. Polkovnikov, Phys. Rev. B **72**, 161201(R) (2005).
38. S. Sachdev, *Quantum Phase Transitions*, (Cambridge University Press, Cambridge, 1999).
39. L.E. Sadler, J.M. Higbie, S.R. Leslie, M. Vengalattore and D.M. Stamper-Kurn, Nature **443**, 312 (2006).
40. G.E. Santoro, R. Martoňák, E. Tosatti and R. Car, Science **295**, 2427 (2002).
41. G.E. Santoro and E. Tosatti, J. Phys. A **39**, R393 (2006).
42. D. Sen, K. Sengupta and S. Mondal, Phys. Rev. Lett. **101**, 016806 (2008).
43. K. Sengupta, S. Powell and S. Sachdev, Phys. Rev. A **69**, 053616 (2004).
44. K. Sengupta, D. Sen and S. Mondal, Phys. Rev. Lett., **100**, 077204 (2008).
45. S. Suzuki and M. Okada, J. Phys. Soc. Jpn. **74**, 1649 (2005).
46. S. Suzuki, H. Nishimori and M. Suzuki, Phys. Rev. E **75**, 051112 (2007).
47. S. Suzuki, J. Stat. Mech., P03032 (2009).
48. C.N. Weiler, T.W. Neely, D.R. Scherer, A.S. Bradley, M.J. Davis and B.P. Anderson, Nature **455**, 948 (2008).
49. W.H. Zurek, Nature (London) **317**, 505 (1985).
50. W.H. Zurek, U. Dorner and P. Zoller, Phys. Rev. Lett. **95**, 105701 (2005).

Chapter 6
Quantum Phase Transition in the Spin Boson Model

S. Florens, D. Venturelli, and R. Narayanan

6.1 Introduction

Quantum phase transitions (QPT) have recently become a widespread topic in the realm of modern condensed matter physics. QPT are phase transformations that occur at the absolute zero of temperature and are triggered by varying a temperature-independent control parameter like pressure, doping concentration, or magnetic field. There are various examples of systems showing quantum critical behavior, which include the antiferromagnetic transition in heavy fermion material like $CeCu_{6-x}Au_x$, that is brought about by changing the Au doping [10]. Another prototypical example of a system exhibiting quantum critical behavior is the quantum Hall effect, wherein a two-dimensional electron gas is tuned, via an externally applied magnetic field, through a quantum critical point (QCP) that intervenes between two quantized Hall plateaux. Other examples of QPT include the ferromagnetic transition in metallic magnets as a function of applied pressure and the superconducting transition in thin films.

Since there are such a wide range of experimentally accessible systems that show quantum critical behavior, it is imperative that we understand QPT at a fundamental level. We shall here endeavor to do just so by giving an introductory account of this fascinating phenomenon. As a striking illustration, we will be comparing and contrasting QPT with the case of more usual thermal (classical) phase transitions (as will be seen later on, thermal phase transitions are also referred to as classical transition, since quantum fluctuations become unimportant in their vicinity). Let us first begin by discussing the ferromagnetic transition, in order to better illustrate the rich phenomenology of phase transitions (both classical and quantum). The route that we

S. Florens (✉)
Institut Néel, CNRS and UJF, 38042 Grenoble, France, `serge.florens@grenoble.cnrs.fr`

D. Venturelli
Institut Néel, CNRS and UJF, 38042 Grenoble, France, `davide.venturelli@grenoble.cnrs.fr`

R. Narayanan
Department of Physics, Indian Institute of Technology, Chennai 600036, India, `rnarayanan@physics.iitm.ac.in`

Florens, S. et al.: *Quantum Phase Transition in the Spin Boson Model.* Lect. Notes Phys. **802**, 145–162 (2010)
DOI 10.1007/978-3-642-11470-0_6

take here to understand the fundamentals of QPTs is as follows: We shall first review
the basic phenomenology of classical (thermal) phase transitions. Then, we shall
illustrate via heuristic arguments how quantum fluctuations can be disregarded in
the vicinity of a thermal phase transition. These arguments also provide clues to the
domain in the phase diagram where one expects quantum fluctuations to dominate.
Also, we shall briefly discuss the question of observability of QPTs. Finally, we
shall end with a discussion of the so-called quantum to classical mapping.

Let us first start with classical (thermal) phase transitions. As a physical system,
say a ferromagnet, approaches its ordering, there is a length scale called the correlation length, ξ, that diverges in a power-law fashion when one comes closer to the
critical point, $\xi \sim |t|^{-\nu}$, so that the system becomes progressively self-similar. Here,
t is dimensionless parameter characterizing the distance to criticality, $t = \frac{T-T_c}{T_c}$ (for
a thermal transition), and T_c is the critical temperature where the phase transformation occurs. Now, the above divergence of the correlation length encapsulates
the information that the fluctuations of the order parameter (say the magnetization)
become spatially long ranged as the system approaches the critical point. Analogous to ξ one can define a timescale ξ_τ that also diverges as a power law as one
approaches a second-order transition. Thus, we have

$$\xi_\tau = \xi^z = |t|^{-\nu z}. \qquad (6.1)$$

The quantity z that controls the divergence of ξ_τ is the so-called dynamical exponent.
Now, associated with this timescale we can define a frequency scale $\omega_c \propto 1/\xi_\tau$ and
through it a corresponding energy $\hbar\omega_c$, which encodes information pertaining to the
energy scale related with order parameter fluctuations. This quantity $\hbar\omega_c$ competes
with $k_B T_c$, the typical energy associated to thermal fluctuations. Now, the question
of importance of quantum fluctuations can be re-cast into a query of which among
these two energy scales prevails. Since $\omega_c \to 0$ as one approaches the critical point,
the energy scale of thermal fluctuations (for any non-zero T_c) always dominates over
the scale $\hbar\omega_c$. In other words for a transition that happens at a finite temperature,
$\hbar\omega_c \ll k_B T_c$. Thus, it can be argued that asymptotically close to a finite temperature transition, it is the thermal fluctuations that are the driving mechanism. This
irrelevance of quantum fluctuations near a thermal phase transition is the reason why
they are given the moniker "classical phase transition."

Now, from our discussion in the previous paragraph it is but obvious that if the
transition were to occur at $T = 0$ (tuned by a non-thermal parameter like doping or
pressure), then the fluctuations that will drive the transition will be wholly quantum
mechanical in origin. It is obvious that one then needs to apply ideas from quantum
statistical mechanics to understand QPT, as pioneered by Hertz in a seminal paper
[6] to tackle the problem of quantum criticality in itinerant magnetic systems. By
using this case of the quantum magnet as a test-bed example, Hertz [6] showed
that any generic d dimensional quantum system can be mapped onto an equivalent
$d + z$ dimensional classical model. This statement is referred to as the quantum
to classical mapping and is of fundamental importance in the field of QPTs. By
using the quantum to classical mapping one can show that the critical behavior of

the quantum model is equivalent to that of a classical model but in z higher dimensions. Although this mapping is believed to be robust for insulating magnets, it was however later shown [1] that Hertz's conclusions were erroneous for a large class of itinerant QPT. This breakdown in fact occurs due to the presence of soft modes in the systems (e.g., the particle-hole excitations in itinerant magnets) other than the order-parameter modes. The presence of these modes induces an effective long-ranged interaction between the order parameter modes, thereby altering the critical behavior [1], as compared to Hertz's original results.

Since QPT occur at zero temperature, it was initially thought that the study of these phase transitions was a mere academic exercise. However, it was soon realized that the presence of a zero-temperature critical point (practically inaccessible) can actually influence the behavior of the system at *finite* temperatures. In other words, at any finite temperature, the critical singularities associated with the QCP are cut by the temperature, so that one observes non-trivial temperature dependence of various observables in a so-called quantum critical regime. The calculation of the quantum critical regime for various models is well beyond the scope of this work. However, the interested reader is directed toward the following papers investigating the effect of non-zero temperatures on QPT in magnetic systems [12, 15]. Also, one will refer to Sect. 6.2.2 for a brief description of the quantum critical regime in the spin boson model (SBM), the specific model of interest in this chapter.

Thus, from the discussion of previous paragraphs, it is obvious that QPT are an extremely interesting physical phenomenon to study. As alluded to before, in this chapter we choose to study a specific toy model example, namely the QPT encountered in the spin boson model (SBM), a variant of Caldeira – Leggett type models [8], wherein a quantum particle is subjected to an external dissipative environment. In the case of the SBM, this quantum "particle" is essentially a two-level system, such as a spin-$\frac{1}{2}$ impurity. While there have been many studies on dissipative quantum models that focused on the effect of decoherence on intermediate timescales, their behavior in the long time limit in the presence of quantum critical points remains relatively unexplored. However, the study of such regimes is extremely important as anomalous low-energy properties emerge due to quantum critical modes. In other words, due to the presence of a QCP, the SBM can display non-trivial dynamics at very long times.

As a more general remark, we note that this model can also be used to study quantum criticality at the level of a single spin-$\frac{1}{2}$ impurity embedded in a correlated system such as Mott insulators [3, 13, 20] or magnetic metals [7, 9]). It also appears as an effective theory for bulk materials themselves (e.g., in quantum spin glasses [17], heavy fermion compounds [16, 19]) via the framework of DMFT.

The SBM is introduced in Sect. 6.2. The various phases of the SBM and the possibility of a QPT between them are discussed in Sect. 6.2.2. In Sect. 6.3, we re-write the SBM by using a Majorana fermion representation for the impurity spins. Section 6.4 is devoted to the derivation of the RG equations by using the Majorana representation presented in Sect. 6.3. In Sect. 6.5, we look at the consequences of the flow equation derived in Sect. 6.3. Section 6.6 is dedicated to the quantum/classical mapping of the SBM to the long-ranged Ising model. Section 6.7 is concerned with the development of a special identity in the SBM model that is used in Sect. 6.8 to derive the so-called Shiba's relation in the case of the sub-ohmic spin boson model. Section 6.9 is dedicated to a discussion on the status of the breakdown of quantum to classical mapping in the SBM, that we use as a conclusion and future outlook regarding quantum phase transitions in dissipative models.

6.2 The Spin Boson Model

As stated earlier, the SBM describes the effect of an external dissipative environment on the quantum mechanical evolution of a two-level system. We will introduce in Sect. 6.2.1 the general properties of the SBM and present in Sect. 6.2.2 its possible phase diagram, obtained on heuristic grounds via an analysis of the various limiting cases.

6.2.1 The Model

The SBM involves a single spin-$\frac{1}{2}$ impurity $\vec{S}$, interacting with a set of bosonic bath variables a_i and $a_i^\dagger$ (in second quantization). The interaction between the bath's oscillator displacement and the spin is controlled via a coupling constant λ. Thus, the SBM Hamiltonian has the general functional form:

$$H = -\Delta S^x + \varepsilon S^z + \lambda S^z \sum_i \left(a_i^\dagger + a_i \right) + \sum_i \omega_i a_i^\dagger a_i. \tag{6.2}$$

Here, in Eq. (6.2), Δ and ε are the transverse and longitudinal magnetic fields, respectively, applied to the quantum spin. A physical sketch for such a SBM, wherein a two-level impurity (when the bias field ϵ is set to zero) is connected to an external environment, is depicted in Fig. 6.1. All that remains to completely specify the model is to endow the bosonic degrees of freedom with a spectrum. This bosonic density of states (DoS) is taken here to be continuous and power-law like and conforms to the functional form:

$$\rho(\omega) \equiv \sum_i \delta(\omega - \omega_i) = \frac{(s+1)\omega^s}{\Lambda^{1+s}} \theta(\omega)\theta(\Lambda - \omega). \tag{6.3}$$

Here, in Eq. (6.3), Λ is a high-energy cutoff. When the exponent s is such that $0 < s < 1$, then the model is said to be in the sub-ohmic regime, the case $s = 1$ is referred to as ohmic, while the case $s > 1$ is called super-ohmic. In fact, as will be seen in the course of this chapter, the quantum critical behavior of the SBM is crucially dependent on the exponent s controlling the behavior of the bath spectrum.

It is also convenient to define a new bosonic variable corresponding to the "local" displacement:

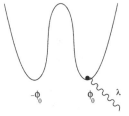

Fig. 6.1 This figure is a pictorial sketch of the SBM. It represents a two-level system, e.g., a particle in two potential minima $\pm\phi_0$, or a spin-$\frac{1}{2}$ impurity, coupled to a bath of harmonic oscillators (such as phonons, nuclear spins) via a coupling constant λ (*wavy line* in this figure)

$$\phi \equiv \sum_i \left(a_i + a_i^\dagger \right), \tag{6.4}$$

which has an associated DoS given by

$$\rho_\phi(\omega) = -\frac{(s+1)|\omega|^s}{\Lambda^{1+s}} \text{sgn}(\omega)\theta(\Lambda^2 - \omega^2). \tag{6.5}$$

To make comparison with existing literature, one can alternatively characterize the bath by means of a spectral function:

$$J(\omega) \equiv \sum_i \pi\lambda^2\delta(\omega - \omega_i) = 2\pi\alpha\omega^s\Lambda^{1-s}\theta(\omega)\theta(\Lambda - \omega). \tag{6.6}$$

Here α is the non-dimensional dissipation strength defined as $\alpha = (s+1)\frac{\lambda^2}{\Lambda^2}$.

6.2.2 The Phases of the SBM

As promised in Sect. 6.2, we will study here the possible phase diagram of the SBM by looking at situations where either one of the two parameters λ or Δ dominates. For instance, let us first consider the case where the dissipative coupling λ is set to zero. The SBM then becomes equivalent to the case of an isolated spin in a transverse magnetic field. It is well known that such a system displays Rabi oscillations. That is, if one were to start with an initial state pointing "up" along the z-direction, then the transverse field Δ periodically drives the system between up and down configurations. This limiting case $\lambda = 0$ is in fact adiabatically related to a whole phase at non-zero λ, dubbed for obvious reasons the delocalized phase, where coherent spin oscillations are expected to occur (at least for small enough λ). We note that the average $\langle S^x \rangle$ is always non-zero as long as the transverse field Δ is finite and thus cannot play the role of an order parameter. However, we can pursue a magnetic analogy by noting that the longitudinal spin average $\langle S^z \rangle$ is identically zero in the delocalized phase, so that we can really relate this portion of the phase diagram to a ground state with zero magnetization. In fact in the alternative regime, i.e., when the dissipation λ dominates over the transverse field Δ, the ground state becomes doubly degenerate, as can be checked on the trivial limiting case $\Delta = 0$, where the z spin component is clearly conserved within the Hamiltonian Eq. (6.3). A simple physical picture emerges, with the system localizing in one of the two minima at $\pm\phi_0$, see Fig. 6.1. Assuming adiabaticity by switching on the transverse field, we arrive to the so-called localized phase, wherein the spontaneous magnetization $\langle S^z \rangle \sim \langle \phi \rangle \equiv M \neq 0$. Now, so far by using heuristic arguments, we have shown that the SBM allows for the existence of a delocalized phase, with $\langle \phi \rangle = 0$, and a localized phase, with $\langle \phi \rangle \neq 0$. Thus, it is quite plausible that a second-order phase transition takes place between the two phases. As we shall see explicitly in Sect. 6.5, there is indeed a second-order quantum localization/delocalization transition for all $0 < s \leq 1$.

The generic phase diagram for the SBM is shown in Fig. 6.2, where a $T = 0$ phase transition, separating localized and delocalized phases, takes place at a critical value α_c of the adimensional dissipation strength. The interesting quantum critical regime emerges above the critical point at finite temperature, where anomalous behavior of all physical quantities is expected.

For instance, the longitudinal spin susceptibility in the quantum critical regime obeys the behavior $\chi_z(T) \sim 1/T^s$, as opposed to the conventional $1/T$ Curie law expected for the whole localized phase. This anomalous power-law behavior is a direct signature of the QCP at α_c and will be demonstrated in the following sections. We note that for the ohmic $s = 1$ case, the conventional treatment for studying the QPT is to map the SBM into an anisotropic Kondo model (AKM) and use previous knowledge on the scaling properties of this well-known Hamiltonian. However, in this chapter, we shall follow a less well-trodden path, namely, performing a

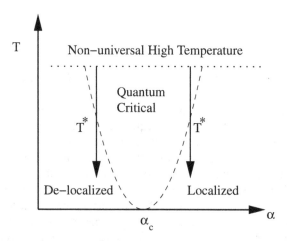

Fig. 6.2 This figure shows the generic phase diagram of the SBM model. Here, a quantum critical point α_c separates localized and delocalized phases, from which a quantum critical region emerges at finite temperature. The scale T^* is the crossover temperature below which various physical observables revert from quantum critical behavior to those associated with the localized or the delocalized phase

renormalization group (RG) calculation directly within the SBM, using a spin representation in terms of Majorana fermions (see Sect. 6.3 for further details). This formalism has the advantage that it can be easily adapted to perform calculations in the sub-ohmic limit, i.e., $0 < s < 1$, see Sect. 6.4.

6.3 The SBM Using the Majorana Representation

In Sect. 6.4 we aim to derive the RG equations for the SBM, based on a perturbative analysis around the localized limit, i.e., $\Delta = 0$. A technical difficulty on this path comes from the fact that the quantum spin-$\frac{1}{2}$ impurity does not follow either bosonic or fermionic commutation relations. Thus, standard calculations based on Wick's theorem cannot be invoked. One of the many ways to avoid this problem is to map the spin-$\frac{1}{2}$ operator onto fermionic degrees of freedom, which can be done in particular using so-called Majorana fermions. The use of this mapping to condensed matter is relatively recent and for a more detailed explanation the readers are referred to [11, 18].

The mapping between the spin-$\frac{1}{2}$ impurity and the Majorana fermions obeys the following correspondence principle:

$$\vec{S} = -\frac{i}{2}\vec{\eta} \times \vec{\eta}.$$
(6.7)

Here, in Eq. (6.7) the η fields represent a triplet of Majorana (real) fermions $\eta \equiv (\eta_1, \eta_2, \eta_3)$ that satisfy the following anticommutation relations: $\{\eta_i, \eta_j\} = \delta_{ij}$. Now, in addition to these Majoranas defined above, one can construct another fermionic field, $\Phi = 2i\eta_1\eta_2\eta_3$, that commutes with the Hamiltonian, and constitutes hence a conserved quantity, with the constraint $\Phi^2 = \frac{1}{2}$. A very useful relation for describing the spin dynamics is given by the correspondence (see [11])

$$\vec{\eta} = 2\Phi\vec{S}.$$
(6.8)

Now, in terms of the Majorana fermions (see Eq. (6.7)), and after the redefinitions $\{S_x \to S_3, S_y \to S_2, S_z \to S_1\}$ which amounts to a $\pi/2$ rotation around the y direction, the Hamiltonian of the SBM can be expressed as

$$H = -i\frac{\Delta}{2}(\eta_1\eta_2 - \eta_2\eta_1) + H_B - i\lambda\phi\eta_2\eta_3.$$
(6.9)

Now in what follows, we will use Eq. (6.9) to perform the perturbative RG analysis. Before we go on to do so, a word of caution regarding the fermionic mapping is in order: Any mapping of the spin-$\frac{1}{2}$ impurity to fermionic operators tends to enlarge the dimensionality of the Hilbert space. How such an enlargement is obviated in the case of the Majorana representation is technical matter that goes well beyond the scope of this chapter, and the reader is directed to [11, 18] for further details.

6.4 Perturbative Renormalization Group in the Localized Regime

In this section, we will perform a perturbative RG treatment starting from the localized phase, i.e., from the limit in which $\Delta = 0$. Our plan of action to derive the RG equations is as follows: We initially start with a model of free spin ($\Delta = 0$, $\alpha = 0$) and then perform a perturbative analysis in both Δ and α, leading to renormalizations of the dissipation α and the transverse field Δ, which depend explicitly on a generic cutoff scale Λ. Following the philosophy of the RG, one aims to compute the renormalized parameters at a lower cutoff scale, Λ', leading to the so-called flow equations.

The key ingredients in developing the perturbation theory are the free Majorana fermion propagator G_η^{free}, as well as the Δ vertex shown in Fig. 6.3, and the λ vertex (first diagram appearing in Fig. 6.4). The free fermion propagator in Matsubara frequency $\omega_n = (2n + 1)\pi T$ at finite temperature T reads $G_\eta^{\text{free}}(i\omega_n) = 1/i\omega_n$. One can then first construct the vertex function Γ_α related to the dissipative coupling λ, shown in Fig. 6.4. The functional form of the vertex function can easily be deduced to be

$$\Gamma_\alpha(\omega, \Lambda) = \frac{\lambda}{\Lambda} + \frac{\lambda}{\Lambda} \Delta^2 G_\eta^{\text{free}}(\omega)^2. \tag{6.10}$$

Here, once again Λ is cutoff scale, set, e.g., by temperature or the bandwidth of the bosonic modes, and ω is a frequency. The above equation can be effectively re-written in terms of an adimensional transverse field $h = 2\Delta/\Lambda$, so that the renormalized dissipation reads

$$\Gamma(\omega, \Lambda) = \frac{\lambda}{\Lambda} \left[1 - \frac{h^2}{4} \exp\left(2 \ln \frac{\Lambda}{\omega} \right) \right]. \tag{6.11}$$

Now, as stated in the introductory part of this section, we re-scale the cutoff Λ to $\Lambda' = \Lambda - d\Lambda$. Under such a re-scaling the vertex function can be re-written as follows:

$$\Gamma(\omega, \Lambda') = \frac{\lambda}{\Lambda'} \left(1 + \frac{d\Lambda}{\Lambda} \right) \left[1 - \frac{h^2}{4} \exp\left(2 \ln \frac{\Lambda}{\omega} \right) + \frac{h^2}{2} \frac{d\Lambda}{\Lambda} \right]. \tag{6.12}$$

The above equation can be re-written in the form of the usual RG β-function by including the frequency-dependent vertex function into the redefinition of the coupling constant. Then in terms of the logarithmic differential $d\ell = -\frac{d\Lambda}{\Lambda}$, Eq. (6.12) can be re-cast into the form

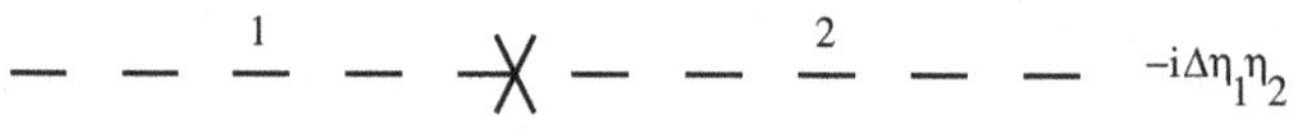

Fig. 6.3 Pictorial representation of the Δ vertex

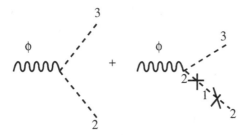

Fig. 6.4 Lowest order diagrams involved in the renormalization of the dissipative term λ

$$\frac{d\lambda}{d\ell} = -\frac{\lambda}{2}h^2, \tag{6.13}$$

and finally more compactly expressed in terms of the dimensionless dissipation $\alpha = 2\lambda^2/\Lambda^2$ as

$$\frac{d\alpha}{d\ell} = -\alpha h^2. \tag{6.14}$$

Now, in a similar vein one can calculate the first corrections to the transverse field Δ, with the diagrams depicted in Fig. 6.5, The technical details of this calculation are very similar to the above calculation, and the final flow equation, written in terms of the scaled magnetic field $h = \Delta/\Lambda$, reads

$$\frac{dh}{d\ell} = (1 - \alpha)h. \tag{6.15}$$

The RG equations that we have so far derived are for the case of the ohmic damping. The derivation of the flow equations in the non-ohmic limit is quite straight forward and can be performed by following the technical details elucidated above. Thus, for the sake of brevity we will not perform these computations here. Instead, we will just quote the results of such an exercise. In the presence of non-ohmic dissipation the α flow equation of Eq. (6.14) gets modified into

$$\frac{d\alpha}{d\ell} = -\alpha h^2 + (1 - s)\alpha. \tag{6.16}$$

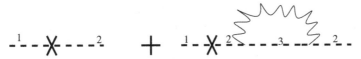

Fig. 6.5 Diagrams that are involved in the renormalization of the transverse field Δ

However, the flow of the magnetic field h retains its functional form given in Eq. (6.15), even in the presence of non-ohmic dissipation.

6.5 Analyzing the RG Flow

In this section we shall discuss the RG flow equations that were derived above. In Sect. 6.5.1, we shall first analyze the β functions for the ohmic case ($s = 1$). Then, in Sect. 6.5.2, we shall show that the super-ohmic case ($s > 1$) is bereft of any critical points. Finally in Sect. 6.5.3, we shall analyze the critical behavior when the bath spectrum is sub-ohmic in character ($0 < s < 1$).

6.5.1 The RG Equations for the Ohmic Case ($s = 1$)

The situation of the ohmic bath spectrum is probably one of the most well-understood case in the study of SBM. This is due to the fact that a linear dispersion of the bath DoS lends itself to an exact mapping to the anisotropic Kondo model (AKM) [4, 8]. Due to this mapping, it is known that the critical dissipation occurs at $\alpha_c = 1$ for small non-zero Δ, and that the phase transition is of the Kosterlitz – Thouless type (infinite order). The flow equations that are found through the mapping to the AKM match the β functions that we obtained by using the Majorana representation (see Eqs. (6.14) and (6.15)). From the structure of these β functions of the ohmic SBM, it is amply clear that the term $-\alpha h^2$ drives the dissipative coupling to zero whenever $\alpha < 1$. However, in the regime $\alpha > 1$, it is now the transverse field term h that is driven to zero, with the dissipative coupling α renormalizing to a finite value. Furthermore, in the limit $\alpha > 1$ we see that the RG equations, Eqs. (6.14) and (6.15), have in fact a line of stable fixed points at zero field, the typical signature of a phase transition of the Kosterlitz-Thouless type. This discussion is encapsulated by Fig. 6.6, which represents the various RG trajectories that are obtained by numerically solving Eqs. (6.14) and (6.15), for various initial values of α and h. From this flow diagram, it is clear that there exists a separatrix such that for any value of α and h that lies below the separatrix, the RG flow terminates at the line of fixed points, whereas if one were to start with initial value of α and h lying above the separatrix, the flow maintains the system in the delocalized phase.

6.5.2 The RG Equations for the Super-ohmic Case ($s > 1$)

In the situation where the bath spectrum is super-ohmic, i.e., $s > 1$, it can be readily argued that the system supports no critical fixed points. This fact can be essentially gleaned from solving the set of equations, Eqs. (6.14) and (6.15) numerically for various initial configurations of α and h, giving the results depicted in Fig. 6.7. One sees that for any initial value of the dissipation and the transverse field, the couplings always flow toward the limit $h = \infty$ and $\alpha = 0$. This implies that for

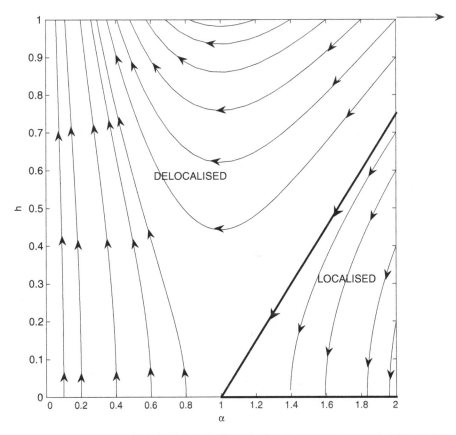

Fig. 6.6 RG flow for the ohmic SBM ($s = 1$). Here, the flow is constructed numerically by giving various initial (bare) values of the coupling constants h and α. See discussion in Sect. 6.5.1 for the interpretation

the super-ohmic case one always ends up in the delocalized phase and no quantum phase transition is allowed.

6.5.3 The RG Equations for the Sub-ohmic Case ($0 < s < 1$)

Now, we turn our attention to the most interesting case, namely the one where the bath spectrum is endowed with a sub-ohmic dispersion, i.e., $0 < s < 1$. Since the mapping of the SBM to the AKM is invalidated in the sub-ohmic regime, the situation of the SBM in the range $0 < s < 1$ was not fully appreciated until recent study from numerical renormalization group (NRG) calculations [2], where a second-order quantum phase transition was explicitly demonstrated for all $0 < s < 1$. At this juncture, it should be noted that this localization/delocalization phase transition found in [2] is missed by the various other analytical treatments of the sub-ohmic

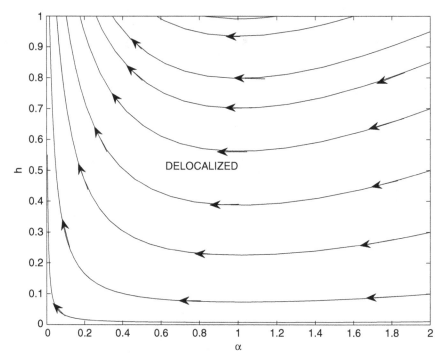

Fig. 6.7 Flow equations of the super-ohmic SBM model ($s > 1$), starting with various initial values of α and h. For further details refer to Sect. 6.5.2

SBM, such as variational ansatz or diagonalizations by unitary transformations, but is correctly predicted by the flow equations derived above. The resulting flow is plotted in Fig. 6.8, with a fixed point occurring at $\alpha_c = 1$ and $h = \sqrt{1-s}$, perturbatively controlled for values of s close to 1.

6.6 Mapping to a Long-Ranged Ising Model

In this section, we shall attempt to derive an effective model for the quantum phase transition discussed previously, based purely in terms of the bosonic mode ϕ. This can be done by representing the spin now in terms of Abrikosov fermions [14] and then integrating the fermionic degrees of freedom perturbatively in λ. By using this route we will see that the SBM can be mapped to a ϕ^4 model with $O(1)$ symmetry and long-ranged interactions in imaginary time. The mapping works as follows:

$$\vec{S} = \sum_{\sigma\sigma'} f_\sigma^\dagger \frac{\vec{\sigma}_{\sigma\sigma'}}{2} f_{\sigma'}. \tag{6.17}$$

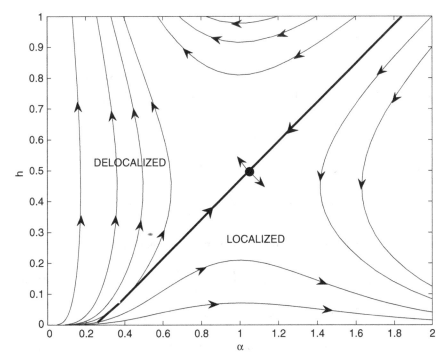

Fig. 6.8 Flow for the sub-ohmic SBM. Once again, the RG trajectories are plotted for various initial values of the transverse field h and the dissipative coupling α. For further details refer Sect. 6.5.3

Here, in Eq. (6.17), the $f_\sigma^\dagger$ field are canonical fermions with an imaginary chemical potential [14], that redefines the Matsubara frequencies $\omega_n \to \omega_n + \pi T/4$, and $\vec{\sigma}$ are the three Pauli matrices. By using Eq. (6.17) to the defining Hamiltonian of the SBM, Eq. (6.2) can be re-written in terms of the following action:

$$S = \int_0^\beta d\tau f^\dagger \left[\partial_\tau - \frac{\Delta}{2}\sigma^x + \frac{\lambda}{2}\sigma^z \sum_i \left(a_i + a_i^\dagger \right) \right] f + \int_0^\beta d\tau \sum_i a_i^\dagger (\partial_\tau + \omega_i) a_i.$$

(6.18)

In the above equation f is a two-component vector whose Hermitian conjugate is given by $f^\dagger \equiv \left(f_\uparrow^\dagger, f_\downarrow^\dagger \right)$. Also in Eq. (6.18), the ∂_τ term is a consequence of time slicing when going into the path integral representation. The philosophy is now to formally integrate out the fermions to get a perturbative expansion in λ of the effective action. This methodology of integrating out the fermions is very similar in spirit to the treatment by Hertz of the itinerant ferromagnet [6] that we have already alluded to in Sect. 6.1. This technical step can be formally performed as the fermionic sector is purely Gaussian, so that the effective theory reads

$$S_{\text{eff}} = \int_0^\beta d\tau \sum_i a_i^\dagger (\partial_\tau + \omega_i) a_i - \text{Tr} \ln \left[\partial_\tau - \frac{\Delta}{2} \sigma^x + \frac{\lambda}{2} \sigma^z \sum_i \left(a_i + a_i^\dagger \right) \right]. \quad (6.19)$$

Defining the "local mode" $\phi = \sum_i \left(a_i^\dagger + a_i \right)$, the bath can be exactly encapsulated by the following Gaussian action, written with the Matsubara frequency $\nu_n = 2n\pi T$:

$$S_{\text{eff}}^{\text{Gauss}} = -\frac{1}{\beta} \sum_{\nu_n} \mathcal{G}_0^{-1}(i\nu_n) \phi(i\nu_n) \phi(-i\nu_n). \quad (6.20)$$

The quantity $\mathcal{G}_0$ in the above equation is given by

$$\mathcal{G}_0(i\nu_n) = \sum_i \left(\frac{1}{i\nu_n - \omega_i} + \frac{1}{-i\nu_n - \omega_i} \right), \quad (6.21)$$

which can be re-expressed in terms of a spectral representation as

$$\mathcal{G}_0(i\nu_n) = \int d\omega \frac{\rho(\omega)}{i\nu_n - \omega}. \quad (6.22)$$

Here, in Eq. (6.22), the bosonic density of states, $\rho(\omega)$, is given by

$$\rho(\omega) = \frac{\omega^s}{\omega_c^{1+s}} \theta(\omega_c + \omega) \theta(\omega_c - \omega) \text{sgn}(\omega). \quad (6.23)$$

The Green's function $\mathcal{G}_0$ can be calculated by substituting the functional form of $\rho(\omega)$ from Eqs. (6.23) into (6.22) and then performing the integration over the frequency variable ω. Once we have performed the integration, the resultant expression can be easily inverted to obtain the functional form for $\mathcal{G}_0^{-1}$ which is given in the low-frequency limit by

$$\mathcal{G}_0^{-1}(i\nu_n) = -\frac{s\omega_c}{2} - \left(\frac{s\omega_c}{2} \right)^2 \frac{\pi |\nu_n|^s}{\sin \frac{\pi s}{2} \omega_c^{s+1}}. \quad (6.24)$$

Thus, substituting the above functional form of the $\mathcal{G}_0^{-1}$ into Eq. (6.20), we see that the Gaussian part of the action for the bath of harmonic oscillators, Eq. (6.19), can be written as

$$S_{\text{eff}}^{\text{Gauss}} = \frac{1}{\beta} \sum_{\nu_n} \left[\frac{s\omega_c}{2} + \left(\frac{s\omega_c}{2} \right)^2 \frac{\pi |\nu_n|^s}{\sin \frac{\pi s}{2} \omega_c^{s+1}} \right] |\phi(\nu_n)|^2. \quad (6.25)$$

Now, that we have taken care of the Gaussian bath term in Eq. (6.19), we turn our attention to the Trln term, which can be Taylor expanded to obtain

$$\sum_{n=1}^\infty \frac{1}{n} \text{Tr} \left(G_0 \sigma^z \phi \frac{\lambda}{2} \right)^{2n}, \quad (6.26)$$

wherein in the above equation, the quantity G_0 is endowed the functional form, $G_0 = \frac{-i\omega_n 1_2 + \frac{\Delta}{2}\sigma^x}{\omega_n^2 + \frac{\Delta^2}{4}}$, where 1_2 is the usual 2×2 identity matrix. Now, we proceed to calculate the traces implicit in Eq. (6.26). The resultant expression, up to order λ^4, is then combined with Eq. (6.25) to obtain

$$S_{\text{eff}} = \int \frac{d\nu}{2\pi} \left(r + A|\nu|^s \right) |\phi(i\nu)|^2 + \int d\tau \, u \, (\phi(\tau))^4. \tag{6.27}$$

Here $r = s\omega_c/2 - \lambda^2/(4\Delta)$ is a mass term that controls the distance to criticality, $u = \lambda^4/(16\Delta^3)$ is the leading interaction term, and $A \propto \omega_c^{s-1}$. We note that this action is equivalent to an Ising model in imaginary time, with interaction decaying as $1/(\tau - \tau')^{1+s}$, as expected from the quantum/classical equivalence [2, 5]. Now, one can use simple power counting arguments to capture the critical behavior of the long-ranged Ising model displayed in Eq. (6.27), thereby also understanding the critical behavior of the underlying microscopic model, Eq. (6.2). By doing a power counting analysis around the Gaussian fixed point one finds that the scale dimension of the $O(\phi^4)$ term is $[u] = 2s - 1$. This implies that for all $s < 1/2$ the scale dimension is negative thus implying that the critical behavior is mean field like. However, for $s > 1/2$ one needs to account for higher loop effects to capture the true critical behavior, leading to non-trivial exponents with respect to the mean field values.

6.7 A Special Identity

In this section, we will derive a special identity that helps us to calculate the fully dressed bosonic propagator in terms of the spin − spin correlator $\chi_z(\tau) = \langle \sigma^z(\tau)\sigma^z(0) \rangle$. We start with an effective action which is a variant of the one that can be obtained from Eq. (6.2). Thus, we have

$$S[\sigma, \phi, J] = S_{\text{Berry}} - \int d\tau \frac{\Delta}{2}\sigma_z(\tau) + \frac{\lambda}{2} \int d\tau \sigma_x \phi(\tau)$$
$$+ \sum_\alpha \int d\tau J^\alpha(\tau)\sigma^\alpha(\tau) + \int d\tau d\tau' \mathcal{G}_0^{-1}(\tau - \tau')\phi(\tau)\phi(\tau'). \tag{6.28}$$

Here, in Eq. (6.28), the term S_{Berry} is the so-called Berry action that encodes the impurity spin commutation relations. This term is not explicitly written down as its functional form relies on spin-coherent states, the discussion of which is beyond the scope of this chapter. Also, in Eq. (6.28), J is a source term for the spin dynamics, and $\mathcal{G}_0$ is again the bare bosonic propagator. The spin − spin correlator can be easily derived from Eq. (6.28), by performing an appropriate functional differentiation of the partition function Z with respect to the source field J. Thus, we have

$$\chi_z(\tau) = \frac{1}{Z} \frac{\delta^2 Z}{\delta J(\tau)\delta J(0)}\Big|_{J=0}. \tag{6.29}$$

Now, in performing the technical calculations inherent in Eq. (6.29), one can re-express the bosonic field ϕ in terms of a new field $\tilde{\phi} = \frac{J}{\lambda} + \phi$. In doing so we use the fact that the partition function Z remains invariant under such a redefinition of the ϕ field. Thus, re-expressing the partition function in terms of the $\tilde{\phi}$ fields and then performing the functional differentiation, we are led to the following relation that connects $\chi_z(\tau)$ to the full bosonic Green's function $\mathcal{G}_\phi(\tau - \tau') = \langle\phi(\tau)\phi(\tau')\rangle$ and the bare bosonic Green's function $\mathcal{G}_0$:

$$\chi_z(\tau) = -\frac{4}{\lambda^2}\mathcal{G}_0^{-1}(\tau) + \frac{4}{\lambda^2}\int d\tau_1 d\tau_2 \mathcal{G}_0^{-1}(\tau_1 - \tau)\mathcal{G}_0^{-1}(\tau_2)\langle\phi(\tau_1)\phi(\tau_2)\rangle. \tag{6.30}$$

By going into the frequency domain representation we can compactly re-write Eq. (6.30) as

$$\chi_z(i\nu_n) = -\frac{4}{\lambda^2}\frac{1}{\mathcal{G}_0(i\nu_n)} + \frac{4}{\lambda^2}\frac{\mathcal{G}_\phi(i\nu_n)}{\mathcal{G}_0(i\nu_n)^2}. \tag{6.31}$$

This identity couples the single particle full bosonic propagator $\mathcal{G}_\phi$ to the spin $-$ spin susceptibility χ_z, naively a four-operator correlation function (see, e.g., the decomposition onto Abrikosov fermions), and shows that both the bosonic field and the longitudinal spin density must become critical altogether at the quantum phase transition. This formula becomes extremely useful in the context of diagrammatic expansions that use the Majorana representation (introduced in Sect. 6.3), because the spin susceptibility is simply related to the single particle Majorana propagators, leading to a very powerful Ward identity. These further theoretical developments go, however, much beyond the scope of this review.

However, in the next section, Sect. 6.8, we shall show the usefulness of the identity derived in this section to recover a well-known result in the context of SBM models, the so-called Shiba's relation.

6.8 Shiba's Relation for the Sub-ohmic Spin Boson Model

In this section we establish the Shiba's relation, usually discussed for the ohmic SBM, in the case of the sub-ohmic model $s < 1$. This relation essentially connects the spin correlations at equilibrium to the zero-frequency spin susceptibility (via the free bath spectrum). To derive this we use the fact that the exact bosonic Green's function, on the real frequency axis, has the following low-frequency form (as can be checked by simple perturbative calculations from the effective action Eq. (6.27))

$$\mathcal{G}_\phi(\nu) = \left(m + a_s^0|\nu|^s + ib_s^0|\nu|^s\text{sgn}(\nu)\right)^{-1}, \tag{6.32}$$

where m is the renormalized mass driving the transition and a_s, b_s are non-critical numerical coefficients. In the limit of small frequencies, the above reduces to

$$\mathcal{G}_\phi(\nu) = \frac{1}{m} - i\frac{b_s^0}{m^2}|\nu|^s\mathrm{sgn}(\nu). \tag{6.33}$$

Now, the identity Eq. (6.31) gives us a relation that connects this full bosonic Green's function $\mathcal{G}_\phi$ to the spin susceptibility. At low frequency, and introducing the bare mass $m_0 = 1/\mathcal{G}_0(0)$, one obviously gets

$$\frac{1}{m} = \frac{1}{m_0} - \frac{\lambda^2}{4m_0^2}\chi_z'(0) \tag{6.34}$$

and

$$\chi_z''(\nu) = -b_s^0\frac{\lambda^2}{4m_0^2}|\nu|^s\mathrm{sgn}(\nu)\left[\chi_z'(0)\right]^2. \tag{6.35}$$

In Eqs. (6.34) and (6.35), the quantities χ_z' and χ_z'' are the real and imaginary parts of the longitudinal spin susceptibility χ_z. The imaginary part of the bare bosonic Green's function reads $\mathcal{G}_0''(\nu) = -\frac{J(|\nu|)}{\lambda^2}\mathrm{sgn}(\nu) = -\frac{b_s^0}{m_0^2}|\nu|^s\mathrm{sgn}(\nu)$. Thus, substituting for b_s^0 in Eq. (6.35), we get

$$\chi_z''(\nu) = \frac{1}{4}J(|\nu|)\mathrm{sgn}(\nu)\left[\chi_z'(0)\right]^2. \tag{6.36}$$

Finally, from the definition that at $T = 0$ the imaginary part of the spin susceptibility is related to the spin correlation function $C(\nu)$ via the simple relation $\chi_z'' = \mathrm{sgn}(\nu)C(\nu)$, we obtain

$$C(\nu) = \frac{1}{4}J(|\nu|)\left[\chi_z'(0)\right]^2, \tag{6.37}$$

which is the generalized Shiba relation for the sub-ohmic spin boson model and is valid in the low-frequency limit for the whole delocalized phase.

6.9 On the Possible Breakdown of Quantum to Classical Mapping

From the general arguments given in the introduction, and the detailed derivation of the classical effective theory Eq. (6.27) for the specific case of the spin boson model, the results of [21] came as a surprise, since critical exponents associated to the spin magnetization $\langle S^z \rangle$ were numerically found by these authors to deviate from the expected mean field result for $0 < s < 1/2$. At present, no consensus has been reached regarding the correctness of this prediction, which was in fact strongly debated afterward [22]. An intriguing possibility is, if true, whether this observation can be reconciled with the ideas of [1] for the spoiling of Hertz's transition in the presence of low-energy modes. In this spirit, the introduction of Majorana modes, where low-lying excitations are naturally associated with the propagation of real fermions, could be a decisive step. Further developments of the Feynman diagrammatics, beyond leading order in perturbation theory, is currently underway for the spin boson model represented by Majorana fermions (Florens et al. unpublished).

Acknowledgments RN thanks Priyanka Mohan for generating some of the figures in this chapter. He also thanks her for discussions and valuable comments.

References

1. D. Belitz, T.R. Kirkpatrick and T. Vojta, Rev. Mod. Phys. **77**, 579 (2005).
2. R. Bulla, N.-H. Tong and M. Vojta, Phys. Rev. Lett. **91**, 170601 (2003).
3. D.G. Clarke, T. Giamarchi and B. I. Shraiman, Phys. Rev. B **48**, 7070 (1993).
4. T.A. Costi, and G. Zarand, Phys. Rev. B, **59**, 12398, (1999).
5. V.J. Emery and A. Luther, Phys. Rev. B **9**, 215 (1974).
6. J.A. Hertz, Phys. Rev. B **14**, 1165 (1976).
7. A.I. Larkin and V.I. Mel'nikov, Sov. Phys. JETP **34**, 656 (1972).
8. A.J. Leggett, S. Chakravarty, A.T. Dorsey, M.P.A. Fisher, A. Garg and W. Zwerger, Rev. Mod. Phys. **59**, 1 (1987).
9. Y. L. Loh, V. Tripathi and M. Turlakov, Phys. Rev. B **71**, 024429 (2005).
10. H. v. Löhneysen, A. Rosch, M. Vojta and P. Wölfle, Rev. Mod. Phys. **79**, 1015 (2007).
11. W. Mao, P. Coleman, C. Hooley and D. Langreth, Phys. Rev. Lett. **91**, 207203 (2003).
12. A.J. Millis, Phys. Rev. B **48**, 7183 (1993).
13. E. Novais, A.H. Castro Neto, L. Borda, I. Affleck and G. Zarand, Phys. Rev. B **72**, 014417 (2005).
14. V. N. Popov and S. A. Fedotov, Sov. Phys. JETP **67**, 535 (1988).
15. S. Sachdev, Phys. Rev. B **55**, 142 (1997)
16. A.M. Sengupta, Phys. Rev. B **61**, 4041 (2000).
17. A.M. Sengupta and A. Georges, Phys. Rev. B **52**, 10295 (1995).
18. A. Shnirman and Y. Makhlin, Phys. Rev. Lett. **91**, 207204 (2003).
19. Q. Si and J. Lleweilun Smith, Phys. Rev. Lett. **77**, 3339 (1996).
20. M. Vojta, C. Buragohain and S. Sachdev, Phys. Rev. B **61**, 15152 (2000).
21. M. Vojta, N. Tong and R. Bulla, Phys. Rev. Lett. **94**, 070604 (2005).
22. A. Winter et al., Phys. Rev. Lett. **102**, 030601 (2009).

Chapter 7
Influence of Local Moment Fluctuations on the Mott Transition

C. Janani, S. Florens, T. Gupta, and R. Narayanan

7.1 Introduction

The Mott metal to insulator transition is a remarkable phenomenon observed in strongly correlated materials, where the localization of electronic waves is driven by on-site electron–electron repulsion (see [7] for a review). Although the appearance of a Mott gap is clearly a charge-related effect, magnetism is expected to play a key role in elucidating the true nature of this phase transition. Indeed, since the Mott insulating state is purely paramagnetic, local moments are well-defined objects between their formation at high temperature (about the local Coulomb interaction) and their ultimate ordering at the Neel temperature. This offers a window for the Mott transition to occur, in which the behavior of these local spin excitations is yet to be clearly understood. The simplest situation lies in case where the low-temperature magnetic ordering is first order, as in Cr-doped V_2O_3. Accordingly magnetic fluctuations should be expected to be weak, so that many predictions can be made from a single-site approach such as the Dynamical Mean Field Theory (DMFT) [4]. In particular, the fact that a low-temperature metallic state leads upon heating to an insulating phase can be understood as a Pomeranchuk effect, where the entropy gain benefits the state with magnetic degeneracy. On the contrary, there are other classes of materials, such as the κ-organics, where the magnetic transition happens to be continuous, so that localized spins will experience strong collective fluctuations. The

C. Janani (✉)
Department of Physics, Indian Institute of Technology, Chennai 600036, India,
janani@physics.iitm.ac.in

S. Florens
Institut Néel and UJF, CNRS, Bp 166, 3082, Grenoble Cedex 9, France,
serge.florens@grenoble.cnrs.fr

T. Gupta
Institute of Mathematical sciences, C. I. T. Campus, Taramani, Chennai 600113, India,
tgupta@imsc.res.in

R. Narayanan
Department of Physics, Indian Institute of Technology, Chennai 600036, India,
rnarayanan@physics.iitm.ac.in

Janani, C. et al.: *Influence of Local Moment Fluctuations on the Mott Transition.* Lect. Notes Phys. **802**, 163–175 (2010)
DOI 10.1007/978-3-642-11470-0_7

most striking experimental finding [9, 11, 12] lies in the progressive disappearance of the Pomeranchuk effect upon approaching the magnetic ordering temperature, so that the Mott transition lines in the pressure–temperature phase diagram bend at low temperature. These qualitative arguments have received recent confirmation from cluster DMFT calculations of the phase diagram of the Hubbard model on a frustrated lattice [14, 17], see also [15]. However, the precise connection between the appearance of low-energy magnetic excitations and deviations from the single-site approach has remained unclear.

For this purpose, we investigate the standard Hubbard–Heisenberg model, treating both the local Coulomb interaction and the spin fluctuations in a local yet dynamical fashion. We derive an effective action in terms of spin-carrying fermionic excitations and slave rotor variables for the charge degree of freedom [3] that we solve within large $\mathcal{N}$ methods, extending previous results [16] on the doped t–J model with random Heisenberg interaction to the half-filled Hubbard model. Our calculations show that the coexistence region of metallic and insulating solutions shrinks by increasing the spin exchange interaction, while the critical lines show deviations in the phase diagram according to a removal of the entropy in the Mott state.

In Sect. 7.2, we review the previously known self-consistent equations for the SU($\mathcal{N}$) Heisenberg spin–liquid model. In Sect. 7.3, we quickly recap how the DMFT equations for the pure Hubbard Hamiltonian are obtained and merge them with results of Sect. 7.2.2, with a derivation of the DMFT equations for the Hubbard–Heisenberg model, and their solution using the slave rotor technique. The results are presented in Sect. 7.4, and our concluding remarks are given in Sect. 7.5. This last section also contains a discussion of open problems that we are currently investigating.

7.2 The Heisenberg Model

In this section, we illustrate how to get a single-site representation of the Heisenberg model. In this respect, we closely mirror the method developed in [2] to obtain the single-site representation. The model is then solved using the large $\mathcal{N}$ method for SU($\mathcal{N}$) spins [18].

7.2.1 Single-Site Representation of the Heisenberg Model

Our starting point in this section is the Heisenberg Hamiltonian which is given by

$$H_{\text{Hei}} = \sum_{ij} J_{ij} \mathbf{S_i} . \mathbf{S_j}. \tag{7.1}$$

Here, the interaction J_{ij} is chosen to be random. The choice of random J_{ij} essentially mimics the oscillatory form of the RKKY interaction. Furthermore, we assume that all sites are connected to all other sites via the medium of the J_{ij}'s. As we shall see later on in this section, the later assumption allows us to derive a single-site representation. Since the J_{ij}'s are chosen to be random, they break translational invariance, which is restored by averaging over the randomness. Integrating out over disorder is done by using the "replica trick" [5]. This trick essentially entails the re-writing of the disorder-averaged free energy as

$$\langle \ln Z \rangle_{\text{dis}} = \lim_{m \to 0} \frac{\langle Z^m \rangle_{\text{dis}} - 1}{m}. \tag{7.2}$$

In Eq. (7.2), Z is the partition function and m is the number of replicas. The averaging over the disorder is done by assigning a probability distribution for the random J_{ij}'s. At this juncture, we could use any non-pathological probability distribution to perform the averaging over the disorder variables. However, for calculational ease, we assume that the random J_{ij} are governed by a probability distribution which is Gaussian. Under the assumption of a Gaussian probability distribution, we can write

$$\langle Z_{\text{Hei}}^m \rangle = \int D[J_{ij}] D[S_i^\alpha(\tau)] \exp \left[-\frac{N}{2J^2} \sum_{ij} J_{ij}^2 \right] \times \exp(-H_{\text{Hei}}). \tag{7.3}$$

In Eq. (7.3), the α's indicate the so-called replica indices, J is the width of the disorder distribution in the Heisenberg exchange term, and N denotes the total number of spins. Due to the Gaussian nature of the disorder distribution, the averaging over the disorder variables in Eq. (7.3) can be easily carried out leading to

$$\langle Z^m \rangle = \int D[S] \left[\exp \left(\frac{J^2}{2N} \int_0^\beta d\tau_1 \int_0^\beta d\tau_2 \sum_{\alpha\gamma,ij} S_i^\alpha(\tau_1).S_j^\alpha(\tau_1) S_i^\gamma(\tau_2).S_j^\gamma(\tau_2) \right) \right]. \tag{7.4}$$

It should be noted that for the sake of convenience, we are working with imaginary time τ at the inverse temperature $\beta = 1/T$ from thereon. As seen from Eq. (7.4), the integration over the random exchange has given rise to an interaction term that is quartic in the spin variable S_i. We now use a Hubbard–Stratonovich (HS) transformation [6] to effectively decouple the quartic spin interaction by means of a bosonic ghost field $Q^{\alpha\gamma}(\tau_1, \tau_2)$. Thus, under the HS transformation Eq. (7.4) can be re-written as

$$\langle Z^m \rangle = \int D[S_i^\alpha(\tau_1)] D[Q^{\alpha\gamma}(\tau_1, \tau_2)]$$
$$\exp \left[-\int_0^\beta d\tau_1 \int_0^\beta d\tau_2 \frac{N}{2J^2} \sum_{\alpha\gamma} (Q^{\alpha\gamma}(\tau_1, \tau_2))^2 \right]$$
$$\times \exp \left[\int_0^\beta d\tau_1 \int_0^\beta d\tau_2 \sum_{\alpha\gamma i} S_i^\alpha(\tau_1).S_i^\gamma(\tau_2) Q^{\alpha\gamma}(\tau_1, \tau_2) \right]. \tag{7.5}$$

From Eq. (7.5), the advantage of performing the HS transformation becomes immediately obvious. Namely, the model has been conveniently re-cast into a single-site
problem, since Eq. (7.5) can be re-written as

$$\langle Z^m \rangle = \int D[Q^{\alpha\gamma}(\tau_1, \tau_2)] \exp(-NF_{\text{Hei}}). \tag{7.6}$$

Here, the effective free energy F_{Hei} is represented by the functional form:

$$F_{\text{Hei}} = \int_0^\beta d\tau_1 \int_0^\beta d\tau_2 \frac{1}{2J^2} \sum_{\alpha\gamma} [Q^{\alpha\gamma}(\tau_1, \tau_2)]^2 - \ln Z[Q]. \tag{7.7}$$

In Eq. (7.7), the quantity $Z[Q]$ is the partition function obtained from the formal
integration over the spin variables and is formally expressed as follows:

$$Z[Q] = Tr \exp\left[\int_0^\beta d\tau_1 \int_0^\beta d\tau_2 \sum_{\alpha,\gamma} \mathbf{S}_i^\alpha(\tau_1).\mathbf{S}_i^\gamma(\tau_2)Q^{\alpha\gamma}(\tau_1, \tau_2)\right]. \tag{7.8}$$

The Tr in Eq. (7.8) indicates a trace over the spin variables $\mathbf{S}_i$. Now, in the limit of
a macroscopic system, i.e., for $N \to \infty$, we can do a saddle point evaluation of the
disorder-averaged partition function depicted in Eq. (7.6). On performing the saddle
point we obtain

$$\frac{\delta F_{\text{Hei}}}{\delta Q^{\nu\eta}(\tau_3, \tau_4)} = \frac{Q^{\nu\eta}(\tau_3, \tau_4)}{J^2} - \langle \mathbf{S}_i^\nu(\tau_3).\mathbf{S}_i^\eta(\tau_4) \rangle = 0. \tag{7.9}$$

Furthermore, we impose the condition that the saddle point solution is translationally invariant in imaginary time, and thus $Q(\tau_3, \tau_4) = Q(\tau_3 - \tau_4)$ depends on the
time difference only. The saddle point equations can then be re-written as a relation
that connects the HS field Q to the spin susceptibility χ. In other terms we have

$$Q(\tau) = J^2 \chi(\tau). \tag{7.10}$$

Here $\chi(\tau)$ in the above equation, Eq. (7.10), is the spin susceptibility per spin component and is given by

$$\chi(\tau) = \frac{1}{N^2 - 1} \langle \mathbf{S}(\tau).\mathbf{S}(\mathbf{0}) \rangle. \tag{7.11}$$

Equation (7.10) can alternatively be viewed as the DMFT self-consistency condition
for the magnetic part of the Hubbard–Heisenberg model that we will discuss in detail
later on in this manuscript.

7.2.2 The Auxiliary Fermion Representation

In Sect. 7.2.1, it was seen that the DMFT self-consistency equation (7.10), connects the bosonic ghost field $Q^{\alpha\gamma}(\tau)$ to a spin susceptibility, Eq. (7.11). To make further progress in evaluating this spin correlator, we take recourse to an auxiliary fermion (Abrikosov fermions) representation [1]. Thus, the $SU(\mathcal{N})$ spins can be represented as

$$\mathbf{S}(\tau) \equiv S_{\sigma\,\sigma'}(\tau) = f_\sigma^\dagger(\tau)f_{\sigma'}(\tau) - \frac{1}{2}\delta_{\sigma\sigma'}. \tag{7.12}$$

Here, the variable σ goes from 1 to $\mathcal{N}$, where $\mathcal{N}$ is the number of flavors of fermions. By utilizing Eq. (7.12), the single-site action for the Heisenberg model can be re-written in terms of the fermionic degrees of freedom as

$$S_{\text{imp}}^{\text{Hei}} = \int_0^\beta d\tau \sum_\sigma f_\sigma^\dagger(\tau)\partial_\tau f_\sigma(\tau) + \int_0^\beta d\tau \int_0^\beta d\tau' \frac{1}{2J^2}(Q(\tau - \tau'))^2$$
$$- \int_0^\beta d\tau \int_0^\beta d\tau' \frac{Q(\tau - \tau')}{\mathcal{N}} \sum_{\sigma\sigma'} f_\sigma^\dagger(\tau)f_{\sigma'}(\tau)f_{\sigma'}^\dagger(\tau')f_\sigma(\tau'). \tag{7.13}$$

The ∂_τ term in the above equation is a consequence of time slicing the Berry phase, as one re-expresses the partition function F_{Hei} of Eq. (7.7) in terms of the effective action $S_{\text{imp}}^{\text{Hei}}$ given by Eq. (7.13). For further details on this time slicing procedure, we refer the readers to [13]. Now, from Eq. (7.13), the slave rotor integral equations are obtained by performing a Hartree–Fock decomposition. This approximation, also referred to as the non-crossing approximation, is vindicated by the large $\mathcal{N}$ limit. Under the Hartree–Fock decomposition, Eq. (7.13) can be re-written in terms of the pseudo-fermion (also called spinon) Green's function, $G_f(\tau - \tau') = -\langle f_\sigma(\tau)f_\sigma^\dagger(\tau')\rangle$ as

$$S_{\text{imp}}^{\text{Hei}} = S[Q] + S_{\text{TS}}[f^\dagger, f] + \int_0^\beta d\tau d\tau'\, Q(\tau - \tau')G_f(\tau - \tau') \sum_{\sigma'} f_{\sigma'}^\dagger(\tau')f_{\sigma'}(\tau). \tag{7.14}$$

Here, in Eq. (7.14), the quantity $S[Q]$ is given by

$$S[Q] = -\int_0^\beta d\tau \int_0^\beta d\tau' \frac{\mathcal{N}Q(\tau - \tau')}{2}(G_f(\tau - \tau'))^2 + \frac{1}{2J^2}\int_0^\beta d\tau \int_0^\beta d\tau'(Q(\tau-\tau'))^2. \tag{7.15}$$

Also, in Eq. (7.14) the term S_{TS} is once again the part of the action that comes via the time slicing operation, (the ∂_τ term in Eq. (7.13)). The action given in Eq. (7.14) can be re-formulated into the form

$$S_{\text{imp}} = S[Q] - \int_0^\beta d\tau' \int_0^\beta d\tau\, G_f^{-1}(\tau - \tau') \sum_\sigma f_{\sigma'}^\dagger(\tau)f_\sigma(\tau). \tag{7.16}$$

Here, in Eq. (7.16), the quantity G_f^{-1} is the fully dressed Green's function. In terms of the fermionic Matsubara frequency ω_n, it can be represented as $G_f^{-1}(i\omega_n) = i\omega_n - \Sigma_f(i\omega_n)$, where the quantity $\Sigma_f(i\omega_n)$ is the self-energy of the f pseudo-fermions. Now, the self-energy $\Sigma(\tau)$ can be easily read off by comparing Eq. (7.14) to its formal counterpart Eq. (7.16), leading us to the following equation:

$$\Sigma_f(\tau) = Q(\tau)G_f(\tau). \tag{7.17}$$

Now, the DMFT self-consistency condition, Eq. (7.10), can be re-expressed in terms of the pseudo-fermion Green's functions as

$$Q(\tau) = J^2 G_f(\tau)^2. \tag{7.18}$$

This in turn implies that the self-energy can be re-expressed as

$$\Sigma_f(\tau) = J^2 G_f(\tau)^3. \tag{7.19}$$

Equation (7.19), together with Dyson equation for G_f^{-1} in terms of the pseudo-fermion self-energy, forms a set of self-consistent equations which can be implemented numerically to be discussed below (see also [18]).

7.3 The Hubbard–Heisenberg Model

We start with the Hubbard–Heisenberg model on a fully connected lattice, with Hamiltonian

$$H = - \sum_{\langle ij \rangle, \sigma} t_{ij} d_{i\sigma}^\dagger d_{j\sigma} - \mu \sum_{i\sigma} d_{i\sigma}^\dagger d_{i\sigma} + U \sum_i d_{i\uparrow}^\dagger d_{i\uparrow} d_{i\downarrow}^\dagger d_{i\downarrow} + H_{\text{Hei}}. \tag{7.20}$$

In the above model, we will take the hopping elements t_{ij} such that it connects the nearest-neighbor sites on an infinite coordination lattice. In Eq. (7.20), the term proportional to U represents the local Coulomb interaction. H_{Hei} is the Heisenberg Hamiltonian with random J_{ij} interaction that we have first encountered in Eq. (7.1). In Sect. 7.2, we have seen how to go to a single impurity for the Heisenberg part of the model in Eq. (7.20). We can follow a procedure based on a similar philosophy for the itinerant part, leading to DMFT equations for the Hubbard part of the Hamiltonian. Here, we will simply write down the effective impurity model for the Hamiltonian shown in Eq. (7.20). For further details, we refer the readers to a review by Georges et al. [4], where the method is well presented. By following the methodology outlined in [4], we can write the effective impurity action as

$$S_{\text{imp}} = S_{\text{Hei}}^{\text{imp}} + S_{\text{Hub}}^{\text{imp}}. \tag{7.21}$$

The first term in Eq. (7.21), $S_{\text{Hei}}^{\text{imp}}$, is the effective single-site impurity action that we have already seen in Eq. (7.13). The second piece of the above equation, $S_{\text{Hub}}^{\text{imp}}$, represents the local impurity effective action of the Hubbard model and is given by

$$S_{\text{Hub}}^{\text{imp}} = \int_0^\beta d\tau \sum_\sigma d_\sigma^\dagger (\partial_\tau - \mu) d_\sigma + U d_\uparrow^\dagger d_\uparrow d_\downarrow^\dagger d_\downarrow +$$
$$\int_0^\beta d\tau \int_0^\beta d\tau' \Delta(\tau - \tau') \sum_\sigma d_\sigma^\dagger(\tau) d_\sigma(\tau'). \quad (7.22)$$

Here, in Eq. (7.22), $\Delta(\tau - \tau')$ is the bath hybridization parameter that has to be determined self-consistently. This gives us one more self-consistency DMFT condition in addition to Eq. (7.10):

$$\Delta(\tau) = t^2 G_d(\tau). \quad (7.23)$$

Here in Eq. (7.23), the quantity $G_d(\tau)$ is the d-electron single-particle Green's function and is given by $G_d(\tau) = -\langle d_\sigma(\tau) d_\sigma^\dagger(0) \rangle$. Now, our aim is to solve the impurity problem for the action given by S_{imp} in Eq. (7.21) in order to obtain the needed correlators $G_d(\tau)$ and $\chi(\tau)$. To proceed toward our stated goal, we adopt the slave rotor representation proposed in [3]. Its main advantage regarding the Hubbard–Heisenberg model is that the f_σ pseudo-fermions, crucial to describe the magnetic fluctuations, are introduced from the outset. The slave rotor representation is then obtained by expressing the physical d-fermions as $d_\sigma^\dagger = f_\sigma^\dagger X$ and $d_\sigma = f_\sigma X^*$, with X a pure phase field, that thus obeys the constraint $|X(\tau)|^2 = 1$. This so-called slave rotor represents the charge fluctuation part, which is central to describe the Mott transition in the Hubbard model. Under such a representation, the Coulomb interaction term $U d_\uparrow^\dagger d_\uparrow d_\downarrow^\dagger d_\downarrow$ can be re-expressed as [3]

$$S_{\text{coul}} = \int_0^\beta d\tau \frac{|\partial_\tau X|^2}{4U} + \int_0^\beta d\tau \Lambda(\tau)[|X|^2 - 1]. \quad (7.24)$$

In Eq. (7.24), $\Lambda(\tau)$ is the Lagrange multiplier term that is used to impose the rotor constraint. Under this re-writing of the Coulomb term the impurity action can be expressed in terms of the rotor fields X and the pseudo-fermion fields $f_\sigma^\dagger$ as

$$S_{\text{imp}} = S_{\text{coul}} + S_{\text{Hei}}^{\text{imp}} + \int_0^\beta d\tau \int_0^\beta d\tau' \Delta(\tau - \tau') \sum_\sigma f_\sigma^\dagger(\tau) X(\tau) f_\sigma(\tau') X(\tau'). \quad (7.25)$$

Here, in Eq. (7.25), S_{coul} is the local Coulomb part of the action (see Eq. (7.24)), written in terms of the X fields. Now, to obtain the slave rotor integral equations, we follow the techniques developed in detail in Sect. 7.2.2. In other words, once again to obtain the integral equations, one follows a Hartree–Fock decomposition of terms in the Eq. (7.25). After performing this approximation, which again is valid in a limit of large $\mathcal{N}$ (one, however, needs for this to include extra quantum numbers for the rotor field, see [3]), we can re-write the expression Eq. (7.25) in terms of the pseudo-particle Green's functions $G_f(\tau)$ and $G_X(\tau)$:

$$
S_{\text{imp}} = S_{\text{coul}} + S_{\text{imp}}^{\text{Hei}} + \int_0^{\beta'} d\tau \int_0^{\beta} d\tau' \left[\Delta(\tau - \tau') \sum_\sigma f_\sigma^\dagger(\tau) f_\sigma(\tau') G_X(\tau - \tau') \right.
$$
$$
+ \int_0^{\beta} d\tau \int_0^{\beta} d\tau' \mathcal{N} \Delta(\tau - \tau') G_f(\tau' - \tau) X(\tau) X^*(\tau'). \tag{7.26}
$$

We can identify the Green's functions G_f and G_X by comparing the above expression with the standard definition for the impurity action, namely

$$
S_{\text{imp}} = \int_0^{\beta} d\tau \int_0^{\beta} d\tau' G_X^{-1}(\tau - \tau') X(\tau) X^*(\tau')
$$
$$
- \int_0^{\beta} d\tau \int_0^{\beta} d\tau' \, G_f^{-1}(\tau - \tau') \sum_\sigma f_\sigma^\dagger(\tau) f_\sigma(\tau'). \tag{7.27}
$$

We have already seen the fully dressed pseudo-fermion Green's function, G_f^{-1}, written in terms of the Matsubara frequency in Sect. 7.2.2. In a similar vein, one can write the fully dressed rotor Green's function G_X^{-1} as a Dyson equation:

$$
G_X^{-1} = \frac{\nu_n^2}{U} + \Lambda - \Sigma_X(i\nu_n), \tag{7.28}
$$

with bosonic Matsubara frequency ν_n. From comparing Eq. (7.26) with its formal counterpart Eq. (7.27), we can write down the expression for the spinon self-energy $\Sigma_f(\tau)$ and the rotor self-energy $\Sigma_X(\tau)$:

$$
\Sigma_f(\tau) = \Delta(\tau) G_X(\tau) + J Q(\tau) G_f(\tau)
$$
$$
\Sigma_X(\tau) = -\mathcal{N} \Delta(-\tau) G_f(\tau). \tag{7.29}
$$

Together with the electron bath $\Delta(\tau) = t^2 G_d(\tau) = t^2 G_f(\tau) G_X(\tau)$ and the spin bath $Q(\tau) = J^2 \chi(\tau) = J^2 G_f(\tau)^2$, the above self-energies form the set of self-consistent equations that we need to solve numerically.

7.4 Numerical Results

In the above section, we have mapped the Hubbard–Heisenberg model onto a self-consistent impurity model given by Eq. (7.21) (the various components in Eq. (7.21) are given by Eq. (7.25) and Eq. (7.14)). This mapping conforms to the self-consistency condition given by Eq. (7.23) and Eq. (7.10) for the Hubbard and Heisenberg parts, respectively, of the impurity action Eq. (7.21). To make further progress, we used the slave rotor impurity solver, which led to Eq. (7.29), so that the DMFT loop can be easily performed using integral equations. To benchmark our results and to make contact with the earlier work of Florens and Georges [3], we first solve in Sect. 7.4.1 the DMFT equations for the case $J = 0$ (pure Hubbard model).

7.4.1 Numerical Results for $J = 0$

To generate the phase diagram we calculated the Green's function G_d from $\beta = 500$ to $\beta = 30$ and varying U for each β (we work here in units of the half-bandwidth $D = 2t = 1$ and take the number of spin species $\mathcal{N} = 3$). We followed two kinds of sweeps in U. A "forward sweep" in which U is varied from a lower (metallic) to a higher (insulating) value and a "backward sweep" in which U is varied from a higher (insulating) to a lower (metallic) value, allowing to assess the coexistence region. Figure 7.1a shows the classic hysteresis behavior wherein there are coexisting metallic and insulating solutions of the DMFT equations. These coexistent phases imply that the Mott transition in the Heisenberg–Hubbard model is of the first-order type when $J = 0$. This result is quantitatively consistent with those found in [3] and slightly deviates from the exact numerical solution of the DMFT equations [4]. As can be seen, the coexistence region exists for a range of interaction values, $U_{c1}(T) \leq U \leq U_{c2}$. Here U_{c1} (respectively U_{c2}) for a given temperature is the value of U at which the Green's function is discontinuous when performing a backward (respectively forward) sweep in U.

The above process is repeated for all values of β, so that the phase diagram for the Hubbard model is thus obtained as in Fig. 7.1b. As one goes to larger temperatures, the width of the coexistence region decreases, finally ending in a critical endpoint. The universality class of the transition at this point is known to be of the liquid–gas type, [10]. The critical endpoint is located at $T = 1/30$, once again consistent with [3]. Also important is the type of Pomeranchuk effect seen in this phase diagram, as the insulating phase is stabilized upon heating, due to its macroscopic degeneracy (for $J = 0$, each lattice site contains a freely fluctuating local moment with $\log 2$ entropy). The local moments in the pure Hubbard model are in fact non-magnetically interacting, owing to the locality of the self-energy in the DMFT approach. This drawback can be avoided by explicitly including a spin–spin interaction. This naturally leads us to investigate in the following section the role of the Heisenberg interaction at $J \neq 0$, which will trigger important spin fluctuations.

7.4.2 Numerical Results for Non-zero J

In this section, we study the effect of turning on a non-zero exchange interaction between the preformed local moments. Let us first discuss a representative case, namely the situation when $J = 0.5$. Once again, for a particular value of β, we

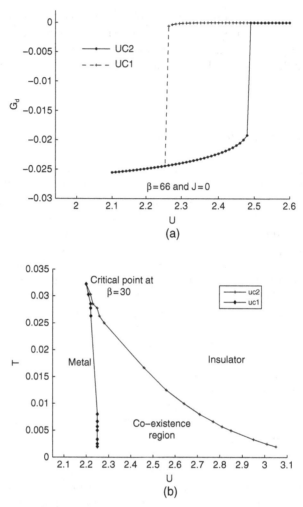

Fig. 7.1 (a) The physical Green's function at imaginary time $\tau = \beta/2$ for the inverse temperature $\beta = 66$ and $J = 0$ as a function of U. (b) The phase diagram in (U,T) of the single-orbital Hubbard model at half-filling for $J = 0$

scan across different values of U in similar manner as described in Sect. 7.4.1. The phase diagram obtained in this way is displayed in Fig. 7.3. We first note that the coexistence regions found in the case $J = 0$ persists, albeit at lower temperatures, so that the critical endpoint occurs for a lower value of T than the corresponding $J = 0$ case (Fig. 7.2b is a testimony for the above fact).

Another interesting facet of the phase diagram is that in the presence of the interaction among the preformed local moments, the critical lines U_{c1} and U_{c2} shift to much lower U values, so that the Heisenberg term clearly stabilizes the insulator, see Fig. 7.3. Indeed, the addition of the exchange interaction must lead to an energy

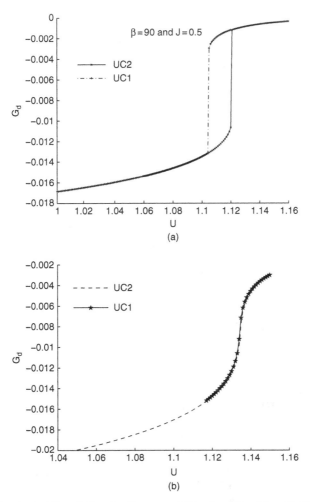

Fig. 7.2 (a) The physical Green's function G_d at $\tau = \beta/2$ for $j = 0.5$ and $\beta = 90$. (b) The physical Green's function G_d at $\tau = \beta/2$ for $j = 0.5$ and $\beta = 66$

gain for the insulator as compared to the metal. Moreover, the narrowing of the coexistence region can also be interpreted by the fact that random interactions tend to quench first-order transitions. Finally, we emphasize again that the Mott insulator with $J = 0$ has a $\log 2$ entropy (as the local moments are non-interacting), so that switching on a non-zero J will reduce the insulator entropy as compared to this free moment value. This clearly comes with an inversion of the critical lines near the Mott endpoint (see Fig. 7.4), as the Pomeranchuk effect is thus weakened. We note, however, that the Pomeranchuk behavior is again observed at very low temperatures, owing to the slow dynamics in Heisenberg spin liquids [16, 18] for which

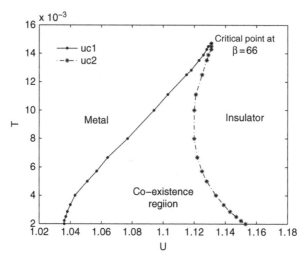

Fig. 7.3 The phase plot for the single-orbital Hubbard–Heisenberg model at half-filling for $J = 0.5$

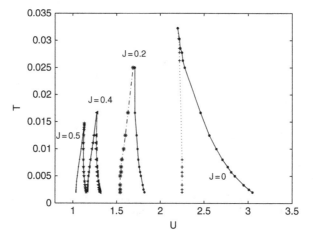

Fig. 7.4 A comparison of (U,T) phase diagrams for different strengths of J

the total entropy is not identically zero. This is in contrast with the study of Park et al. [17], in which a model with short-range magnetic interactions (solved within cluster DMFT) shows that the inversion of the critical line is not altered down to zero temperature.

Finally, it is believed that the critical behavior at the critical endpoint remains in the liquid–gas universality class. The construction of the explicit free energy is an interesting point that will be discussed in detail in an upcoming publication [8].

7.5 Conclusions, Future Outlook, and Open Problems

In this manuscript, we have looked at the effect of local moments fluctuations on the Mott transition. We have benchmarked our results by first investigating the case of zero Heisenberg exchange $J = 0$ and comparing with those of [3]. Accordingly with the DMFT scenario for the Mott transition in the pure Hubbard model, we recovered the first-order metal–insulator transition at low temperatures. Turning on the magnetic interaction J, the coexistence region was shown to shrink toward a stabilization of the insulator. Concomitantly, entropic arguments were presented for the inversion of the critical lines near the Mott endpoint.

There are some other interesting open problems that we are currently focusing on. For example, it would be very worthwhile to study various thermodynamic, magnetic, and transport properties as one scans across the various parts of the phase diagram with increasing spin exchange. The nature of the zero temperature Mott transition, which now happens between a Fermi liquid metal and a spin liquid insulator, remains also to be elucidated. These investigations will be the subject matter of an upcoming publication [8].

References

1. A. Auerbach *Interacting Electrons and Quantum Magnetism* (Springer-Verlag, New York, 1994).
2. A. J. Bray and M. A. Moore J. Phys. C: Solid St. Phys., **13**, L655 (1980).
3. S. Florens and A. Georges, Phys. Rev. B **66**, 165111 (2002).
4. A. Georges, G. Kotliar, W. Krauth and M. J. Rozenberg, Rev. Mod. Phys. **68**, 13 (1996).
5. G. Grinstein, *Fundamental Problems in Statistical Mechanics VI*, E.G.D Cohen (ed.) (Elsevier, New York, 1950).
6. J. A. Hertz, Phys. Rev.B **14**, 1165 (1976).
7. M. Imada, A. Fujimori and Y. Tokura, Rev. Mod. Phys. **70**, 4 (1998).
8. C. Janani, S. Florens, T. Gupta and R. Narayanan, to be published.
9. F. Kagawa, T. Itou, K. Miyagawa, and K. Kanoda, Phys. Rev. B **69**, 064511 (2004).
10. G. Kotliar, E. Lange and M. J. Rozenberg, Phys. Rev. Lett. **84**, 22 (2000).
11. S. Lefebvre, P. Wzietek, S. Brown, C. Bourbonnais, D. Jérome, C. Mézière, M. Fourmigué, and P. Batail, Phys. Rev. Lett. **85**, 5420 (2000).
12. P. Limelette, P. Wzietek, S. Florens, A. Georges, T.A. Costi, C. Pasquier, D. Jérome, C. Mézière, and P. Batail, Phys. Rev. Lett. **91**, 016401 (2003).
13. J.W. Negele and H. Orland "Quantum Many-Particle Systems", *Frontiers in Physics*, (Addison-Wesley, California and New York, 1988).
14. T. Ohashi, T. Momoi, H. Tsunetsugu, and N. Kawakami, Phys. Rev. Lett **100**, 076402 (2008).
15. S. Onoda and N. Nagaosa, J. Phys. Soc. Jap. **72**, 2445 (2003).
16. O. Parcollet and A. Georges, Phys. Rev. B **59**, 5341 (1999).
17. H. Park, K. Haule, and G. Kotliar, Phys. Rev. Lett. **101**, 186403 (2008); Soc. Jap. **72**, 2445 (2003).
18. S. Sachdev and J. Ye, Phys. Rev, Lett. **70**, 3339 (1993).

Chapter 8
Signatures of Quantum Phase Transitions via Quantum Information Theoretic Measures

I. Bose and A. Tribedi

8.1 Introduction

Quantum phase transitions (QPTs) in many body systems occur at $T = 0$ brought about by tuning a non-thermal parameter, e.g. pressure, chemical composition or external magnetic field [1, 2]. In a QPT, the ground state wave function undergoes qualitative changes at the transition point. The transition is driven by quantum fluctuations whereas ordinary phase transitions occurring at nonzero temperatures are driven by thermal fluctuations. Like a thermal phase transition, a QPT can be first order, second order or higher order. The thermal critical point, associated with a second-order phase transition, is characterized by the presence of thermal fluctuations on all length scales resulting in a divergent correlation length. The free energy and the thermodynamic functions develop singularities as temperature $T \rightarrow T_c$, the critical temperature. At the quantum critical point (QCP), quantum fluctuations occur on all length scales leading to a divergent correlation length. The ground state and related physical quantities become non-analytic as the tuning parameter g tends to the critical value g_c. The influence of QPTs extends into the finite T part of the phase diagram so that experimental detection of QPTs is possible.

Recent years have witnessed a surge in research activity involving a close interaction between the subjects of quantum information science and many body condensed matter physics [3–5]. The key concept in quantum information is that of entanglement, a truly unique feature of quantum mechanical systems. Entanglement implies non-local correlations between quantum particles which do not have a classical counterpart. Two entangled particles have joint properties rather than individual identities and the underlying correlations extend over macroscopic distances. If two quantum systems A and B are entangled, the values of certain physical properties

I. Bose (✉)
Department of Physics, Bose Institute, 93/1, A.P.C. Road, Kolkata 700009, India,
indrani@bosemain.boseinst.ac.in

A. Tribedi
Department of Physics, Bose Institute, 93/1, A.P.C. Road, Kolkata 700009, India,
amittribedi@gmail.com

Bose, I., Tribedi, A.: *Signatures of Quantum Phase Transitions via Quantum Information Theoretic Measures*. Lect. Notes Phys. **802**, 177–200 (2010)
DOI 10.1007/978-3-642-11470-0_8   © Springer-Verlag Berlin Heidelberg 2010

of A are correlated with the values of the same properties of B. The properties can remain correlated even if the systems A and B are spatially separated leading to the well-known description "spooky action at a distance". Schrödinger commented on the nature of entanglement, "I would not call that *one* but rather *the* characteristic trait of quantum mechanics, the one that enforces its entire departure from classical lines of thought". The peculiarities of the concept of entanglement have been anal-ysed over the years but only recently the realization has come that entanglement has a practical aspect as well. Entanglement, an essential resource in quantum computa-tion and communication applications, can be generated, destroyed, manipulated as well as quantified. There is now a large body of work on identifying different types of entanglement and developing appropriate measures to quantify them [6, 7].

Entanglement measures provide an additional characterization of many body states in condensed matter, e.g. the states of an interacting spin system. Since the ground state wave function changes qualitatively in a QPT, it is of significant interest to probe how the genuine quantum aspect of the wave function, namely entanglement, changes as the transition point is traversed. The questions that natu-rally arise are whether entanglement between the subsystems of a quantum system extends over macroscopic distances, as ordinary correlations do, in the vicinity of the critical point and whether it carries other distinctive signatures of the transition. In recent years, QPTs have been extensively studied in spin systems using well-known quantum information theoretic measures like entanglement, fidelity, reduced fidelity, fidelity susceptibility Fidelity, a concept borrowed from quantum informa-tion theory, is defined as the overlap modulus between ground states correspond-ing to slightly different Hamiltonian parameters. The fidelity typically drops in an abrupt manner at a QCP indicating a dramatic change in the nature of the ground state wave function. This is accompanied by a divergence of the fidelity susceptibil-ity [16–23]. A number of entanglement measures have now been identified which develop special features close to the QCP. In this chapter, we focus on QPTs in a two-leg antiferromagnetic (AFM) Heisenberg spin ladder (Fig. 8.1) in an external magnetic field [24]. The Hamiltonian describing the model is given by

$$\mathcal{H} = \sum_{j=1}^{L} [J_{||}(\mathbf{S}_{1,j}.\mathbf{S}_{1,j+1} + \mathbf{S}_{2,j}.\mathbf{S}_{2,j+1}) + J_{\perp}\mathbf{S}_{1,j}.\mathbf{S}_{2,j}] - H \sum_{j=1}^{L} (S_{1,j}^z + S_{2,j}^z), \quad (8.1)$$

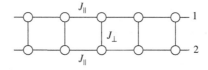

Fig. 8.1 A two-chain ladder with rung and intra-chain nearest-neighbour exchange couplings of strengths $J_\perp$ and $J_{||}$, respectively. The indices **1** and **2** label the two chains of the ladder

where the indices 1 and 2 distinguish the lower and upper legs of the ladder and j labels the rungs. The spins have magnitude $\frac{1}{2}$ ($|\vec{S_j}| = \frac{1}{2}$) and interact via the AFM Heisenberg exchange interaction. The intra-chain and rung exchange couplings are of strengths $J_{\parallel}$ and $J_{\perp}$, respectively. The total number of rungs is L and periodic boundary conditions are assumed. The factor $g\mu_B$ (g is the Landé splitting factor and μ_B the Bohr magneton) is absorbed in H. If $J_{\perp} = 0$, the ladder decouples into two non-interacting spin-$\frac{1}{2}$ Heisenberg chains with no gap to spin excitations. For any arbitrary $J_{\perp} \neq 0$, the excitation spectrum acquires a gap (spin gap). In the strong coupling limit, $J_{\perp} >> J_{\parallel}$, a simple physical picture of the ground state and the origin of the spin gap can be given. The spins along the rungs predominantly form singlets in the ground state. A spin excitation is created by replacing a singlet by a triplet which propagates along the ladder due to the intra-chain exchange interaction. In first-order perturbation theory, the spin gap Δ is given by $\Delta \approx J_{\perp} - J_{\parallel}$ separating the lowest excited state from the dimerized ground state.

QPTs have been observed experimentally in strong coupling organic ladder compounds like $Cu_2(C_5H_{12}N_2)_2Cl_4$ [25], $(C_5H_{12}N)_2CuBr_4$ [26] and $(5IAP)_2CuBr_4$. $2H_2O$ [27] by the tuning of an external magnetic field. The ladder systems have two QCPs at $H = H_{c1}$ and $H = H_{c2}$ [25–28]. At $T = 0$ and for $0 < H < H_{c1}$, the ladder is in the spin gap phase. In the presence of the magnetic field, there is a Zeeman splitting of the triplet ($S = 1$) excitation spectrum with the $S^z = 1$ component having the lowest energy. The spin gap is now $\Delta - H$. At $H = H_{c1} = \Delta$, the gap closes and a QPT occurs to the Luttinger liquid (LL) phase characterized by a gapless excitation spectrum. At the upper critical field $H = H_{c2}$, there is another QPT to the fully polarized ferromagnetic (FM) state. The excitation spectrum in this state is again gapped. The magnetization data exhibit universal scaling behaviour in the vicinity of H_{c1} and H_{c2}, consistent with theoretical predictions. In the gapless regime $H_{c1} < H < H_{c2}$, the ladder Hamiltonian can be mapped onto an XXZ chain Hamiltonian the thermodynamic properties of which can be calculated exactly using the Bethe Ansatz (BA). The theoretically computed magnetization versus magnetic field curve is in excellent agreement with the experimental data. QPTs can be observed in the organic ladder compounds as the magnitudes of the critical fields are experimentally accessible. In the following sections, we first introduce the appropriate quantum information theoretic measures like entanglement and fidelity which provide signatures of QPTs in various condensed matter, specially spin systems. A few well-known spin models are discussed as an illustration. We finally describe the results obtained by us on entanglement and fidelity signatures of QPTs in organic ladder compounds.

8.2 Entanglement and Fidelity Measures Probing QPTs

We consider a system of interacting spins of magnitude $\frac{1}{2}$. Each spin thus serves as a qubit (the fundamental unit of quantum information) with two possible states: $|\uparrow\rangle$ and $|\downarrow\rangle$. A state of the system is said to be entangled if it cannot be written as a product of individual spin states. A well-known example is the singlet state, $\frac{1}{\sqrt{2}}(|\uparrow\downarrow\rangle - |\downarrow\uparrow\rangle)$, of two spin-$\frac{1}{2}$ particles. On the other hand, the state $\frac{1}{\sqrt{2}}(|\downarrow\downarrow\rangle + |\downarrow\uparrow\rangle)$ is a separable, i.e. unentangled state, since it can be written as a product of individual spin states, $\frac{1}{\sqrt{2}}|\downarrow\rangle(|\downarrow\rangle + |\uparrow\rangle)$. A quantum system is in a pure state when the state can be described by a wave function $|\psi\rangle$. The wave function may be a single key or a linear combination of basis states. A mixed state is a statistical mixture of pure states represented by the density operator

$$\rho = \sum_k p_k |\psi_k\rangle \langle\psi_k|, \tag{8.2}$$

where p_k is the probability that the system is in the pure state $|\psi_k\rangle$ with $\sum_k p_k = 1$. In the case of a pure state, $\rho = |\psi\rangle\langle\psi|$. The thermal density matrix provides an example of a mixed state with $p_k = \frac{e^{-\beta E_k}}{Z}$, the Boltzmann factor normalized by the partition function Z. It is easy to show that

$$Tr(\rho) = 1 \qquad \text{in all cases}$$
$$Tr(\rho^2) = 1 \qquad \text{for pure states.} \tag{8.3}$$
$$Tr(\rho^2) < 1 \qquad \text{for mixed states}$$

Also ρ is a positive operator with matrix elements ≥ 0. The mean value of an observable represented by an operator A is

$$\langle A\rangle = Tr(\rho A). \tag{8.4}$$

A pure state of N spins is separable if the wave function $|\psi\rangle$ describing the pure state can be written in a product form:

$$|\psi\rangle = |\phi_1\rangle |\phi_2\rangle |\phi_3\rangle |\phi_4\rangle .... |\phi_N\rangle, \tag{8.5}$$

where the ϕ_i's are individual spin functions. The state $|\psi\rangle$ is entangled if it cannot be written in the product form. A mixed state (Eq. (8.2)) is separable if it can be expressed in a form such that each $|\psi_k\rangle$ is in a product form. If no such decomposition is possible, the mixed state is entangled. The ground states of interacting many body systems are, in general, entangled.

Entanglement can be of various types, e.g. bipartite, multipartite, zero temperature, finite temperature [3–7]. Bipartite (multipartite) entanglement refers to the entanglement between the subsystems of a quantum system when the number of such subsystems is two (more than two). The problem of developing appropriate quantification measures for the various types of entanglement constitutes an active field of research. We now discuss a few of the entanglement measures which are in wide use. These mostly include bipartite entanglement measures for systems of qubits, each qubit having two possible states. Examples of a qubit are a spin-$\frac{1}{2}$ particle, a photon with two possible polarization states, a trapped ion with two atomic states, etc. For a bipartite system in a pure state, one can define the reduced density operator describing a subsystem as

$$\rho_A = Tr_B \, \rho = Tr_B \, |\psi\rangle \langle\psi|, \tag{8.6}$$

where A, B denote the subsystems. $\rho = |\psi\rangle \langle\psi|$ is the pure state density matrix and the partial trace, Tr_B, is taken over the states of the subsystem B. The reduced density operator ρ_B can be defined in a similar manner. As an example, consider a pure state of four spins, $|\psi\rangle = \frac{1}{\sqrt{2}}(|\uparrow\uparrow\downarrow\downarrow\rangle - |\downarrow\downarrow\uparrow\uparrow\rangle)$. The subsystems A and B consist of the pairs of spins $(1, 2)$ and $(3, 4)$, respectively. The reduced density matrix $\rho_{12} = \frac{1}{2}|\uparrow\uparrow\rangle \langle\uparrow\uparrow| + \frac{1}{2}|\downarrow\downarrow\rangle \langle\downarrow\downarrow|$. If the pure state ρ is separable, the reduced density operators ρ_A and ρ_B are also separable since the original state is in a factorized form. If the pure state is entangled, the subsystems are in mixed states. For pure states, a good measure of entanglement is given by the amount of "mixedness" of the states of the subsystems. The quantification is provided by the von Neumann entropy of either of the subsystems:

$$S = -Tr(\rho_A log \, \rho_A) = -Tr(\rho_B log \, \rho_B), \tag{8.7}$$

where the logarithm has base 2 as is standard in quantum information theory. The entropy S is always non-negative, i.e. $S \geq 0$ and can be taken to be a measure of the uncertainty about the quantum state of the system. The value of S is zero when the subsystems are in pure states, i.e. we have full knowledge of the states of the subsystems. In the case of spin-$\frac{1}{2}$ particles or a collection of qubits, the maximum value of S is 1.

In the case of bipartite entanglement measures , the bipartitioning of the system can be achieved in several ways. We define the following two measures appropriate for spin systems defined on a lattice. The single-site von Neumann entropy, a measure of the entanglement of a single spin with the rest of the system, is given by

$$S(i) = -Tr \, \rho(i) \, log_2 \, \rho(i), \tag{8.8}$$

where $\rho(i)$ is the single-site reduced density matrix [9]. The two-site von Neumann entropy is given by

$$S(i,j) = -Tr \, \rho(i,j) \, log_2 \, \rho(i,j), \tag{8.9}$$

where $\rho(i,j)$ is the reduced density matrix for a subsystem of two spins located at sites i and j. The reduced density matrix is obtained from the full density matrix by tracing out the spins other than the ones at sites i and j. For a pure state $|\psi\rangle$, one can express the von Neumann entropy S (Eq. (7)) [29] as

$$S = \varepsilon(C(\psi)), \tag{8.10}$$

where C, a quantity known as concurrence, is given by

$$C(\psi) = \left| \left\langle \psi | \tilde{\psi} \right\rangle \right|. \tag{8.11}$$

The wave function $\left| \tilde{\psi} \right\rangle$ is obtained via a spin flip transformation on state $|\psi\rangle$. In the case of a single spin state,

$$\left| \tilde{\psi} \right\rangle = \sigma_y \left| \psi^* \right\rangle. \tag{8.12}$$

For a system of n spins, the above transformation is applied to each individual spin. The function $\varepsilon(C)$ in Eq. (8.10) is given by

$$\varepsilon(C) = -\frac{(1+\sqrt{1-C^2})}{2} \log_2 \frac{(1+\sqrt{1-C^2})}{2} - \frac{(1-\sqrt{1-C^2})}{2} \log_2 \frac{(1-\sqrt{1-C^2})}{2}. \tag{8.13}$$

$\varepsilon(C)$ is a monotonic function of C and varies from 0 to 1 as C changes from 0 to 1. Concurrence is thus a measure of entanglement in its own right. Consider the state $|\psi\rangle$ to be the singlet state of two spin-$\frac{1}{2}$ particles, i.e. $|\psi\rangle = \frac{1}{\sqrt{2}}(|\uparrow\downarrow\rangle - |\downarrow\uparrow\rangle)$. The concurrence in this case has the maximum value, $\left| \left\langle \psi | \tilde{\psi} \right\rangle \right| = 1$. If $|\psi\rangle = |\uparrow\downarrow\rangle$ (a separable state), the concurrence $C = 0$. The case of mixed states is more difficult to treat. Given a density matrix ρ of a bipartite system composed of two subsystems A and B, one can consider all possible decompositions of ρ into pure states

$$\rho = \sum_i p_i |\psi_i\rangle \langle \psi_i|, \tag{8.14}$$

where p_i's denote the probabilities. For each pure state $|\psi_i\rangle$, the entanglement is measured by the reduced von Neumann entropy $S(\psi_i)$ of either of the subsystems (Eq. (8.7)). The entanglement of formation of the mixed state ρ is defined to be the average entanglement of the constituent pure states, minimized over all decompositions of ρ:

$$E(\rho) = \min \sum_i p_i S(\psi_i). \tag{8.15}$$

In general, the minimization is extremely difficult to carry out. The minimization problem can, however, be solved analytically for pairs of qubits (two-level systems) [29, 30]. The closed-form expansion of $E(\rho)$ is given by $E(\rho) = \varepsilon(C(\rho))$ with the function $\varepsilon(C)$ as defined in Eq. (8.13). The concurrence C is computed in the following manner. Let $\rho(i,j)$, the reduced density matrix for a pair of spin-$\frac{1}{2}$ particles (qubits), be represented in the standard basis $\{|\uparrow\uparrow\rangle, |\uparrow\downarrow\rangle, |\downarrow\uparrow\rangle, |\downarrow\downarrow\rangle\}$. One can define the spin-reversed density matrix as $\tilde{\rho} = (\sigma_y \otimes \sigma_y) \rho^* (\sigma_y \otimes \sigma_y)$. The concurrence C is given by

$$C = \max(\lambda_1 - \lambda_2 - \lambda_3 - \lambda_4, 0), \tag{8.16}$$

where λ_i's are the square roots of the eigenvalues of the matrix $\rho \tilde{\rho}$ in descending order. Again, $C = 0$ implies an unentangled state, whereas $C = 1$ corresponds to maximum entanglement.

We now define the fidelity measure F which can be used to probe QPTs. F is given by the modulus of the overlap of the ground state wave functions $|\psi_0(\lambda)\rangle$ and $|\psi_0(\lambda + \delta\lambda)\rangle$ for closely spaced Hamiltonian parameter values λ and $\lambda + \delta\lambda$ [16]:

$$F(\lambda, \lambda + \delta\lambda) = |\langle\psi_0(\lambda)|\psi_0(\lambda + \delta\lambda)\rangle|. \tag{8.17}$$

Equation (8.17) gives a definition of the global fidelity. The reduced fidelity, RF, refers to a subsystem and is defined to be the overlap between the reduced density matrices $\rho \equiv \rho(h)$ and $\tilde{\rho} \equiv \rho(h + \delta)$ of the ground states $|\phi_0(h)\rangle$ and $|\phi_0(h + \delta)\rangle$, h and $h + \delta$ being two closely spaced Hamiltonian parameter values [16, 20, 31–35]. The RF is

$$F_R(h, h + \delta) = Tr \sqrt{\rho^{\frac{1}{2}} \tilde{\rho} \rho^{\frac{1}{2}}}. \tag{8.18}$$

The fidelity susceptibility, $\chi_F(\lambda)$, is given by the second derivative of the fidelity $F(\lambda, \lambda + \delta\lambda)$. Equation (8.17) with respect to $\delta\lambda$:

$$\chi_F(\lambda) = \lim_{\delta\lambda \to 0} \frac{-2 \ln F}{\delta\lambda^2} = -\frac{\partial^2 F}{\partial(\delta\lambda)^2} \tag{8.19}$$

One can similarly define the reduced fidelity susceptibility.

8.3 QPTs in Model Spin Systems

The prototypical model for the study of QPTs is the transverse Ising model (TIM) in one dimension (1D) [1, 2]. It is a special case of the more anisotropic XY model, the Hamiltonian of which can be written as follows:

$$H = -\lambda \sum_{i=1}^{N} [(1 + \gamma) S_i^x S_{i+1}^x + (1 - \gamma) S_i^y S_{i+1}^y] - \sum_{i=1}^{N} S_i^z. \tag{8.20}$$

The anisotropy parameter $\gamma = 0$ corresponds to the isotropic XX model. The TIM is obtained for $\gamma = 1$. In the interval $0 < \gamma \leq 1$, the model belongs to the Ising universality class. In the thermodynamic limit $N \to \infty$, a QPT occurs at the critical point $\lambda_c = 1$. The magnetization in the x-direction, $\langle S^x \rangle$, serves as the order parameter of the transition with $\langle S^x \rangle \neq 0$ for $\lambda > \lambda_c$ and is zero for $\lambda \leq \lambda_c$. The magnetization along the z-direction, $\langle S^z \rangle$, is different from zero for any value of λ. We now discuss briefly the nature of the TIM ground state for limiting values of the tuning parameter λ. When λ is close to zero value, the ground state, $|\psi_g\rangle$, is a product of spins pointing in the positive z-direction:

$$\left|\psi_g\right\rangle_{\lambda\to 0} \approx \ldots \left|\uparrow\right\rangle_i \left|\uparrow\right\rangle_{i+1} \ldots . \tag{8.21}$$

In the limit $\lambda \to \infty$, the ground state has the form

$$\left|\psi_g^+\right\rangle_{\lambda\to\infty} \approx \ldots \left|\to\right\rangle_j \left|\to\right\rangle_{j+1} \ldots, \tag{8.22}$$

where the spin state $\left|\to\right\rangle \equiv \frac{1}{\sqrt{2}}(\left|\uparrow\right\rangle + \left|\downarrow\right\rangle)$ represents a spin pointing in the positive x-direction. The ground state is actually doubly degenerate and the second ground state is given by

$$\left|\psi_g^-\right\rangle_{\lambda\to\infty} \approx \ldots \left|\leftarrow\right\rangle_j \left|\leftarrow\right\rangle_{j+1} \ldots, \tag{8.23}$$

with the spin state $\left|\leftarrow\right\rangle \equiv \frac{1}{\sqrt{2}}(\left|\uparrow\right\rangle - \left|\downarrow\right\rangle)$ representing a spin pointing in the negative x-direction. The two states, $\left|\psi_g^+\right\rangle_{\lambda\to\infty}$ and $\left|\psi_g^-\right\rangle_{\lambda\to\infty}$, are connected to each other by a global phase flip, $U_{PF} = \Pi_{j=1}^N \sigma_j^z$ [2, 9]:

$$\left|\psi_g^-\right\rangle_{\lambda\to\infty} \equiv U_{PF} \left|\psi_g^+\right\rangle_{\lambda\to\infty} . \tag{8.24}$$

The $\lambda = 0$ ground state is invariant under the global spin flip operation. The changes in the ground state as one moves away from the limiting cases can be computed using perturbation theory. One can show that the essential character of the $\lambda = 0$ ground state is retained in the range of λ values $0 < \lambda < \lambda_c$, i.e. the ground state is invariant under global phase flip with $\langle S^x \rangle = 0$. Similarly, the features of the $\lambda \to \infty$ ground state, e.g. the ground state degeneracy under global phase flip, are retained in the range $\lambda_c < \lambda < \infty$. At the quantum critical point $\lambda = \lambda_c = 1$, a transition in the fundamental nature of the ground state takes place. The symmetry under global phase flip breaks at the QCP and as a result the system develops a nonzero magnetization, $\langle S^x \rangle \neq 0$, in the x-direction. $\langle S^x \rangle$ thus plays the role of an order parameter in distinguishing between two phases: the paramagnetic phase for $\lambda < \lambda_c$ ($\langle S^x \rangle = 0$) and the ferromagnetic phase $\lambda > \lambda_c$ ($\langle S^x \rangle \neq 0$). The order parameter is in principle observable as the system chooses one or the other of the two degenerate ground states due to the presence of small external perturbations. A straightforward calculation of the magnetization $\langle S^x \rangle$ is difficult but one can obtain $\langle S^x \rangle$ from the large-r limit of the correlation function $\left\langle S_j^x S_{j+r}^x \right\rangle$ [2, 9]:

$$\begin{aligned} \langle S^x \rangle &= 0 & \lambda \le 1 \\ &\sim (1 - \lambda^{-2})^{\frac{1}{8}} & \lambda > 1 \end{aligned} . \tag{8.25}$$

We now focus on the entanglement properties of the TIM ground state as the QCP is approached. In the initial studies, the focus was on bipartite entanglement as measured by concurrence [8, 9]. The concurrence tends to zero in both the limits

$\lambda \ll 1$ and $\lambda \gg 1$ as the ground states in these cases approach the product form. Also, the concurrence $C(n)$, a measure of pairwise entanglement for spins separated by n lattice constants, is zero for $n > 2$.

The nearest-neighbour (n.n.) concurrence, $C(1)$, turns out to be a smooth function of λ and reaches its maximum value close to the QCP. The general conclusion is that the pairwise entanglement measured by concurrence does not become long-ranged as $\lambda \to \lambda_c$ in contrast to the usual correlation length ξ which diverges as $\xi \sim |\lambda - \lambda_c|^{-\nu}$ with $\nu = 1$. The next-nearest-neighbour (n.n.n.) concurrence , $C(2)$, becomes maximum at $\lambda = \lambda_c$. The critical properties are best captured by the derivatives of the concurrence with respect to λ. In the thermodynamic limit $N \to \infty$, $\partial_\lambda C(1)$ diverges on approaching the critical point as [8, 9]

$$\partial_\lambda C(1) \sim \frac{8}{3\pi^2} ln|\lambda - \lambda_c| \qquad (8.26)$$

Equation (8.26) provides a quantitative measure of non-local correlations in the critical region. The study of precursors of the critical behaviour in finite systems is based on finite-size scaling. According to the scaling theory, the concurrence, a function of both the system size and the tuning parameter, depends only on the combination $N^{\frac{1}{\nu}}(\lambda - \lambda_m)$ in the vicinity of the transition point with ν being the correlation length exponent and λ_m the position of the minimum of $\partial_\lambda C(1)$. In the case of log divergence, the scaling ansatz is of the form [8]

$$\partial_\lambda C(1)(N, \lambda) - \partial_\lambda C(1)(N, \lambda_0) \sim Q[N^{\frac{1}{\nu}}(\lambda - \lambda_m)] - Q[N^{\frac{1}{\nu}}(\lambda_0 - \lambda_m)], \quad (8.27)$$

where λ_0 is some non-critical value and $Q(x) \sim Q(\infty) ln x$ for large x. Figure 8.2 shows the plots of $\partial_\lambda C(1)$ versus λ for different values of N, the system size. There is no divergence for finite N and λ_m scales as $\lambda_m \sim \lambda_c + N^{-1.86}$ (left inset of Fig. 8.2). The first derivative diverges logarithmically with increasing system size as [8]

$$\partial_\lambda C(1)|_{\lambda_m} = -0.2702 \ln N + \text{const.} \qquad (8.28)$$

The scaling ansatz for logarithmic singularities stipulates that the ratio of the prefactors ($\frac{8}{3\pi^2}$ and 0.2702) of the logarithms in Eqs. (8.26) and (8.28) should be equal to the correlation length exponent ν. Since, $\frac{8}{3\pi^2} \sim 0.2702$, the value of ν is 1 which is the same as the value obtained from the solution of the Ising model. Thus, although the concurrence does not become long ranged as $\lambda \to \lambda_c$, it exhibits scaling behaviour typical of critical point transitions. In the case of the anisotropic XY model, $0 < \gamma < 1$, universality in the critical properties of entanglement has been verified [8]. The universality class is that of the TIM. The scaling ansatz holds true with the critical exponent $\nu = 1$, consistent with the universality hypothesis. The range of entanglement, ξ_E, is not, however, universal and tends to infinity as γ tends to zero. The total concurrence $\sum_n C(n)$ is a weakly increasing function of γ with

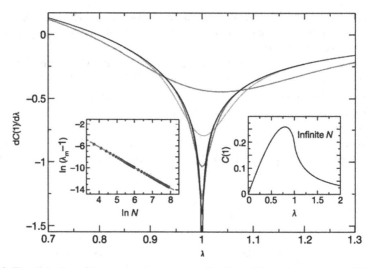

Fig. 8.2 The derivative of the n.n. concurrence as a function of the tuning parameter λ in the 1D TIM. The curves correspond to different lattice sizes with the minimum getting more pronounced as the system size is increased. The position of the minimum changes as system size N changes (*left inset*) and a logarithmic divergence occurs for an infinite system ($N \to \infty$). The *right inset* shows the variation of the n.n. concurrence with λ in an infinite system. The maximum of the plot has no relation with the critical properties of the Ising model [From 8]

$0.15 < \sum_n C(n) < 0.2$ for $0.1 < \gamma \le 1$, i.e. there is no significant change in the total entanglement content of the ground state [8].

The vanishing of the concurrence does not preclude the existence of other types of entanglement between distant spins. Vidal et al. [10] analysed the entanglement between a block of L adjacent spins with the rest of the chain via the von Neumann entropy of the reduced density matrix (Eq. (8.7)) of a subsystem. The block of L spins and the rest of the spins in the chain constitute, respectively, the subsystems A and B. The entropy is expected to saturate once the length L exceeds the correlation length ξ. A different situation prevails at the QCP where the correlation length diverges. For the TIM, the computed von Neumann entropy $S^{(L)}$ is given by

$$S^{(L)} = \frac{c + \bar{c}}{6} \, log_2 L + k, \qquad (8.29)$$

so that $S^{(L)}$ diverges logarithmically with the length of the block. The coefficients $c, \bar{c}$ correspond to the central charges of conformal field theory and specify the universality class of the critical point transition. Verstraete et al. [36] introduced the concept of localizable entanglement which is a measure of the maximum amount of entanglement that can be localized between two spins by performing local operations on the other spins of the chain. They defined an entanglement length ξ_E which diverges, i.e. becomes long ranged at the QCP. Bose and Chattopadhyay [37] considered some simple spin models for which the ground states are known exactly. The models exhibit first-order QPTs brought about by tuning an external magnetic field. The entanglement measure, concurrence, was shown to change discontinuously at the transition points. Later studies of first-order QPTs provided more examples of such discontinuities [38, 39]. Wu et al. [11] developed a general theory on the connection between bipartite entanglement measures and QPTs. The theory is valid for QPTs characterized by non-analyticities in the derivatives of the ground state energy. A first-order QPT involves a finite discontinuity in the first derivative of the ground state energy. A second-order QPT is similarly characterized by a finite discontinuity or divergence in the second derivative of the ground state energy with the first derivative being continuous. These characterizations arise from the $T = 0$ limits of the first-and second-order thermodynamic phase transitions in which free energy substitutes for the ground state energy. Under some general assumptions, Wu et al. [11] proved a theorem that a discontinuity in a bipartite entanglement measure like concurrence and negativity is both necessary and sufficient to signal a first-order QPT. Similarly, a discontinuity or a divergence in the first derivative of the bipartite entanglement measure (assuming that the measure itself is continuous) is both a necessary and sufficient indicator of a second-order QPT. The earlier observation of the divergence of the first derivative of the n.n. concurrence $\partial_\lambda C(1)$ (Eq. (8.26)) at a QCP is consistent with the statement of the theorem. To understand the link better, one should analyse the expressions for the derivatives of the ground state energy. If ε is the energy per particle, then

$$\partial_\lambda \varepsilon = \frac{1}{N} \sum_{ij} Tr[(\partial_\lambda U(i,j)) \rho(i,j)], \tag{8.30}$$

$$\partial_\lambda^2 \varepsilon = \frac{1}{N} \sum_{ij} \{Tr[(\partial_\lambda^2 U(i,j)) \rho(i,j)] + Tr[(\partial_\lambda U(i,j)) \partial_\lambda \rho(i,j)]\}, \tag{8.31}$$

where $\rho(i,j)$ is the reduced two-particle density operator and $U(i,j)$ is the sum of all the one and two body terms of the Hamiltonian corresponding to the particles at sites i and j. Assume that $U(i,j)$ is a smooth function of the parameters of the Hamiltonian and that $\rho(i,j)$ is finite at the critical point. In that case, a first-order QPT has its origin in the discontinuity of one or more of the $\rho(i,j)$'s at the QCP. Similarly, in the case of a second-order QPT, the discontinuity or singularity of $\frac{\partial^2 \varepsilon}{\partial \lambda^2}$ occurs due to the discontinuity or divergence of one or more of the $\frac{\partial \rho(i,j)}{\partial \lambda}$'s at the QCP. Typical bipartite entanglement measures like concurrence and negativity are linear functions of the elements of $\rho(i,j)$, so that the discontinuity/divergence in $\rho(i,j)/\partial_\lambda \rho(i,j)$ translates into the discontinuity/divergence of the entanglement measure or its derivative.

The single-site von Neumann entropy defined in Eq. (8.8) has been shown to be a good indicator of QPTs. The same is true for the two-site von Neumann entropy $S(i,j)$ (Eq. (8.9)). In a translationally invariant system $S(i,j)$ depends only on the distance $n = |j - i|$. As pointed out in [13], the spins that are entangled with one or both the spins at sites i and j contribute to S. Let us consider the case of the $S = \frac{1}{2}$

anisotropic XY model in a transverse magnetic field the Hamiltonian of which is given in Eq. (8.20). As already mentioned, the model, away from the isotropic limit, belongs to the universality class of the TIM. One can show that $S(i,j)$ has a simple dependence on the spin correlation functions in the large n limit. Away from the critical point, $S(i,j)$ is found to saturate over a length scale ξ_E as n increases. Near the QCP, one obtains [13]

$$S(i,j) - S(\infty) \sim n^{-1} e^{-\frac{n}{\xi_E}}. \tag{8.32}$$

ξ_E is termed the entanglement length as it sets the scale over which the entanglement measure decays. As the QCP is approached, ξ_E diverges with the critical exponent ν. One can further show that $\xi_E = \frac{\xi_C}{2}$, where ξ_C is the usual correlation length. The entanglement measure $S(i,j)$ thus captures the long-range correlations associated with a QCP. At the critical point itself, $S(i,j) - S(\infty)$ has a power law decay, i.e. $S(i,j) - S(\infty) \sim n^{-\frac{1}{2}}$ [13]. Also, in the limit of large n, the first derivative of $S(i,j)$ with respect to the tuning parameter develops a λ-like cusp at the critical point. The universality and a finite-size scaling of the entanglement have also been demonstrated. We have so far discussed how bipartite entanglement measures provide signatures of QPTs. Oliveira et al. [12] have proposed a generalized global entanglement (GGE) measure $G(2, n)$ that quantifies multipartite entanglement (ME). $G(2, n)$ for a translationally symmetric system is given by

$$G(2, n) = \frac{d}{d-1}[1 - \sum_{l,m=1}^{d^2} |[\rho(j, j+n)]_{lm}|^2], \tag{8.33}$$

where $\rho(j, j+n)$ is the reduced density matrix of dimension d. The factor 2 in $G(2, n)$ indicate that the reduced density matrix is that for a pair of particles. Oliveira et al. [12] have shown that the GGE signals a critical point location as well as the order of a QPT. A discontinuity in $G(2, n)$ signals a first-order QPT, brought about by a discontinuity in one or more of the elements $[\rho(j, j+n)]_{lm}$ of the reduced density matrix. In the case of a second-order QPT, a discontinuity or divergence in the first derivative of $G(2, n)$ with respect to the tuning parameter occurs due to a discontinuity or divergence in the first derivatives of one or more elements of the reduced density matrix. In the case of the XY $S = \frac{1}{2}$ spin chain in a transverse magnetic field, the GGE measure shows a diverging entanglement length ξ_E as the QCP is approached. As in the case of the two-site von Neumann entropy, $\xi_E = \frac{\xi_C}{2}$ and both the length scales diverge near the QCP with the same critical exponent. The GGE measure is furthermore maximal at the critical point. $G(2, n)$ is thus more versatile than concurrence in signalling QPTs [12]. We have mentioned in the introduction that fidelity and fidelity susceptibility also serve as indicators of QPTs. The case of the XY spin chain in a transverse magnetic field is illustrated in Fig. 8.3 [18]. The Hamiltonian has the same form as in Eq. (8.20) but the parameter λ appears in the transverse magnetic field term. The fidelity $F(\lambda, \gamma)$ is a function of two parame-

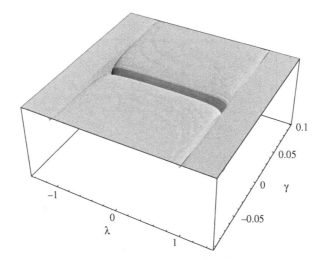

Fig. 8.3 Fidelity $F(\lambda, \gamma)$ in the XY spin chain in a transverse magnetic field versus the anisotropy parameter γ and the magnetic field parameter λ. Quantum phase transitions are indicated by the drops in F. Ising-like transitions occur at $\lambda = \pm 1$ and the anisotropy transition at $\gamma = 0$ [From 18]

ters λ and γ. The Ising transitions occur at $\lambda = \pm 1$ signalled by the drops in F. The anisotropy transition occurs at $\gamma = 0$. In using the fidelity measure to study QPTs, one obvious advantage is that no prior knowledge of the order parameter and symmetry breaking is required. Recently, general scaling analyses of the fidelity susceptibility have been proposed [16]. The critical exponents appearing in the scaling formulations define universality classes as in the usual critical phenomena.

Chen et al. [40] have shown that there is an intrinsic relation between ground state fidelity and QPTs. The Hamiltonian of a quantum many body system undergoing QPTs has the general form

$$H(\lambda) = H_0 + \lambda H_1, \tag{8.34}$$

where H_1 describes the driving part of the Hamiltonian and λ is the tuning parameter. Let the eigenstates and eigenvalues of $H(\lambda)$ be denoted by $\psi_n(\lambda)$ and $E_n(\lambda)$. A first-order QPT is brought about by ground state level crossing. For $\lambda < \lambda_c$, the ground state energy $E_0(\lambda) < E_1(\lambda)$, the energy of the first excited state. At $\lambda = \lambda_c$, a role reversal takes place so that $E_1(\lambda) < E_0(\lambda)$ for $\lambda > \lambda_c$. The crossing of the energy levels (Fig. 8.4) is responsible for the discontinuity in the first derivative of the ground state energy at the transition point $\lambda = \lambda_c$. Because of the change in the nature of the ground state, the fidelity F (Eq. (8.17)) has a sharp drop at the transition point. The sudden drop in F has the same physical origin as the discontinuity in the first derivative of the ground state energy. In the case of a second-order QPT, there is no ground state level crossing so that the ground state energy $E_g(\lambda) = E_0(\lambda)$ for all values of λ. The expression for the second derivative of the ground state energy is given by [40]

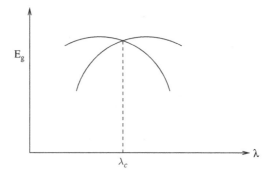

Fig. 8.4 A crossing of ground state energy levels in a first-order quantum phase transition. The first derivative of the ground state energy $\frac{dE_g}{d\lambda}$ is discontinuous at the transition point $\lambda = \lambda_c$

$$\frac{\partial^2 E_0(\lambda)}{\partial \lambda^2} = \sum_{n \neq 0} \frac{2 \,|\, \langle \psi_n(\lambda)| \, H_1 \, |\psi_0(\lambda)\rangle \,|^2}{E_0(\lambda) - E_n(\lambda)}. \tag{8.35}$$

The fidelity susceptibility $\chi_F(\lambda)$ can be written as

$$\chi_F(\lambda) = \sum_{n \neq 0} \frac{|\, \langle \psi_n(\lambda)| \, H_1 \, |\psi_0(\lambda)\rangle \,|^2}{[E_0(\lambda) - E_n(\lambda)]^2}. \tag{8.36}$$

Comparing Eqs. (8.34) and (8.35), one finds that in both the cases singularities arise due to the vanishing of the energy gap in the thermodynamic limit. This is a generic feature of critical point transitions [1, 2]. In the case of the TIM, the critical point $\lambda_c = 1$ separates the FM ($\langle S^x \rangle \neq 0$) and paramagnetic ($\langle S^x \rangle = 0$) phases. In both the phases, the first excited state is separated by an energy gap from the ground state. The energy gap goes continuously to zero as the QCP is approached with an appropriate critical exponent. The second derivative of the ground state energy and the fidelity susceptibility thus are equivalent indicators of QPTs.

8.4 Signatures of QPTs in a Spin Ladder

We have already described the spin ladder (Fig. 8.1) in Sect. 8.1. AFM spin ladders are examples of interacting many body systems which exhibit a range of novel phenomena [41, 42]. An n-chain spin ladder consists of n chains coupled by rungs, the simplest example being a two-chain ladder with $n = 2$. The study of ladders as prototypical many body systems became important after the discovery of high temperature superconductivity in the strongly correlated cuprate materials. The dominant electronic and magnetic properties of the cuprates are associated with the CuO_2 planes which have the structure of a square lattice [43]. The treatment of strong correlation is less rigorous in two dimensions (2D) than in 1D. Ladder models, with structure interpolating between $1D$ and $2D$, serve as ideal candidates to address issues related to strong correlation and also to investigate how electronic and magnetic properties change as one progresses from the chain to the plane. In undoped ladder models, each site of the ladder is occupied by a spin (usually of magnitude $\frac{1}{2}$) and the spins interact via the AFM Heisenberg exchange interaction. In doped ladder models, some of the spins are replaced by positively charged holes which are mobile. The Hamiltonian describing the doped systems are the t-J and Hubbard ladder models [41–43]. With the discovery of a large number of materials having a ladder-like structure, the study of ladders has acquired considerable importance. The materials exhibit a range of phenomena including superconductivity in hole-doped systems, the 'odd–even' effect in which the excitation spectrum of an n-chain ladder is gapped (gapless) if n is even (odd) and quantum phase transitions (QPTs) tuned by an external magnetic field [41–43]. Many of the experimental observations were motivated by theoretical predictions, superconductivity being a prime example [44–46].

Figure 8.5 shows the temperature T versus magnetic field H phase diagram of a two-chain spin ladder, the Hamiltonian of which is given in Eq. (8.1) [28]. As already mentioned in Sect. 8.1, there are two QCPs at $H = H_{c1}$ and $H = H_{c2}$. We now discuss how the different entanglement and fidelity measures, already defined, provide signatures of the QPTs in the ladder system. One notes that the z-component $S_{tot}^z = \sum_{j=1}^{L} (S_{1,j}^z + S_{2,j}^z)$ of the total spin is a conserved quantity. Using this fact, the Hamiltonian can be diagonalized for different values of L with the help of the numerical diagonalization package TITPACK [47]. This enables one to compute the variation of the fidelity $F(H, H + \delta)$ (Eq. (8.17)) as a function of H with $\delta = .001$. The inset of Fig. 8.6 shows sharp drops in F at $H_{c1}^L = \Delta_L$, where Δ_L is the energy gap. A polynomial fitting of the Δ_L versus $\frac{1}{L}$ data points yields $\Delta_L \approx 11.8416 +$

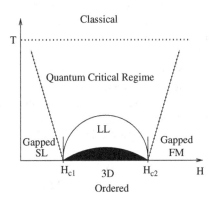

Fig. 8.5 Temperature T versus magnetic field H phase diagram of a two-chain spin ladder, the Hamiltonian of which is given in Eq. (8.1). The two critical points at $H = H_{c1}$ and $H = H_{c2}$ separate three distinct phases: a spin gap (spin liquid SL) phase, a Luttinger liquid (LL) phase and a gapped ferromagnetically polarized phase. The *shaded region* represents a 3D-ordered phase which occurs at sufficiently low T due to the weak interactions between isolated ladders

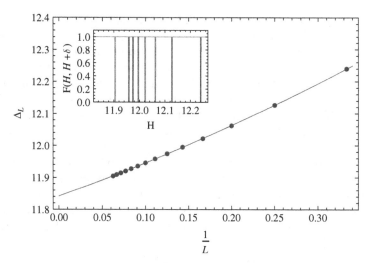

Fig. 8.6 Plot of Δ_L, the spin gap, versus $\frac{1}{L}$ from numerical diagonalization data of the ladder Hamiltonian (Eq. (8.1)), L being the number of rungs in the ladder; *(inset)* ground state fidelity versus magnetic field H for $L = 3, 4, ..., 16$

$0.9739\left(\frac{1}{L}\right) + 0.6621\left(\frac{1}{L}\right)^2$. On extrapolating to the thermodynamic limit, one obtains $H_{c1} = \Delta_{\infty} \approx 11.8416$. The estimate is close to the value $H_{c1} \approx J_{\perp} - J_{\parallel}$ obtained in first-order perturbation theory [28]. The values of the coupling strengths $J_{\perp}$ and $J_{\parallel}$ are $J_{\perp} = 13\,K$ and $J_{\parallel} = 1.15\,K$, the values for the AFM ladder compound $(5IAP)_2CuBr_4.2H_2O$ with $J_{\perp} \gg J_{\parallel}$ [27]. The magnitude of the upper critical point H_{c2} can be calculated exactly as $H_{c2} = J_{\perp} + 2J_{\parallel}$ taking into account the fact that the fully polarized FM ground state ($H > H_{c2}$) becomes unstable when the lowest energy of the spin wave excitations falls below the energy of the polarized state. The numerical diagonalization data reproduce the exact value of H_{c2}. The inset of Fig. 8.7 shows the variation of the first derivative of the magnetization $m(H)$ with respect to H versus H in the thermodynamic limit adopting the extrapolation precedure outlined in [48]. The magnetization $m(H)$ is the average magnetization per site and because of translational invariance $m(H) = \langle S_i^z \rangle$. The inset of Fig. 8.7 shows that the derivative $\frac{dm}{dH}$ tends to diverge as $H \rightarrow H_{c1}$ and H_{c2}. This is consistent with the existence of a square root singularity in $m(H)$ in the vicinity of the quantum critical points H_{c1} and H_{c2} [28, 49]. The single-site reduced density matrix $\rho(i)$ can be written in terms of the spin expectation value $\langle S_i^z \rangle$ as [33]

$$\rho(i) = \begin{pmatrix} \frac{1}{2} + \langle S_i^z \rangle & 0 \\ 0 & \frac{1}{2} - \langle S_i^z \rangle \end{pmatrix} \qquad (8.37)$$

in the $|\uparrow\rangle, |\downarrow\rangle$ basis. From Eq. (8.8), the single-site von Neumann entropy is

$$S(i) = -\sum_i \lambda_i \log_2 \lambda_i, \qquad (8.38)$$

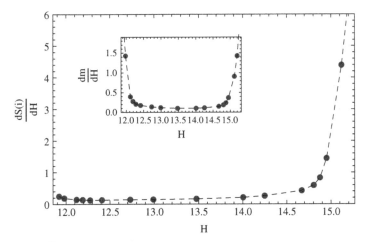

Fig. 8.7 Plot of $\frac{dS(i)}{dH}$ versus H (using numerical diagonalization data); (*inset*) variation of $\frac{dm}{dH}$ with H

where the λ_i's are the two diagonal elements of $\rho(i)$. Figure 8.7 shows the variation of $\frac{dS(i)}{dH}$ with H. Unlike $\frac{dm}{dH}$, $\frac{dS(i)}{dH}$ tends to diverge only at the upper critical point H_{c2} and has a finite value as $H \rightarrow H_{c1}$, the lower critical point. The values of H_{c1} and H_{c2} are $H_{c1} = 11.8416\,K$ and $H_{c2} = 15.3\,K$ as obtained from numerical diagonalization data.

The strongly coupled ladder model in high magnetic field can be mapped onto a 1D XXZ AFM Heisenberg chain with an effective Hamiltonian [28, 50]:

$$\mathcal{H}_{\it eff} = J_{||} \sum_{j=1}^{L} [\tilde{S}_j^x \tilde{S}_{j+1}^x + \tilde{S}_j^y \tilde{S}_{j+1}^y + \frac{1}{2}\tilde{S}_j^z \tilde{S}_{j+1}^z] - \tilde{H} \sum_{j=1}^{L} \tilde{S}_j^z, \qquad (8.39)$$

where $\tilde{H} = H - J_\perp - \frac{J_{||}}{2}$ is an effective magnetic field and $\tilde{S}_j^\alpha$'s ($\alpha = x, y, z$) are pseudo spin-$\frac{1}{2}$ operators which can be expressed in terms of the original spin operators. There are several exact results known for the XXZ spin-$\frac{1}{2}$ chain in a magnetic field [51, 52]. In particular, the zero temperature magnetization $m(H)$ close to the quantum critical points is given by the expressions (we use the symbol H instead of $\tilde{H}$)

$$m(H) \sim \frac{\sqrt{2}}{\pi} \sqrt{(H - H_{c1})/J_{||}}, \quad H > H_{c1}, \qquad (8.40)$$

$$m(H) \sim 1 - \frac{\sqrt{2}}{\pi} \sqrt{(H_{c2} - H)/J_{||}}, \quad H < H_{c2}. \qquad (8.41)$$

Similar expressions are obtained in the case of an integrable spin ladder model with the help of the thermodynamic BA [49]. Using the analytic expressions of $m(H)$ in

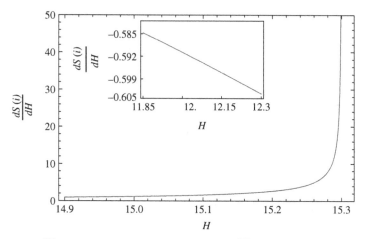

Fig. 8.8 Plot of $\frac{dS(i)}{dH}$ versus H near $H = H_{c2}$; (*inset*) plot of $\frac{dS(i)}{dH}$ versus H near H_{c1}

Eqs. (8.40) and (8.41), the first derivative of single-site von Neumann entropy with respect to magnetic field H, $\frac{dS(i)}{dH}$, can be calculated analytically from Eqs. (8.37) and (8.38). Again, the derivative diverges near H_{c2} (Fig. 8.8) but not as the quantum critical point H_{c1} is approached, consistent with the numerical results. The values of H_{c1} and H_{c2} are $H_{c1} = J_\perp - J_\parallel = 11.85\ K$ and $H_{c2} = J_\perp + 2J_\parallel = 15.3\ K$. The estimate of H_{c1} is from first-order perturbation theory.

The correlation functions of the $S = \frac{1}{2}$ XXZ chain in a magnetic field are known [53] in the gapless phase $H_{c1} < H < H_{c2}$. In terms of the original spin operators, these are given by

$$\langle S_1^z(r)S_1^z(0)\rangle = \frac{m^2}{4} + \frac{1}{r^2} + \cos(2\pi mr)\left(\frac{1}{r}\right)^{2K}, \tag{8.42}$$

$$\langle S_1^+(r)S_1^-(0)\rangle = \cos[\pi(1 - 2m)r]\left(\frac{1}{r}\right)^{\frac{2K+1}{2K}} + \cos(\pi r)\left(\frac{1}{r}\right)^{\frac{1}{2K}}, \tag{8.43}$$

where K is the Luttinger liquid exponent. For simplicity, some prefactors (constants) in the terms appearing in Eqs. (8.42) and (8.43) have been dropped. The expressions for the correlation functions are utilized to study the variation of the two-site entanglement $S(i, j)$ and concurrence $C_{i,i+1}$ with respect to the magnetic field. These quantities can be computed from the two-site reduced density matrix $\rho(i, j)$ which, in terms of the spin expectation values and correlation functions, is given by [54]

$$\rho(i,j) = \begin{pmatrix} \frac{1}{4} + \langle S_i^z\rangle + \langle S_i^z S_j^z\rangle & 0 & 0 & 0 \\ 0 & \frac{1}{4} - \langle S_i^z S_j^z\rangle & \langle S_i^x S_j^x\rangle + \langle S_i^y S_j^y\rangle & 0 \\ 0 & \langle S_i^x S_j^x\rangle + \langle S_i^y S_j^y\rangle & \frac{1}{4} - \langle S_i^z S_j^z\rangle & 0 \\ 0 & 0 & 0 & \frac{1}{4} - \langle S_i^z\rangle + \langle S_i^z S_j^z\rangle \end{pmatrix}. \tag{8.44}$$

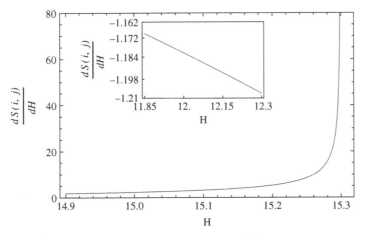

Fig. 8.9 Plot of $\frac{dS(i,j)}{dH}$ versus H near $H = H_{c2}$; (*inset*) plot of $\frac{dS(i,j)}{dH}$ versus H near H_{c1}

The two-site von Neumann entropy (Eq. (8.9)) can be written as

$$S(i,j) = -\sum_i \epsilon_i \, log_2 \, \epsilon_i, \tag{8.45}$$

where ϵ_i's are the eigenvalues of $\rho(i,j)$. Using Eqs. (8.40)–(8.44), the first derivative of $S(i,j)$ with respect to H can be calculated near both the QCPs (Fig. 8.9). The derivative diverges as $H \to H_{c2}$ but approaches a finite value as $H \to H_{c1}$. The n.n. concurrence (Eq. (8.16)) can be written as [9, 29, 30, 55, 56]

$$C_{i,i+1} = 2 \, \mathrm{Max}[0, \, |\rho_{23}(i, i+1)| - \sqrt{\rho_{11}(i, i+1)\rho_{44}(i, i+1)}]. \tag{8.46}$$

Figure 8.10 shows the derivative of $C_{i,i+1}$ with respect to H versus H. The derivative, as in the case of one-site and two-site entanglement measures, diverges as $H \to H_{c2}$ but has a finite value as $H \to H_{c1}$. As discussed in [11], the first derivatives of one or more elements of the reduced density matrix $\rho(i,j)$ with respect to the tuning parameter are expected to diverge at the QCPs. In the case of the ladder, one can verify that this is the case as $H \to H_{c1}$ and H_{c2} with the divergent contributions coming from $\rho_{11}(i,j)$ and $\rho_{44}(i,j)$. The theorem in [11] that a discontinuity or a divergence of the first derivative of concurrence signals a second-order QPT holds true provided certain conditions are satisfied. One of these (Condition (*b*) in [11]) stipulates that the discontinuous or divergent derivatives $\frac{\partial \rho(i,j)}{\partial \lambda}$ do not either all vanish accidentally or cancel with other terms in the expression for the first derivative of concurrence. One finds that precisely this condition is violated in the case of the two-chain ladder as $H \to H_{c1}$. The first derivative $\frac{dC_{i,i+1}}{dH}$ involves terms containing the factor $m(H)\frac{dm(H)}{dH}$ which leads to a cancelation of singularities as $H \to H_{c1}$ (see Eq. (8.40)) so that the theorem is no longer valid. A similar cancelation of singularities does not occur as $H \to H_{c2}$ (see Eq. (8.41)) so that $\frac{dC_{i,i+1}}{dH}$ signals correctly the

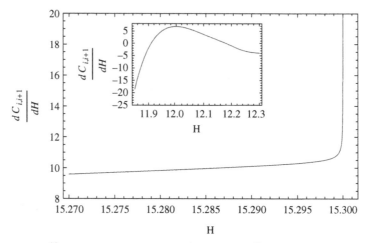

Fig. 8.10 Plot of $\frac{dC_{i,i+1}}{dH}$ versus H near $H = H_{c2}$; (*inset*) plot of $\frac{dC_{i,i+1}}{dH}$ versus H near H_{c1}

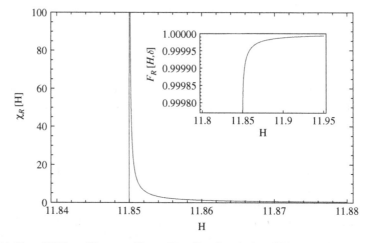

Fig. 8.11 Plot of RFS $\chi_R(H)$ versus H near $H = H_{c1}$; (*inset*) plot of RF $F_R(H, H + \delta)$ versus H near $H = H_{c1}$

occurrence of a QPT. In the cases of single-site and two-site entanglement measures, one can show that a cancelation of singularities occurs in the expressions for the derivatives as $H \rightarrow H_{c1}$. We now discuss whether fidelity-related measures exhibit special features at the QCPs H_{c1} and H_{c2}. The one-site RF $F_R(H, H + \delta)$, defined by Eqs. (8.11) and (8.37), drops sharply at the two QCPs (insets of Figs. 8.11 and 8.12). One can define the reduced fidelity susceptibility (RFS) to be

$$\chi_R(H) = \lim_{\delta \to 0} \frac{-2 \ln \mathcal{F}_R(H, H + \delta)}{\delta^2}. \tag{8.47}$$

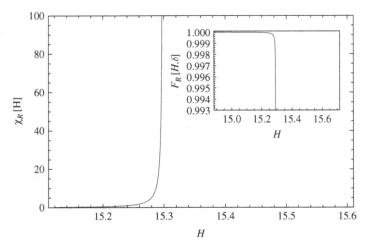

Fig. 8.12 Plot of RFS $\chi_R(H)$ versus H near $H = H_{c2}$; (*inset*) plot of RF $F_R(H, H + \delta)$ versus H near $H = H_{c2}$

As shown in Figs. 8.11 and 8.12, the RFS diverges at both the QCPs. The result is in contrast with what is observed in the case of entanglement measures, where a special feature develops only in the vicinity of the critical point H_{c2}. The calculations of the RF and the RFS are possible because they involve only local measures. A calculation of the global fidelity would not have been possible lacking a knowledge of the true many body ground state.

8.5 Concluding Remarks

In this chapter, we have discussed how different quantum information theoretic measures like entanglement and fidelity provide signatures of QPTs in interacting spin systems. The spin models specifically considered include the spin-$\frac{1}{2}$ anisotropic XY chain, the 1D transverse Ising model which is a special case of the XY model and a two-chain n spin ladder in an external magnetic field. Extensive studies have been carried out on other spin models like the Heisenberg *XXZ*, *XXX*, the Lipkin–Meshkov–Glick model, dimerized spin chains, spin-1 models. The results pertaining to these models are described in detail in [3–5]. In the case of the spin ladder, we have shown that fidelity-related measures (RF and RFS) are better indicators of QPTs than bipartite entanglement measures like concurrence and one/two-site reduced von Neumann entropy. The failure of the latter measures to signal the QPT at $H = H_{c1}$ can be attributed to the square root singularities in magnetization (Eqs. (8.40) and (8.41)). Such singularities are generic to other AFM systems with a spin gap like the spin-1 chain in a magnetic field [57–59]. Thus, the behaviour reported for the spin ladder may be a general feature of a class of gapped 1D AFM systems.

In the course of exploring the links between quantum information and condensed matter, it is pertinent to ask whether quantum information theory provides useful insight on condensed matter systems. The information theoretic measures confer a novel characterization on many body states. There are non-trivial differences between classical and non-local quantum correlations. In the TIM, the classical correlation length diverges as $\lambda \to \lambda_c$ but the non-local correlation measured by concurrence is short-ranged. Recently, it has been shown that for a certain class of multiparity states, the multiparity quantum and classical correlations are independent of each other [60], in fact the former can exist in the absence of the latter. The role of multipartite entanglement at a QCP is yet to be fully elucidated though there are results to show that the pairwise entanglement decreases dramatically at the QCP in relation to multipartite entanglement [61, 62]. In certain phase transitions, entanglement has been shown to play the role of a quantum order parameter [63]. The phase transitions considered are those between a normal conductor and a superconductor in the BCS and η-pairing models and from a Mott-insulator to a superfluid phase in the Bose–Hubbard model. The order parameters of these transitions are directly related to the amounts of entanglement present in the systems. The fidelity approach in the study of QPTs has the advantage that no a priori knowledge of the order parameter, symmetry and the type of QPTs are required. The Landau theory of second-order phase transitions operates in the framework of spontaneous symmetry breaking with the correlation functions of local order parameters playing a crucial role. The transition brought about by symmetry breaking is between a disordered and an ordered phase. However, QPTs like topological phase transitions occur between two disordered phases without any symmetry breaking. The fidelity approach is particularly suitable for the study of such QPTs [16]. Using the nuclear magnetic resonance technique, experimental detection of QPTs in terms of the fidelity has been achieved [64].

Detection of entanglement can be made with the help of an entanglement witness (EW) which is an observable, the expectation value of which is positive in separable and negative in entangled states [6, 65, 66]. Thermodynamic observables like internal energy, magnetization and susceptibility have been proposed as EWs [3, 67–70]. There is now experimental evidence that entanglement can affect the macroscopic properties of solids like specific heat and magnetic susceptibility [71, 72]. Christensen et al. [73] have carried out comprehensive neutron scattering measurements on a square lattice $S = \frac{1}{2}$ Heisenberg AFM. The data reveal a non-spin wave continuum of low-lying excitations and strong quantum effects. Analysis of the data suggests entanglement of spins at short distances. Quantum magnets serve as good laboratories for entangled states and with technologies developing for the implementation of condensed matter Hamiltonians on optical lattices [4, 74], there are exciting prospects ahead to obtain experimental signatures of entanglement in the properties of condensed matter systems including those in the vicinity of QPTs.

Acknowledgments A. T. is supported by the Council of Scientific and Industrial Research, India, under Grant No. 9/15 (306)/ 2004-EMR-I.

References

1. S. Sachdev, Science **288**, 475 (2000).
2. S. Sachdev, *Quantum Phase Transitions* (Cambridge University Press, Cambridge, 1999).
3. L. Amico, R. Fazio, A. Osterloh and V. Vedral, Rev. Mod. Phys. **80**, 517 (2008).
4. M. Lewenstein, A. Sanpera, V. Ahufinger, B. Damski, A. Sen and U. Sen, Adv. Phys. **56**, 2 (2007).
5. R. Fazio and H. van der Zant, Phys. Rep. **355**, 235 (2001).
6. O. Gühne and G. Tóth, Phys. Rep. **474**, 1 (2009).
7. M. Plenio and V. Vedral, Contemp. Phys. **39**, 431 (1998).
8. A. Osterloh, L. Amico, G. Falci, and R. Fazio, Nature (London) **416**, 608 (2002).
9. T.J. Osborne and M.A. Nielsen, Phys. Rev. A **66**, 032110 (2002).
10. G. Vidal, J.I. Latorre, E. Rico, and A. Kitaev, Phys. Rev. Lett. **90**, 227902 (2003).
11. L.-A. Wu, M.S. Sarandy and D.A. Lidar, Phys. Rev. Lett. **93**, 250404 (2004).
12. T.R. Oliveira, G. Rigolin, M.C. de Oliveira and E. Miranda, Phys. Rev. Lett. **97**, 170401 (2006).
13. H.-D. Chen, J. Phys. A **40**, 10215 (2007).
14. A. Tribedi and I. Bose, Phys. Rev. A **75**, 042304 (2007).
15. A. Tribedi and I. Bose, Phys. Rev. A **77**, 032307 (2008).
16. S -J. Gu, arxiv: 0811.3127v1.
17. H.T. Quan, Z. Song, X. F. Liu, P. Zanardi, and C. P. Sun, Phys. Rev. Lett. **96**, 140604 (2006).
18. P. Zanardi and N. Paunković, Phys. Rev. E **74**, 031123 (2006).
19. M. Cozzini, R. Ionicioiu and P. Zanardi, Phys. Rev. B **76**, 104420 (2007).
20. H.-Q. Zhou, e-print arXiv:0704.2945.
21. P. Zanardi, M. Cozzini and P. Giorda, J. Stat. Mech.: Theory Exp. L02002 (2007).
22. P. Zanardi, H.T. Quan, X. Wang and C.P. Sun, Phys. Rev. A **75**, 032109 (2007).
23. S. Chen, L. Wang, S.J. Gu and Y. Wang, Phys. Rev. E **76**, 061108 (2007).
24. A. Tribedi and I. Bose, Phys. Rev. A **79**, 012331 (2009).
25. G. Chaboussant, P.A. Crowell, L. P. Lévy, O. Piovesana, A. Madouri and D. Mailly, Phys. Rev. B **55**, 3046 (1997).
26. B.C. Watson et al., Phys. Rev. Lett. **86**, 5168 (2001).
27. C.P. Landee, M.M. Turnbull, C. Galeriu, J. Giantsidis and F.M. Woodward, Phys. Rev. B **63**, 100402 (2001).
28. G. Chaboussant et al., Eur. Phys. J. B **6**, 167 (1998).
29. K.M. O'Connor and W.K. Wootters, Phys. Rev. A **63**, 052302 (2001)
30. W.K. Wootters, Phys. Rev. Lett. **80**, 2245 (1998).
31. N. Paunković, P.D. Sacramento, P. Nogueira, V.R. Vieira and V.K. Dugaev, Phys. Rev. A **77**, 052302 (2008).
32. H.-M. Kwok, C.-S. Ho and S.- J. Gu, Phys. Rev. A **78**, 062302 (2008).
33. J. Ma, L. Xu, H. Xiong and X. Wang, Phys. Rev. E **78**, 051126 (2008).
34. J. Ma, L. Xu and X. Wang, arXiv:0808.1816.
35. H.-N. Xiong, J. Ma, Z. Sun and X. Wang, Phys. Rev. B **79**, 174425 (2009).
36. F. Verstraete, M. Popp and J.I. Cirac, Phys. Rev. Lett. **92**, 227902 (2004).
37. I. Bose and E. Chattopadhyay, Phys. Rev. A **66**, 062320 (2002).
38. F. C. Alcaraz, A. Saguia and M. S. Sarandy, Phys. Rev. A **70**, 032333 (2004).
39. J. Vidal, R. Mosseri and J. Dukelsky, Phys. Rev. A **69**, 054101 (2004).
40. S. Chen, L. Wang, Y. Hao and Y. Wang, Phys. Rev. A **77**, 032111 (2008).
41. E. Dagotto and T.M. Rice, Science **271**, 618 (1996).
42. E. Dagotto, Rep. Prog. Phys. **62**, 1525 (1999).
43. E. Dagotto, Rev. Mod. Phys. **66**, 763 (1994)
44. E. Dagotto, J. Riera and D. Scalapino, Phys. Rev. B **45**, 5744 (1992).
45. S. Gopalan, T.M. Rice and M. Sigrist, Phys. Rev. B **49**, 8901 (1994).

46. I. Bose and S. Gayen, Phys. Rev. B **48**, 10653 (1993).
47. H. Nishimori, AIP Conf. Proc. **248**, 269 (1992).
48. T. Sakai and M. Takahashi, Phys. Rev. B **43**, 13383 (1991).
49. M.T. Batchelor, X.W. Guan, N. Oelkers and Z. Tsuboi, Adv. Phys. **56**, 465 (2007).
50. F. Mila, Eur. Phys. J. B **6**, 201 (1998).
51. C.N. Yang and C.P. Yang, Phys. Rev. **150**, 327 (1966).
52. F.D.M. Haldane, Phys. Rev. Lett. **47**, 1840 (1981).
53. T. Giamarchi and A.M. Tsvelik, Phys. Rev. B **59**, 11398 (1999).
54. U. Glaser, H. Büttner and H. Fehske, Phys. Rev. A **68**, 032318 (2003).
55. M.C. Arnesen, S. Bose and V. Vedral, Phys. Rev. Lett **87**, 017901 (2001)
56. D. Gunlycke, V. M. Kendon, V. Vedral and S. Bose, Phys. Rev. A **64**, 042302 (2001).
57. R. Chitra and T. Giamarchi, Phys. Rev. B **55**, 5816 (1997).
58. H. J. Schulz, Phys. Rev. B **22**, 5274 (1980).
59. I. Affleck, Phys. Rev. B **43**, 3215 (1991).
60. D. Kaszlikowski, A. Sen (De), U. Sen, V. Vedral and A. Winter, Phys. Rev. Lett. **101**, 070502 (2008).
61. T. Roscilde et al., Phys. Rev. Lett. **93**, 167203 (2004).
62. T. Roscilde et al., Phys. Rev. Lett. **97**, 147208 (2005).
63. F.G.S.L. Brandão, New J. Phys. **7**, 254 (2005).
64. J. Zhang, X. Peng, N. Rajendran and D. Suter, Phys. Rev. Lett. **100**, 100501 (2008).
65. G. Tóth, Phys. Rev. A **71**, 010301(R) (2005).
66. M.R. Dowling, A.C. Doherty and S.D. Barlett, Phys. Rev. A **70**, 062113 (2004).
67. Č. Brukner, V. Vedral and A. Zeilinger, Phys. Rev. A **73**, 012110 (2006).
68. M. Wieśniak, V. Vedral and Č. Brukner, New J. Phys. **7**, 258 (2005).
69. I. Bose and A. Tribedi, Phys. Rev. A **72**, 022314 (2005).
70. T. Veŕtesi and E. Bene, Phys. Rev. A **73**, 134404 (2006).
71. S. Ghosh, T.F. Rosenbaum, G. Aeppli and S.N. Coppersmith, Nature (London) **425**, 48 (2003)
72. V. Vedral. Nature **425**, 28 (2003).
73. N. B. Christensen et al., Proc. Natl. Acad. Sci. **104**, 15264 (2007).
74. V. Vedral, Nature **453**, 1004 (2008).

Chapter 9
How Entangled Is a Many-Electron State?

V. Subrahmanyam

9.1 Introduction

Quantum entanglement is perceived as a resource for quantum communication and information processing and has emerged over the last few years as a major research area in various diverse fields such as physics, mathematics, chemistry, electrical engineering, and computer science [1, 2]. Entanglement in a quantum state is a signature of quantum correlations between different spatial parts of the system. Consider a bipartite system, consisting of two parts A and B, in a pure state. Let $\|A_i >, i = 1, N$, be a set of orthogonal basis states for the subsystem A, and similarly $|B_i\rangle, i = 1, M$, be an orthogonal basis for the subsystem B. Let the total system be described by a pure state,

$$|\psi\rangle_{AB} = \alpha_{13}|A_1\rangle|B_3\rangle + \alpha_{29}|A_2\rangle|B_9\rangle. \tag{9.1}$$

We have taken a simple superposition of two orthogonal components for simplicity here. In general, we can have NM terms, which is the dimension of the Hilbert space for the total system. The von Neumann entropy of this state can be calculated from $-\text{Tr}\rho \log \rho$, where $\rho = |\psi\rangle\langle\psi|$ is the density matrix associated with the above state. The entropy associated with a pure state is zero, as one of the eigenvalues is unity, with the rest of the eigenvalues of ρ being zero. However, the state of the subsystem A may be described by a mixed state density matrix, indicating a nonzero entropy. The reduced density matrix that describes the subsystem A is calculated by tracing over the degrees of freedom of the subsystem B, given as

$$\rho_A = Tr_B\rho = |\alpha_{13}|^2|A_1\rangle\langle A_1| + |\alpha_{29}|^2|A_2\rangle\langle A_2|. \tag{9.2}$$

It is clear that the above describes a mixed state, indicating a nonzero entropy, if both the coefficients are nonzero. This indicates an entanglement between the two parts of the system. If the subsystem A is described by a pure state density matrix,

V. Subrahmanyam (✉)
Department of Physics, Indian Institute of Technology, Kanpur 208016, India, vmani@iitk.ac.in

Subrahmanyam, V.: *How Entangled Is a Many-Electron State?*. Lect. Notes Phys. **802**, 201–214 (2010)
DOI 10.1007/978-3-642-11470-0_9

and hence the subsystem entropy is zero, the two parts A and B are not entangled. The above example illustrates that entanglement comes about purely due to quantum superpositions, hence the correlations that exist between the two parts are essentially quantum correlations. For instance, if the total system is in a pure state, with only one of the coefficients in Eq. (9.1) nonzero, the state is factorizable. Then tracing over B degrees has no effect on the part A state, which remains in a pure state.

Consider a local unitary transform of the state in Eq. (9.1), given as

$$|\psi'_{AB}\rangle = U_A|\psi_{AB}\rangle, \tag{9.3}$$

where the local unitary transformation U_A affects only subsystem A. It is clear that the process of partial tracing can be done exactly as before, giving us the reduced density matrix for A as

$$\rho'_A = U_A\rho_A U_A^\dagger, \tag{9.4}$$

in terms of the reduced density matrix we got from the total state $|\psi\rangle$. Since a unitary transform does not change the eigenvalues, though it could change the basis states, both ρ_A and ρ'_A have the same eigenvalues, and hence the same entropy, indicating $|\psi'\rangle$ and $|\psi\rangle$ have the same entanglement. One important aspect of this entanglement between subsystems A and B is that it can be changed by a unitary transformation U_{AB}, which itself cannot be factorized. The transformed state now is given by

$$|\phi_{AB}\rangle = U_{AB}|\psi_{AB}\rangle, \tag{9.5}$$

and the unitary transformation has an effect on both the subsystem degrees of freedom. Since U_{AB} cannot be written as a direct product, after a partial trace over B, the reduced density matrix will have different coefficients, implying a change in the eigenvalues, and hence the entropy of the subsystem A may differ in the state $|\phi\rangle$ from that of $|\psi\rangle$.

There are a number of measures of quantum entanglement in spin systems that have been widely investigated which all have the above property that the measure is invariant over local unitary transformations. Some of the popular measures of entanglement are the von Neumann entropy of subsytem that has been discussed above, the concurrence measure [3], which describes the entanglement between two spin-1/2 particles or qubits in a pure or a mixed state, the global entanglement measure [4, 5], which describes the average site mixedness or how impure a given local one-spin density matrix is. There are many more measures of entanglement that have been studied in the context of spin systems. In the context of many-electron states, there have not been such an intense activity. We will study these measures in the context of well-known spin or electronic states.

Let us consider N_e electrons either localized or hopping around on N lattice sites. We can choose a basis state for any site as given by $|0\rangle, |\uparrow\rangle, |\downarrow\rangle, |\uparrow\downarrow\rangle$, which corresponds to no occupancy (or a hole), up-spin electron occupancy, down-spin electron occupancy, and double occupancy, respectively. ts is characterized by 2^N amplitudes, corresponding to as many basis states. A general many-electron state can be written as

$$|\psi> = \sum_{\{a_i,..a_N\}} \phi_{a_1,a_2,...,a_N} |a_1, a_2, ..., a_N >, \qquad (9.6)$$

where a_i labels the basis states of site i. The above pure state is characterized by 4^N complex-number amplitudes for the basis states. To represent the state using a classical memory device, an exponentially large storage is needed, for example, for $N = 30$ a storage capacity of $4^{30} \sim 10^{18}$ would be needed, i.e., a billion 1-Gb classical chips. Conversely, so many numbers that need a storage easily saturating the global storage capacity on earth can be stored in a quantum memory chip with a mere 30 memory locations. In fact, this mammoth memory capacity of quantum systems, if at all could be harnessed, is central to envisage and develop quantum-bit-based computers.

The details as to whether interactions are there and what is the nature of interactions determine the exact many-electron state we like to study. Since the entanglement properties can be discussed entirely through the many-body state itself, we will refer to the interactions only through the structure of the state in this chapter. There are four states per site, viz., unoccupied site or hole state, occupied with a up-spin electron, occupied with a down-spin electron, and a doubly occupied site. The number of states can decrease depending on the interaction strength. For instance, for a large on-site repulsive interaction (as in Hubbard model), the low-energy states would be confined to states with no double occupied sites, effectively making the number of states per site to be three. These are strongly correlated states. Similarly, if we work with the case of half-filling and strong correlation, $N_e = N$, then the number of states per site becomes two, with each site occupied by either a spin-up or a spin-down electron. These are spin-only states.

9.2 Spin States

Let us briefly discuss the case of spin-only states, by keeping the site basis states $|\uparrow>, |\downarrow>$ only and dropping the unoccupied sites and doubly occupied sites. A pure state of N spin-1/2 objects, as shown in Eq. (9.1), is characterized by 2^N amplitudes, corresponding to as many basis states. All the information about a subsystem is carried by the reduced density matrix (RDM) by eliminating the rest of the degrees of freedom. For instance, a two-site RDM for sites l, m is given by a partial trace of the density matrix of the N-qubit state, $\rho = |\psi\rangle\langle\psi|$, as

$$\rho_{lm} = Tr'\rho, \qquad (9.7)$$

where the prime indicates tracing over all the spins except at sites l and m. There are N one-site RDMs, $^N C_2$ two-site RDMs, and so on to make up the total number of

RDM's to be $\sum_l^N C_l = 2^N$, which is exactly the number of independent amplitudes that characterize the pure states of N sites. The matrix elements of each RDM are in general complicated functions of 2^N amplitudes. No two RDMs have exactly the same information of the original pure states. Conversely, no two RDMs will give exactly the same entanglement property of the full state. Consider $\rho_1, \rho_2, \rho_{12}$, corresponding to single-site RDMs of sites 1 and 2, and the two-site RDM of the sites 1 and 2. The entanglement of site 1 and the rest is straightforward to calculate from the eigenvalues of ρ_1, and similarly for the site 2. Now, ρ_{12} can give an independent information, not captured by either ρ_1 or ρ_2, viz., how much is the two-party entanglement between the sites, viz., concurrence. Similarly, an l-site RDM will give an independent information as to how much a true l-party entanglement is there between the l sites in question. However, these higher partial density matrices are quite difficult to work with, beyond a two-site density matrix, that too if only it is four dimensional.

The single-site RDM of site i can be written as

$$\rho_i = \begin{pmatrix} \langle \frac{1}{2} + S_i^z \rangle & \langle S_i^+ \rangle \\ \langle S_i^+ \rangle & \langle \frac{1}{2} - S_i^z \rangle \end{pmatrix}. \tag{9.8}$$

In the above the expectation values are with respect to the original N-spin parent state, from which the RDM has been derived, i.e., $\langle S_i^z \rangle = \langle \psi | S_i^z | \psi \rangle$. The entanglement ε_i of ith site with the rest of the system is calculated from the single-site RDM, and it can be specified by

$$\varepsilon_i = 2(1 - Tr\rho_i^2). \tag{9.9}$$

This measure of entanglement is also known as linear entropy. The prefactor is so chosen to make the maximum entanglement for a site unity. The maximum value the above can take for the spin-only states is unity, corresponding to a maximal entanglement between the site and the rest of the system when both the eigenvalues of the RDMs are equal to 1/2. If the site has no entanglement with the rest of the system, then the above is zero. Now, how entanglement is distributed over various parts of the system can be inferred by a global measure using the single-site RDMs, viz., the global entanglement calculated by averaging the single-site entanglement, given by

$$\varepsilon = \frac{1}{N} \sum_i \varepsilon_i. \tag{9.10}$$

Though the above measure is calculated from local single-site density matrices, the process of averaging over all the sites is capable of capturing the global features, and thus this measure is called the global entanglement measure.

We expect the entanglement to be an extensive quantity like magnetization, or density of electrons. If a large number of sites are individually entangled with the rest of the system, we get a nonzero ε for the given many-spin state. However, it should be noted that a dimer state with neighboring spins forming local singlets will also have $\varepsilon = 1$, though the state has no global entanglement sharing. In other words, one can rule out global entanglement sharing if ε is quite small, but conversely $\varepsilon \sim 1$ may not necessarily imply a globally entangled many-body state.

A two-site RDM contains two different entanglements, viz., the entanglement between the subsystem (containing the two sites) and the rest, and the entanglement between the two sites themselves. In general the reduced density matrix, R_{ij} represents a mixed state for the pair of sites labeled (i,j). The von Neumann entropy calculated from the eigenvalues r_n of R_{ij} as $-\sum r_n \log_2 r_n$ quantifies entanglement of this pair with the rest of the qubits. The concurrence measure [3] has the important information as to how these qubits are entangled among themselves and is given as

$$C_{ij} = \max\ (0, \lambda_1^{1/2} - \lambda_2^{1/2} - \lambda_3^{1/2} - \lambda_4^{1/2}). \tag{9.11}$$

In the above λ_i are the eigenvalues in decreasing order of the matrix $R\hat{R}$, where $\hat{R}$ is the time-reversed matrix, $\hat{R} = \sigma_y \times \sigma_y R^* \sigma_y \times \sigma_y$. In general it is quite messy to calculate the above concurrence. However, for a state with a fixed number of down spins (that is conserving the total z-component of the spin), it can be written in terms of the diagonal and off-diagonal spin correlation functions. The two-site RDM of sites i and j, after a partial trace over other site spin degrees of freedom, is given by

$$R_{ij} = \begin{pmatrix} u_{ij} & 0 & 0 & 0 \\ 0 & x_{ij} & z_{ij} & 0 \\ 0 & z_{ij}^* & y_{ij} & 0 \\ 0 & 0 & 0 & v_{ij} \end{pmatrix}. \tag{9.12}$$

The various matrix elements in the above are given in terms of the spin–spin correlation functions, the coefficient z_{ij} is given in terms of the off-diagonal correlation function between the two sites, and the other quantities are related to the diagonal correlation function given as

$$u_{ij} = \left\langle \left(\frac{1}{2} - S_i^z\right)\left(\frac{1}{2} - S_j^z\right)\right\rangle, v_{ij} = \left\langle \left(\frac{1}{2} + S_i^z\right)\left(\frac{1}{2} + S_j^z\right)\right\rangle,$$

$$x_{ij} = \left\langle \left(\frac{1}{2} - S_i^z\right)\left(\frac{1}{2} + S_j^z\right)\right\rangle, y_{ij} = \left\langle \left(\frac{1}{2} + S_i^z\right)\left(\frac{1}{2} - S_j^z\right)\right\rangle,$$

$$z_{ij} = \langle S_j^+ S_i^- \rangle. \tag{9.13}$$

In the above the angular brackets denote an expectation value in the original state. Here, the concurrence for the two sites, calculated by the method outlined above, has a simple form [6]:

$$C_{ij} = 2\ \max\ (0, |z_{ij}| - \sqrt{u_{ij}v_{ij}}). \tag{9.14}$$

As can easily be seen from above the concurrence measure uses both diagonal and off-diagonal correlation functions. Whether or not two sites have a nonzero concurrence is not at all intuitive, given a specific state with known correlation functions. For a long-ranged concurrence, the necessary condition is the existence of off-diagonal long-range order (ODLRO) . If there is no off-diagonal long-range order (ODLRO), $z_{ij} \rightarrow 0$, as $|\mathbf{r}_i - \mathbf{r}_j| \rightarrow \infty$, the concurrence of two sites far apart would go to zero. The existence of ODLRO is a necessary condition for a long-ranged concurrence. However, even with a ODLRO, long-ranged concurrence may be absent, if $|z_{ij}| < \sqrt{u_{ij}v_{ij}}$.

Let us consider a few examples to illustrate the measures we have discussed above, viz., the global entanglement measure ε and the pair-wise concurrence. Consider the singlet state between two sites, labeled A and B, given as

$$|\text{Singlet}\rangle = \frac{1}{\sqrt{2}}|\uparrow_A \downarrow_B - \downarrow_A \uparrow_B\rangle. \tag{9.15}$$

It is straightforward to calculate the concurrence, and we have $C_{AB} = 1$. Similarly, it is easy to see that the reduced density matrices for A and B are proportional to the identity matrix, implying $\varepsilon = 1$. Now, consider the GHZ or cat state between three sites given as

$$|\text{GHZ}\rangle = \frac{1}{\sqrt{3}}|\uparrow_A \uparrow_B \uparrow_C + \downarrow_A \downarrow_B \downarrow_C\rangle. \tag{9.16}$$

Here again, the single-site RDMs are proportional to the identity matrix, and hence $\varepsilon = 1$, which implies the presence of entanglement. However, if we examine pair-wise concurrences, we find that there is no entanglement between any two pairs, $C_{AB} = C_{BC} = C_{CA} = 0$. This is a classic example of an entanglement that exists between three parties, with no bipartite concurrences. That is, the two-site RDM, ρ_{AB}, represents a mixed state, implying the subsystem $A - B$ is entangled with C, but there is no entanglement between A and B.

Let us now examine states of many spins. As the number of spins increases, the variety of states, and hence the variety of entanglements (multipartite type), increase many fold. In general, the entanglement will depend on various diagonal and off-diagonal multi-spin correlation functions. For a nondegenerate translationally invariant state with the total spin $S = 0$, the concurrence can be shown to be related to $\Gamma_{ij} = \langle S_i^z S_j^z \rangle$, the diagonal correlation function alone [7]. For $S = 0$ states, the total spin expectation value is zero, which means $\langle S^z \rangle = 0$ and similarly the other components. Translational invariance implies that $\langle S_i^z \rangle$ is independent of the site label and is equal to $\langle S^z \rangle / N$, which is zero in this state. Similarly, the diagonal and off-diagonal spin–spin correlation functions between the sites i and j obey the equality for this state,

$$\langle S_i^z S_j^z \rangle = \langle S_i^x S_j^x \rangle = \langle S_i^y S_j^y \rangle. \tag{9.17}$$

This implies that the matrix elements in Eq. (9.11) are given now in terms of the diagonal correlation function alone, and we have

$$u_{ij} = v_{ij} = \frac{1}{4} + \Gamma_{ij}, \quad z_{ij} = \Gamma_{ij}. \tag{9.18}$$

Now, the pair concurrence is completely determined by the diagonal spin correlation function, we have

$$
\begin{aligned}
C_{ij} &= 0, \quad \text{for } \Gamma_{ij} > 0 \\
&= 0, \quad \text{for } \Gamma_{ij} < 0, |\Gamma_{ij}| < \frac{1}{12} \\
&= 6(|\Gamma_{ij}| - \frac{1}{12}), \quad \text{for } \Gamma_{ij} < 0, |\Gamma_{ij}| > \frac{1}{12}.
\end{aligned} \tag{9.19}
$$

Thus, from the diagonal correlation function, we can estimate the bipartite concurrence in spin states. For a positive Γ_{ij} the concurrence is zero, which corresponds to a pair of sites belonging to the same sublattice on a bipartite lattice. In the ground state of Heisenberg antiferromagnets (which correspond to $S = 0$ states here), two spins belonging to the same sublattice have a tendency to be parallel to each other, in which case the entanglement is diminished. For a negative Γ_{ij}, which corresponds to a pair of spins belonging to two different sublattices, the concurrence can be nonzero, subject to the inequality shown above. It is shown that only nearest-neighbor concurrence is nonzero for the ground state of Heisenberg antiferromagnet in one dimension and the square latttice, and all concurrences are zero for triangular and Kagome lattice [7]. As the number of neighbors increases, pair concurrences decrease, and more multipartite entanglements become more significant. Now, since $\langle S_i^z \rangle = 0$, along with the other components, for every site in a $S = 0$ state, the single-site RDM is proportional to identity matrix again. This implies that $\varepsilon = 1$ for all such states.

The entanglement distribution among various parts of the spin system can be sensed by examining an averaging over the system. For instance, the average of two-party concurrences (calculated from two-party RDM's) C_{ave} can give indication as to which states share bipartite entanglement among so many pairs. A state with its entanglement distributed over larger number of subparts is better for entanglement sharing. We will discuss this aspect briefly in the context of one-magnon states.

Let us consider N spins sitting in a cubic lattice confiuguration. The state with all the spins up is a direct product state, and hence has no entanglement structure. There are N states with exactly one down spin; these are one-magnon states. One can consider any linear combination of these one-magnon states and study the entanglement distribution. It is shown that time-reversal symmetric states (corresponding amplitudes are real) tend to have less sharing of the entanglement than the states which are not symmetric under time reversal [8]. Let us consider a spin-wave state (the wave function of the single down spin is given by a plane wave characterized by a wave vector $\mathbf{k}$) of N spins sitting on a three-dimensional cubic lattice, given as

$$|\mathbf{k}\rangle = \frac{1}{\sqrt{N}} \sum_i e^{i\mathbf{k}\cdot\mathbf{r}_i}|\mathbf{r}_i\rangle, \tag{9.20}$$

where $\mathbf{r}_i$ is the location of the down spins in the ith basis state. It is straightforward to show that the concurrence between any two sites is the same and equal to $2/N$, and hence the average pair concurrence is

$$\mathcal{C}_{ave} = \frac{2}{N}. \tag{9.21}$$

In the case of an arbitrary one-magnon state, given as

$$|\phi\rangle = \sum_i \alpha_i |\mathbf{r}_i\rangle, \tag{9.22}$$

where α_i is the amplitude for the i'th configuration with the down spin at the location $\mathbf{r}_i$, the pair concurrence two sites i and j is given by $C_{ij} = 2|\alpha_i\alpha_j|$. Now the average pair concurrence will depend on the particular linear combination that is considered. For a state with random amplitudes with all the numbers are either real (corresponding to the case of time-reversal invariant states) or complex numbers (corresponding to no time reversal invariance), it has been shown [8] that

$$\mathcal{C}_{ave} = \frac{4}{\pi N} \quad \text{Time-reversal invariant} \tag{9.23}$$

$$= \frac{\pi}{2N} \quad \text{No time reversal.} \tag{9.24}$$

In the above, the real (complex) amplitudes are drawn randomly from a Gaussian orthogonal (unitary) ensemble, which are used routinely in the study of particle trajectories in chaotic dynamical systems. The use of the random matrix distributions makes it easy to find the above result for the average concurrence.

It is straightforward to calculate for these one-magnon states the global entanglement measure ε from the single-site RDMs. Since these states are translationally invariant, all the single-site RDM are same, with eigenvalues $1/N, 1-1/N$, and thus we have

$$\varepsilon = \frac{2(N-1)}{N^2}. \tag{9.25}$$

An interesting state with all pair concurrences are zero but with a nonzero global entanglement is the GHZ state or the cat state, which is an equal superposition of all spins up and all spins down.

Let us consider the dimer state of N spins, consisting of $N/2$ pair of nearest-neighbor singlets, given as

$$|Dimer\rangle = |Singlet\rangle_{12}|Singlet\rangle_{34}. \tag{9.26}$$

In this state, the pair concurrences $C_{12} = C_{34} \ldots = 1$, for pairs sharing a singlet, otherwise the pair concurrence is zero. Of the total $N(N-1)/2$ pairs possible, $N/2$ pairs are maximally entangled, and the rest of the pairs unentangled. Thus, the average pair concurrence for this state is

$$C_{ave} = \frac{1}{N-1} \quad \text{Dimer state.} \tag{9.27}$$

In this dimer state too, the single-site RDM is proportional to the identity matrix, which implies the global entanglement is maximal here, $\varepsilon = 1$. Compared with the one-magnon states, the dimer state has $N/2$ down spins and hence possibility of maximal entanglement between $N/2$ pairs of spins. But, the one-magnon states (especially the spin wave states discussed above) share entanglement better, with larger average concurrence.

9.3 Many-Electron States

In the context of many-electron states even two-site RMDs are difficult to work with and unravel the pair-wise entanglement , as the RDM is now 16-dimensional since no occupancy and double occupancy can occur. The well-studied concurrence measure we discussed in the previous section has a simple way of calculation [3], which only works if maximum nonzero eigenvalue of the two-site RDM is 4. For many-electron states, the two-site concurrence is quite difficult to work with.

However, the global entanglement measure, which involves only single-site RDMs, is easily generalized for this situation. So we will use a generalization of the global entanglement measure to study different many-electron states. We will study spatially uniform many-electron states, and hence the mixedness of a given site reduced density matrix can easily pull out the average entanglement property of the many-body state. The number of nonzero eigenvalues of ρ_i, the single-site RDM of site i, can be up to 4, and hence the maximal mixedness of ρ_i or maximal entanglement of the site i with the rest of the system is achieved if all the eigenvalues are same, viz., each eigenvalue in that case would be equal to 1/4. Then, a suitable way of definition of the global entanglement here is given by

$$e = \frac{4}{3N} \sum_i (1 - Tr\rho_i^2). \tag{9.28}$$

The sum in the above is over all sites of the system. Obviously, the entanglement measure would depend on the electron-filling factor, number of up (down) spin electrons, apart from the electron correlation functions, in general. The prefactor is chosen so that the maximal entanglement for a lattice of N sites, where each site is maximally entangled with the rest of the sites, is given by $e = 1$. This global entanglement measure has been investigated for the many-electron states, viz., strongly correlated states, superconducting states recently [9]. Below, we will give the main features of the global entanglement in this context.

Let us consider a quantum state of $N_e = nN$ electrons, where n is the electron density. The state is further characterized by the densities of up and down spin electrons, with $N_\uparrow = n_\uparrow N$ and $N_\downarrow = n_\downarrow N$ as the up- and down-spin electron numbers, respectively. That is, we consider a quantum state with conserved densities of up- and down-spin electrons. As we shall see below, an entanglement hierarchy, characterized by the maximum entanglement, is possible. Because of the conserved quantities, the single-site RDM of a given site has a diagonal structure, using the site-basis states as unoccupied, doubly occupied, up-spin-only occupied, down-spin-only occupied states, given as

$$
\rho = \begin{pmatrix} 1-n+d & 0 & 0 & 0 \\ 0 & d & 0 & 0 \\ 0 & 0 & n_\uparrow - d & 0 \\ 0 & 0 & 0 & n_\downarrow - d \end{pmatrix}.
\tag{9.29}
$$

In the above, d is the probability of double occupancy, that is, the number of doubly occupied sites $D = Nd$,

$$
d \equiv \frac{D}{N} = \frac{1}{N} < \sum_i n_{i\uparrow} n_{i\downarrow} >,
\tag{9.30}
$$

where the sum is over all sites in the lattice. The angular brackets denote an expectation value in the original many-electron state, $n_{i\uparrow}$ is the number operator for the up-spin electrons at site i. Since, the local density matrix is diagonal, the global entanglement ε can be easily calculated for a given many-body state, which reduces to the calculation of double occupancy in the state. In general, the calculation of d, the double occupancy is quite involved for a many-body state with generic electron correlations, as the product of operators in the above equations cannot be decoupled and averaged. There are two simple limits where it is easy to do, namely for a metallic state, with no correlation between up and down electrons, where the above product simply factorizes, and each factor is determined by the corresponding up or down spin electron density. The other simple case is the limit of very strong correlations, which make double occupancy impossible, and hence $d = 0$.

There are three types of states in terms of maximum possible nonzero eigenvalues, and hence the maximum possible entanglement within the space of the particular type of states. The first type consists of spin-only states, which we discussed in the previous section, with no double occupancy and no holes (the case of half-filling). Since there are no holes, the number of electrons should equal the number of sites and no site should be doubly occupied, viz., $N_e = N$, $d = 0$. Essentially there are two physical states per site. There are only two nonzero eigenvalues, let them be λ and $1 - \lambda$. Now the eigenvalues depend on the up (down) spin electron density. Since we are working with spatially uniform states, the global entanglement for these states is given by

$$e_1 = \frac{8}{3}\lambda(1 - \lambda).$$ (9.31)

To maximize entanglement, both eigenvalues tend to be equal. The global maximum for this type of states is obtained for $\lambda = 1/2$. From the structure of single-site density matrix it is clear that this corresponds to states with $n_\uparrow = n_\downarrow = n/2$. It is straightforward to calculate the maximal entanglement for these states, we have

$$e_1(n_\uparrow = \frac{1}{2} = n_\downarrow) = \frac{2}{3}.$$ (9.32)

The states with maximal entanglement here are similar to $S = 0$ spin-only states we discussed in the previous section. The low-energy states of Hubbard model with large on-site correlations, and at half-filling, also correspond to these states.

The second type of states has unoccupied sites (holes), implying it is away from half-filling, but still no doubly occupied sites. Double occupancy is forbidden due to strong on-site correlations. These states correspond to strongly correlated states, for example, the low-energy states of large-U Hubbard model , with the number of electrons less than the number of lattice sites. Here, since the single-site manifold is three-dimensional, there can be three nonzero eigenvalues possible. Let the eigenvalues of the single-site RDM be $\lambda_1, \lambda_2, \lambda_3 = 1 - \lambda_1 - \lambda_2$. The entanglement measure for these states is given by

$$e_2 = \frac{8}{3}(\lambda_1 + \lambda_2 - \lambda_1\lambda_2 - \lambda_1^2 - \lambda_2^2).$$ (9.33)

Again, the maximum entanglement occurs if all the eigenvalues are equal, $\lambda_1 = \lambda_2 = \lambda_3 = 1/3$. This corresponds to states with $n_\uparrow = n_\downarrow = 1/3$. The maximal entanglement for this type of states then works out to be

$$e_2(n_\uparrow = n_\downarrow = \frac{1}{3}) = \frac{8}{9}.$$ (9.34)

The maximal global entanglement for these states is larger than the spin-only states we considered earlier. The increase in the entanglement is due to the presence of holes, which increases the number of degrees of freedom per site.

The third type of states can have doubly occupied sites as well as unoccupied sites, giving states per site the maximum number of degrees of freedom available. We expect that global entanglement should increase again, compared to the second type of states considered above. Now, the single-site RDM can have four nonzero eigenvalues. Let the eigenvalues be $\lambda_1, \lambda_2, \lambda_3, \lambda_4 = 1 - \lambda_1 - \lambda_2 - \lambda_3$. The entanglement measure for these states is given by

$$e_2 = \frac{8}{3}(\lambda_1 + \lambda_2 + \lambda_3 - \lambda_1\lambda_2 - \lambda_2\lambda_3 - \lambda_1\lambda_3 - \lambda_1^2 - \lambda_2^2 - \lambda_3^2).$$ (9.35)

Again, we maximize the entanglement for these states, the maximal entanglement occurs if all the eigenvalues are equal to 1/4 now. The global maximum corresponds to states with $d = 1/4, n_\uparrow = n_\downarrow = 1/2$. That is, at half-filling, these states have optimized double occupancy along with unoccupied sites. The true global maximum entanglement occurs for these states,

$$e(n_\uparrow = n_\downarrow = \frac{1}{2}, d = \frac{1}{4}) = 1. \tag{9.36}$$

These states correspond to metallic states, with no spin correlations.

We will discuss now specific states with correlations, and by how much the entanglement would decrease as the double occupancy is increased. Again, we will not refer to actual interactions which are responsible for introducing the correlations. A metallic state is constructed by filling up the single-particle states (plane wave states for translationally invariant states) by the up spins and down spins. For a metallic state, or uncorrelated spin state, it is straightforward to calculate the double occupancy d_0. The product of number operators decouples as there are no correlations between the up and down spin electrons, we have

$$d_0 = n_\uparrow n_\downarrow. \tag{9.37}$$

Now, it is easy to optimize the global entanglement, and we get $e = 1$ for $n_\uparrow = n_\downarrow = 1/2, d_0 = 1/4$. It should be noted that there are a large number of states with different energies with maximal entanglement, some of them could be metallic ground states and some other could be excited states. That is, the entanglement structure need not depend on the energy of the state!

For a general many-electron state it is difficult to calculate the double occupancy in the state, and hence the calculation of entanglement is not straightforward. The entanglement is calculated for the Gutzwiller projection states in one dimension as a function of the projection parameter [9]. However, we can easily examine how the entanglement changes as a function of the double occupancy d. A given value of d may correspond to low-lying energy states of the Hubbard model . For the ground state of Hubbard model with on-site interaction strength U, it is easy to establish that

$$d(U) = d_0(1 - \delta), \quad 0 \leq \delta(U, n) \leq 1. \tag{9.38}$$

In the above we have introduced a new quantity δ which is determined by which particular correlated state we consider. In principle calculating δ is quite tough, even if the state is known exactly, like the ground state in one-dimension. Two simple situations are as follows: (a) for no interactions, corresponding to metallic states, $\delta = 0$; (b) for infinite strength of interactions, corresponding to very strong correlation, no double occupancy $\delta = 1$. That is, the double occupancy will be less than that of the uncorrelated metallic state. This is because of the on-site repulsive interactions, the energy cost tends to discourage double occupancy. Since equalizing the number of down spins and the number of up spins raises the entanglement, we will set $n_\uparrow = n_\downarrow = n/2$. The global entanglement in the state for the case of half-filling $(n = 1)$ is given by

$$e(n = 1, \delta) = 1 - \frac{\delta^2}{3}. \tag{9.39}$$

This means for any nonzero interaction strength, $\delta > 0$, the global entanglement decreases. Away from half-filling, the global entanglement depends on δ as

$$e(n, \delta) = e(n, 0) + \frac{2n^2(1 - n)^2}{3}\delta - \frac{n^4}{3}\delta^2, \tag{9.40}$$

where $e(n, 0) = \frac{8n}{3}(1 + \frac{n^2}{2})(1 - \frac{n}{4}) - 2n^2$. Now, there is a possibility of more entanglement for the correlated case than for the uncorrelated case $(\delta = 0)$, depending on the coefficients of the linear term, viz., if $2(1 - 1/n)^2 > \delta$, then $e(n, \delta) \rangle e(n, 0)$. The condition on the filling fraction and the double occupancy for the correlated state to have more entanglement than the uncorrelated state is given by

$$n > \left(1 - \sqrt{\frac{\delta}{2}}\right)^{-1} \quad \text{or} \quad d > n - \frac{n^2}{4} - \frac{1}{2}. \tag{9.41}$$

In the above range, however, the actual entanglement is small compared to the global maximum of $e = 1$ (which happens for $n = 1, \delta = 0$). The tendency of inhibiting double occupancy, due to the repulsive interactions, decreases the global entanglement. One could look at excited states of Hubbard model, which have less inhibition of double occupancy, at the cost of having higher energy, then the global entanglement can be greater for the correlated case.

Also, one could look at situations where double occupancy is encouraged, as it may happen with attractive interactions. The best known example is the superconducting states, with in-built ingredients that favor entanglement, viz., the off-diagonal long-range order and the double occupancy. The superconducting states have been examined recently [9] from this perspective. The main difference that arises from the preceding discussion is that the single-site RDM has off-diagonal matrix elements here, namely the matrix element between the unoccupied site and a doubly occupied site is nonzero, as this state does not conserve the number of superconducting pairs! This corresponds to a nonzero matrix element ζ, the top row and second column in the matrix structure shown in Eq. (9.12). This is related to the superconducting order parameter, the energy gap Δ_0, through

$$\zeta = \frac{3n\Delta_0}{4E_F}\sinh^{-1}\frac{\hbar\omega_D}{\Delta_0}, \tag{9.42}$$

where E_F is the Fermi energy that depends on the filling fraction, and ω_D is the Debye energy or an appropriate energy scale over which the possible attractive interaction makes Cooper pairs stable, making the superconducting state favorable compared to the normal metallic state. It is easy to calculate the double occupancy for this state as

$$d = \frac{n^2}{4} + \zeta^2. \tag{9.43}$$

The double occupancy is now more than that in the metallic state, $d_0 = n^2/4$. But as we argued earlier from the eigenvalues of the RDM, the entanglement is maximized if all the eigenvalues are the same, implying the state with a nonzero ζ will have less entanglement than the metallic state with $\zeta = 0$.

Though the superconducting state will have less global entanglement than the uncorrelated state, there is a new type of entanglement that can arise in the super-conducting case, namely a single-site concurrence . The single-site RDM can be viewed as bipartite state of up and down spins by a simple mapping that makes the basis states to product states. Now, we can go through Wootters' procedure for finding the concurrence between the two parts, namely the up and down spins at any site. It has been shown that the single-site concurrence can be nonzero, ζ, satisfies the condition

$$\zeta(1 + \zeta) \geq \frac{n}{2}\left(1 - \frac{n}{2}\right). \tag{9.44}$$

Thus, the superconducting states are quite interesting from the variety of entan-glements that they can exhibit (see [9] for a plot of the global entanglement and the single-site concurrence as a function of the superconducting order parameter). These states have near-maximal global entanglement (but definitely less global entangle-ment than the uncorrelated metallic states) and can have a nonzero single-site con-currence.

References

1. M.A. Nielsen and I.L. Chuang, *Quantum Computation and Quantum Information* (Cambridge University Press, Cambridge, 2000).
2. G. Benenti, G. Casati, and G. Strini, *Principles of Quantum Computation and Information* (World Scientific, Singapore, 2007).
3. W.K. Wootters, Phys. Rev. Lett. **80**, 2245 (1998).
4. D.A. Meyer and N.R. Wallach, J. Math. Phys. **43**, 4273 (2002).
5. G.K. Brennen, Quan. Inf. Comput. **3**, 619 (2003).
6. P. Zanardi, Phys. Rev. **A65**, 042101 (2002).
7. V. Subrahmanyam, Phys. Rev. **A69**, 022311 (2004).
8. A. Lakshminarayan and V. Subrahmanyam, Phys. Rev. **A67**, 052304 (2003).
9. V. Subrahmanyam, arXiv: 0905.3441 (2009).

Chapter 10
Roles of Quantum Fluctuation in Frustrated Systems – Order by Disorder and Reentrant Phase Transition

S. Tanaka, M. Hirano, and S. Miyashita

10.1 Introduction

Frustration in systems shows many interesting equilibrium and dynamical properties [1–6]. In frustrated classical systems, there is no ground state where all the interactions are satisfied energetically. Because the ground states in regularly frustrated classical systems have many energetically *unsatisfied* interactions, there are many degenerated ground states and the residual entropy is larger than that of unfrustrated systems. Figure 10.1 shows the ground states of three Ising spins on antiferromagnetic triangle cluster. The closed and open circles in Fig. 10.1 denote up and down spins, respectively. The crosses in Fig. 10.1 represent energetically unsatisfied interactions.

Fully frustrated classical systems have macroscopically degenerated ground states. Figure 10.2 shows the examples of the ground states of triangular Ising antiferromagnet. The residual entropy of the triangular Ising antiferromagnets is $0.323k_B N$, where N is the number of spins. This is about 46.6% of the total entropy ($Nk_B \log 2$) [7–11].

Another typical example of fully frustrated classical system is the kagome Ising antiferromagnet. Kagome lattice has the so-called corner-sharing structure, whereas triangular lattice has the edge-sharing structure. Kagome Ising antiferromagnet has also many degenerated ground states. The examples of the ground state of the kagome Ising antiferromagnet are shown in Fig. 10.3. The residual entropy of the kagome Ising antiferromagnets is $0.502k_B N$, which is about 72.4% of the total entropy [12].

S. Tanaka (✉)
Institute for Solid State Physics, University of Tokyo, 5-1-5 Kashiwanoha Kashiwa-shi, Chiba, Japan, shu-t@issp.u-tokyo.ac.jp

M. Hirano
Department of Physics, University of Tokyo, 7-3-1, Hongo, Bunkyo-ku, Tokyo, Japan, hirano@spin.phys.s.u-tokyo.ac.jp

S. Miyashita
Department of Physics, University of Tokyo, 7-3-1, Hongo, Bunkyo-ku, Tokyo, Japan; CREST, JST, 4-1-8 Honcho Kawaguchi, Saitama 332-0012, Japan, miya@spin.phys.s.u-tokyo.ac.jp

Tanaka, S. et al.: *Roles of Quantum Fluctuation in Frustated System – Order by Disorder and Reentrant Phase Transition*. Lect. Notes Phys. **802**, 215–234 (2010)
DOI 10.1007/978-3-642-11470-0_10 © Springer-Verlag Berlin Heidelberg 2010

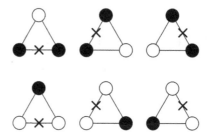

Fig. 10.1 The ground states of three Ising spins on antiferromagnetic triangle cluster. The *closed* and *open circles* denote up and down spins, respectively. The *crosses* denote energetically unsatisfied interactions

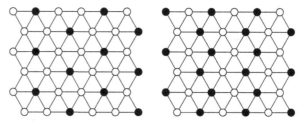

Fig. 10.2 Examples of the ground state of the triangular Ising antiferromagnet. The *closed* and *open circles* denote up and down spins, respectively

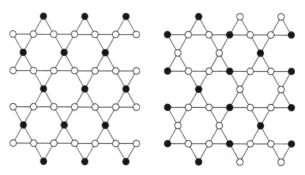

Fig. 10.3 Examples of the ground state of the kagome Ising antiferromagnet. The *closed* and *open circles* denote up and down spins, respectively

Because there are not only many degenerated ground states but also many degenerated excited states in classical frustrated systems, ordering due to thermal fluctuation can be realized. This nature is called "order by disorder" which was first proposed by Villain et al. [13]. They have studied the pure generalized "domino" model, where the ferromagnetic long-range order appears at low temperature but not at zero temperature. The origin of this nature is the effect of frustration. Now, many examples of order by thermal disorder in regularly frustrated systems are wellknown

[14–22]. Ordering due to quantum fluctuation can also be realized [23–26]. For example, Villain model [27, 28] is the typical fully frustrated Ising spin system and there are many degenerated ground states in this model. If we apply weak transverse field, amplitude of wavefunction of the commensurate structure becomes dominant in the ground state [26]. Order by quantum disorder also takes place in the quantum dimer model [29–31]. In quantum dimer model, the exotic phase transition appears. Quantum dimer model has one parameter v/t, where v and t represent the on-site potential and the kinetic energy, respectively. For example, the model with $v/t = 0$ corresponds to an Ising antiferromagnet with transverse field on its dual lattice. The model with $v/t = 1$ corresponds to a classical dimer model and this point is called "Rokhsar–Kivelson point" [32].

Another interesting ordering nature due to thermal fluctuation in frustrated systems is reentrant phase transition where the order parameter behaves non-monotonic as a function of control parameters [13, 33–51]. For example, as the temperature decreases, paramagnetic $\rightarrow$ antiferromagnetic $\rightarrow$ paramagnetic $\rightarrow$ ferromagnetic phases appear. In this case, the antiferromagnetic order parameter behaves non-monotonically as a function of temperature. It is well known that the reentrant phase transition comes from entropy effect. The reentrant phase transition also appears in many electric materials [52, 53].

Dynamical properties, e.g., the slow relaxation, in the frustrated systems are also interesting topics. Slow relaxation appears in not only random systems but also in regularly frustrated systems [21, 22, 54]. To obtain the ground states by numerical methods, we should consider the algorithms which avoid the slow relaxation. Well-known algorithms are the simulated annealing [55, 56] and the exchange method [57–60], in other words, parallel tempering method. In these methods, we can obtain the stable state by controlling the temperature, i.e., thermal fluctuation. On the other hand, another algorithm to obtain the ground state, which uses the quantum fluctuation instead of the thermal fluctuation, has been developed during recent years. This algorithm is called quantum annealing [26, 61–73]. The realizations of the quantum annealing are classified into roughly three categories. The first one is the stochastic method, for example, quantum Monte Carlo simulation. As we all know, because efficient algorithms of Monte Carlo simulation have been developed [74–78], the quantum annealing method has also been developed by combining the novel algorithm of Monte Carlo method [79]. The second one is the deterministic method, for example, real-time dynamics by solving Schrödinger equation [61, 71], mean-field annealing [80, 81], and quantum Bayes inference [73]. The last one is experiment. The artificial lattice such as optical lattice can simulate quantum systems by recent experimental development [82–85].

In this chapter, we focus on roles of quantum fluctuation in frustrated Ising systems comparing with roles of thermal fluctuation. We show effects of thermal fluctuation and of quantum fluctuation in triangular Ising antiferromagnets in Sect. 10.2. In Sect. 10.3, we consider the classical reentrant phase transition in decorated bond systems and non-monotonic behavior of the correlation function due to quantum fluctuation. We also study the dynamics of the frustrated Ising spin system in Sect. 10.3. We conclude this chapter in Sect. 10.4.

10.2 Triangular Lattice

In this section, we consider the effects of the thermal fluctuation and of the quantum fluctuation in triangular Ising antiferromagnets. The Hamiltonian is given by

$$\mathcal{H} = J \sum_{\langle i,j \rangle} \sigma_i^z \sigma_j^z - \Gamma \sum_i \sigma_i^x, \quad (J > 0, \Gamma > 0),\qquad(10.1)$$

where σ_i^α represents the α-element of the Pauli matrix of $S = 1/2$ at ith site:

$$\sigma^z = \begin{pmatrix} 1 & 0 \\ 0 & -1 \end{pmatrix}, \quad \sigma^x = \begin{pmatrix} 0 & 1 \\ 1 & 0 \end{pmatrix}.\qquad(10.2)$$

Now, we take J as the energy unit. First, we consider the classical case without transverse field. In this classical system, there are many degenerated ground states as stated in Sect. 10.1. Although the long-range order does not exist even at zero temperature, the correlation function shows the power law decay [86]

$$\left\langle \sigma_i^z \sigma_j^z \right\rangle \propto r_{ij}^{-1/2},\qquad(10.3)$$

where r_{ij} is the distance between sites i and j. This indicates that the state is at a kind of critical point and sensitive to external perturbations [87–91]. Now we consider the correlation function at finite temperatures. For simplicity, we focus on the correlation function along the direction of the a-axis (see Fig. 10.4);

$$C(n) = \left\langle \sigma_{\mathbf{r}}^z \sigma_{\mathbf{r}+n\mathbf{a}}^z \right\rangle.\qquad(10.4)$$

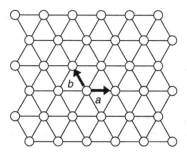

Fig. 10.4 Triangular lattice. a-axis and b-axis are defined as depicted by *thick lines*

Figure 10.5a shows $C(n)$ at several temperatures in the case of $N = 30^2$ with a periodic boundary condition, where N is the number of spins. The correlation function takes large positive values at every three sites at low temperature. Because $C(3)$ is the maximum positive value, we focus on the thermal effect of $C(3)$ as shown in Fig. 10.5b. $C(3)$ behaves monotonic as a function of the temperature.

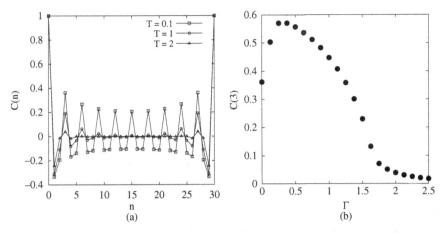

Fig. 10.5 The correlation function in the classical triangular Ising antiferromagnet. (**a**) The correlation function along the a-axis $C(n)$ as a function of distance n. The *squares*, the *circles*, and the *triangles* indicate the case of $T = 0.1$, $T = 1$, and $T = 2$, respectively. *Lines* between points are drawn to highlight the trends. (**b**) $C(3)$ as a function of the temperature

To consider the effect of the quantum fluctuation, we study triangular Ising antiferromagnet with a transverse field by quantum Monte Carlo simulation [92, 93].

Figure 10.6 shows $C(n)$ at several transverse fields at the temperature, $T = 0.1$, in the case of $N = 30^2$ with a periodic boundary condition.

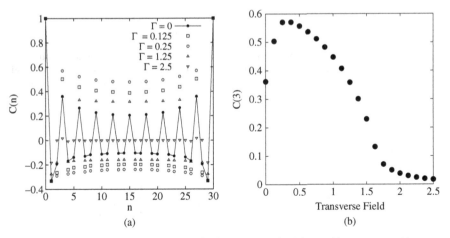

Fig. 10.6 The correlation function at $T = 0.1$ in the triangular Ising antiferromagnet with transverse field. (**a**) $C(n)$ as a function of distance n. The *closed circles*, the *open squares*, the *open circles*, the *triangles*, and the *inverted triangles* denote $\Gamma = 0$ (classical), $\Gamma = 0.125$, $\Gamma = 0.25$, $\Gamma = 1.25$, and $\Gamma = 2.5$, respectively. *Lines* between points are drawn to highlight the trends. (**b**) $C(3)$ as a function of transverse field

At weak transverse field, $C(n)$ at distances in multiples of three are enhanced by the transverse field. This behavior is quantum-field-assisted ordering nature. The correlation function is suppressed at large transverse field. In Fig. 10.6b, we depict the transverse field dependence of $C(3)$.

10.3 Reentrant Phase Transition

As explained above, reentrant phase transitions often take place in frustrated systems. In this section, we consider the classical (i.e., thermal-induced) reentrant phase transition and quantum (i.e., transverse-field-induced) reentrant behavior in a decorated bond system depicted in Fig. 10.7a, where the circles and the triangles denote the system spins σ_i $(i = 1, 2)$ and the decorated spins σ_i $(i = 3, \ldots, N_d + 2)$, respectively. We consider the square lattice system with decorated bond depicted in Figure 10.7b. The critical value of ferromagnetic Ising model on square lattice is exactly given by $J/T_c = K_c = \frac{1}{2} \log \left(1 + \sqrt{2}\right)$, where J and T_c denote the magnitude of ferromagnetic interaction and the critical temperature, respectively [94].

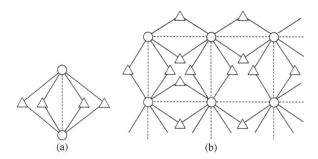

Fig. 10.7 Decorated bond system. The *circles* and the *triangles* denote the system spins and the decorated spins, respectively. The number of the decoration spins is N_d. The *solid lines* and the *dotted line* denote the ferromagnetic coupling J and the antiferromagnetic coupling $N_d J/2$, respectively

The Hamiltonian of this system is given by

$$\mathcal{H} = \frac{N_d J}{2} \sigma_1^z \sigma_2^z - J \sum_{i=3}^{N_d+2} \left(\sigma_1^z + \sigma_2^z\right) \sigma_i^z - \Gamma \left(\sigma_1^x + \sigma_2^x + \sum_{i=3}^{N_d+2} \sigma_i^x\right), \quad (J, \Gamma > 0),$$

(10.5)

where N_d is the number of the decoration spins and σ_i^α represents the α-component of Pauli matrix of $S = 1/2$ at the site i. Now we take J as the energy unit. The solid lines and the dotted line denote the ferromagnetic coupling $-J$ and the antiferromagnetic coupling $N_d J/2$, respectively.

First of all, we consider the case that the transverse field is zero (i.e., classical case). The correlation function between the system spins behaves in a non-monotonic way as a function of temperature. We can calculate exactly the effective coupling between the system spins by tracing out the degree of freedom of the decoration spins:

$$\text{Tr}_{\{\sigma_i=\pm1;i=3,\cdots,N_d+2\}}e^{-\beta\mathcal{H}} = Ae^{-\beta J_{\text{eff}}\sigma_1^z\sigma_2^z}, \tag{10.6}$$

$$K_{\text{eff}} = \beta J_{\text{eff}} = \tfrac{N_d}{2}\log\cosh(2\beta J) - \tfrac{N_d}{2}\beta J. \tag{10.7}$$

The correlation function of the system spins is given by

$$\langle\sigma_1^z\sigma_2^z\rangle = \tanh K_{\text{eff}}. \tag{10.8}$$

Now we call K_{eff} "effective coupling." The correlation function and the effective coupling as a function of temperature behave non-monotonic as shown in Fig. 10.8. If K_{eff} is larger than the critical value of the effective coupling K_c, the ferromagnetic long-range order appears in this system. In the same way, if K_{eff} is smaller than $-K_c$, the antiferromagnetic long-range order appears. Now, we set the number of decoration spins $N_d = 8$. Square of the magnetization m^2 and that of the staggered magnetization m_{st}^2 as a function of temperature are depicted in Fig. 10.9.

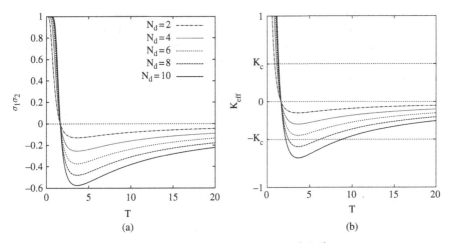

Fig. 10.8 (a) The correlation function between the system spins $\langle\sigma_1^z\sigma_2^z\rangle$ as a function of temperature. (b) The effective coupling K_{eff} as a function of temperature. K_c indicates the critical value of the ferromagnetic Ising model on two-dimensional square lattice

The staggered susceptibility m_{st}^2 shows non-monotonic on the temperature, and this model exhibits the reentrant phase transition . Other models where the reentrant phase transition takes place are studied by a number of researchers [52, 53, 33–38, 13, 39–51].

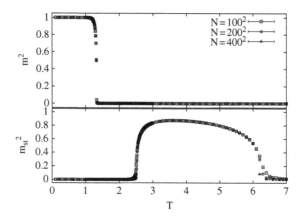

Fig. 10.9 The uniform susceptibility m^2 and the staggered susceptibility m_{st}^2 in the decorated bond system as a function of temperature in the case of $N_d = 8$

Next, we consider the case that the transverse field is finite. The correlation function between system spins is calculated by exact diagonalization. Figure 10.10a shows the correlation function in the ground state as a function of transverse field. The correlation function also behaves non-monotonic as a function of transverse field. This fact indicates a similarity between the thermal fluctuation effect and the quantum fluctuation effect.

The value of transverse field at which the correlation function becomes zero in Fig. 10.10b depends on the number of the decoration spins. In Fig. 10.8a, on the other hand, we find the correlation function becomes zero at a temperature independently of N_d. This fact shows a difference between the thermal fluctuation effect and the quantum fluctuation effect.

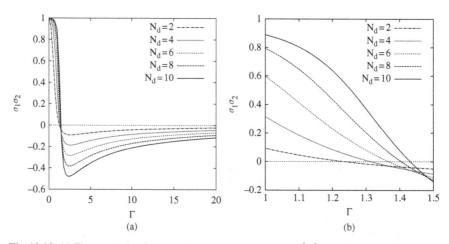

Fig. 10.10 (a) The correlation function between the system spins $\sigma_1^z \sigma_2^z$ as the function of transverse field. (b) Magnified figure of (a)

10.3.1 Classical Dynamics

We considered the equilibrium properties of the frustrated decorated bond system. In this section, we will study dynamical behaviors of the correlation function between the system spins $\sigma_1^z \sigma_2^z$ of the model depicted in Fig. 10.7a. Now we define that N is equal to $N_d + 2$. First, we consider stochastic dynamics of the correlation function in the classical system ($\Gamma = 0$):

$$C(t) = \sum_{\{\sigma_i\}=\pm 1} P(\{\sigma_i\}, t)\, \sigma_1 \sigma_2, \qquad (10.9)$$

where $P(\{\sigma_i\}; t)$ is the probability distribution at time t. We adopt the Glauber Ising model for the time evolution:

$$
\frac{\partial P(\sigma_1, \cdots, \sigma_i, \cdots, \sigma_N; t)}{\partial t} = -\sum_i P(\sigma_1, \cdots, \sigma_i, \cdots, \sigma_N; t)\, w_{\sigma_i \to -\sigma_i}
$$
$$
+ \sum_i P(\sigma_1, \cdots, -\sigma_i, \cdots, \sigma_N; t) w_{-\sigma_i \to \sigma_i}, \quad (10.10)
$$

where the transition probability from σ_i to $-\sigma_i$, $w_{\sigma_i \to -\sigma_i}$ is given by

$$
w_{\sigma_i \to -\sigma_i} = \frac{P_{eq}(\sigma_1, \cdots, -\sigma_i, \cdots, \sigma_N)}{P_{eq}(\sigma_1, \cdots, \sigma_i, \cdots, \sigma_N) + P_{eq}(\sigma_1, \cdots, -\sigma_i, \cdots, \sigma_N)}, \qquad (10.11)
$$

where $P_{eq}(\sigma_1, \ldots, \sigma_i, \ldots, \sigma_N)$ is the equilibrium probability distribution:

$$
P_{eq}(\sigma_1, \ldots, \sigma_i, \ldots, \sigma_N) = \frac{e^{-\beta \mathcal{H}(\{\sigma_i\})}}{Z}, \qquad (10.12)
$$
$$
Z = \mathrm{Tr}\, e^{-\beta \mathcal{H}(\{\sigma_i\})} \qquad (10.13)
$$

Here, we use the probability vector $\mathbf{P}(t)$:

$$
\mathbf{P}(t) = \begin{pmatrix} P(++, \ldots, +; t) \\ P(-+, \ldots, +; t) \\ P(+-, \ldots, +; t) \\ P(--, \ldots, +; t) \\ \vdots \\ P(--, \ldots, -; t) \end{pmatrix}. \qquad (10.14)
$$

The time evolution is given by $\mathbf{P}(t + \Delta t) = \mathcal{L}\mathbf{P}(t)$, where $\mathcal{L}$ is a $2^N \times 2^N$ matrix. Their elements are expressed by

$$\mathcal{L}_{ij} = \frac{1}{N} w_{j \to i} \Delta t \quad \text{(for } j \neq i\text{)},\tag{10.15}$$

$$\mathcal{L}_{ii} = 1 - \sum_{j \neq i} \mathcal{L}_{ji}.\tag{10.16}$$

Now we suddenly decrease the temperature from $T = \infty$ to $T = 1$ and study the spin dynamics by iterating the time evolution operator $\mathcal{L}$ with $\Delta t = 1$. If we decrease temperature gradually (i.e., adiabatically), the correlation function behaves non-monotonic, because the equilibrium value of the correlation function is non-monotonic as a function of temperature. However, it is nontrivial whether the time evolution of correlation function behaves non-monotonic when the temperature suddenly decreases. Figure 10.11 shows the time evolution of the correlation function between system spins.

The initial condition is set to be the equilibrium probability distribution at $T = \infty$, in other words, uniform distribution: $\mathbf{P}(t = 0) = {}^{t}\left(\frac{1}{2^N}, \ldots, \frac{1}{2^N}\right)$. From Fig. 10.11, we can find that the correlation function evolves non-monotonic. Then, we may say that the "effective temperature" of the system gradually decrease even when the temperature suddenly decreases. In a system of simple two Ising spins with ferromagnetic interaction, a monotonic decay occurs when the initial condition is set at $\mathbf{P}(t = 0) = {}^{t}\left(\frac{1}{2}, \frac{1}{2}\right)$. This fact shows the difference between the decorated bond systems and their effective coupling model. This difference comes from an entropy effect of the decoration spins.

To investigate the microscopic mechanism of this non-monotonic behavior, we study the eigenvalue of the time evolution operator $\mathcal{L}$. For simplicity, we consider the case of $N_d = 1$. Although $\mathcal{L}$ is not symmetrized, $\mathcal{L}$ can be symmetrized in this form of

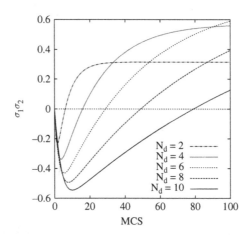

Fig. 10.11 Stochastic dynamics of the correlation function between system spins

$$\tilde{\mathcal{L}} \equiv P_{eq}^{-1/2} \mathcal{L} P_{eq}^{1/2},$$ (10.17)

where $P_{eq}^{1/2}$ is the diagonal matrix:

$$\left(P_{eq}^{1/2}\right)_{ii} = \sqrt{P_{eq}(i)} = \sqrt{\frac{e^{-\beta E_i}}{Z}}.$$ (10.18)

Then $(\tilde{\mathcal{L}})_{ij}$ is given by

$$\left(\tilde{\mathcal{L}}\right)_{ij} = \frac{1}{N} \frac{\sqrt{e^{-\beta(E_i+E_j)}}}{e^{-\beta E_i} + e^{-\beta E_j}} \Delta t = \left(\tilde{\mathcal{L}}\right)_{ji},$$ (10.19)

$$\left(\tilde{\mathcal{L}}\right)_{ii} = \mathcal{L}_{ii},$$ (10.20)

where λ_k is the corresponding eigenvalue of the eigenvector $\mathbf{v}_k$ of $\tilde{\mathcal{L}}$,

$$\left(P_{eq}^{-1/2} \mathcal{L} P_{eq}^{1/2}\right) \mathbf{v}_k = \tilde{\mathcal{L}} \mathbf{v}_k = \lambda_k \mathbf{v}_k.$$ (10.21)

No generality is lost by assuming that

$$1 = \lambda_1 > \lambda_2 \geq \cdots \geq \lambda_{2^N} > 0.$$ (10.22)

For a finite t iterations, $\mathcal{L}^t$ is positive matrix. Then, the Perron–Frobenius theorem guarantees that the maximum of the eigenvalues is not degenerated. Moreover, $\mathcal{L}$ is a probabilistic matrix, the maximum of the eigenvalues is $\lambda_1 = 1$ and $\mathbf{u}_1$ is the equilibrium distribution at T. Then, the right-handed eigenvectors of $\mathcal{L}$ is proportional to $P_{eq}^{1/2} \mathbf{v}_k$. Taking into account the normalization of the probability distribution $\sum_i P_{eq}(i) = 1$, we obtain

$$\mathbf{u}_1 = \frac{P_{eq}^{1/2} \mathbf{v}_1}{\sqrt{{}^t\mathbf{v}_1 P_{eq} \mathbf{v}_1}}.$$ (10.23)

In this way, we can calculate the right-handed eigenvalues and the right-handed eigenvectors of the time evolution operator $\mathcal{L}$:

$$\mathcal{L} \mathbf{u}_k = \lambda_k \mathbf{u}_k.$$ (10.24)

The initial state can be expanded by $\{\mathbf{u}_k\}$

$$\mathbf{P}(0) = \mathbf{u}_1 + \sum_{k=2}^{2^N} c_k \mathbf{u}_k.$$ (10.25)

After t time evolution, in other words, t Monte Carlo step (MCS) by iterating the operator $\mathcal{L}$, the state evolves as

$$\mathbf{P}(t) = \mathcal{L}^t \mathbf{P}(0) = \mathbf{u}_1 + \sum_{k=2}^{2^N} c_k \lambda_k^t \mathbf{u}_k. \tag{10.26}$$

Then, the time evolution of the correlation function is given by

$$C(t) = \sum_{k=1}^{2^N} c_k(t) = \sum_{k=1}^{2^N} c_k \lambda_k^t \sum_{\{\sigma_i\}=\pm 1} u_k(\{\sigma_i\}) \sigma_1 \sigma_2. \tag{10.27}$$

Figure 10.12a shows the eigenvalues of the time evolution operator $\mathcal{L}$ and the coefficients $\{c_k\}$.

From Fig. 10.12a, we find that the relevant coefficients are c_1, c_4, and c_7. Then it is sufficient to consider the dynamics of the fourth and the seventh modes. Time evolution of the fourth mode, seventh mode, and the correlation function are shown in Fig. 10.12b. From Fig. 10.12b, we find that the non-monotonic behavior comes from the relaxation of the fourth mode.

Now we consider the relaxation process from the viewpoint of spin configuration space. In Fig. 10.13, the configurations $1, 2, 3, 4, \ldots$, and 8 denote $(\sigma_1, \sigma_2, \sigma_3) = (+++), (-++), (+-+), (--+), \ldots$, and $(---)$, respectively. In configurations $1, 4, 5$, and 8, the system spins σ_1 and σ_2 correlate ferromagnetically, and the configurations $2, 3, 7$, and 8 antiferromagnetically. Figure 10.13a shows the time evolution of the probability distribution in the configuration space. At $t = 0$, the

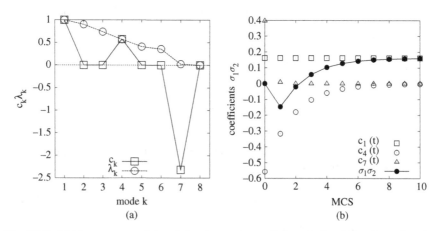

Fig. 10.12 (a) The *squares* and the *circles* denote the coefficients $\{c_k\}$ and the eigenvalues λ_k, respectively. The relevant modes are first, fourth, and seventh modes. *Lines* between points are drawn to highlight the trends ($N_d = 1$). (b) The *open squares*, the *open circles*, the *open triangles*, and the *closed circles* denote the time evolution of the first mode, the fourth mode, the seventh mode, and the correlation function, respectively ($N_d = 1$)

probability distribution is uniform, the amplitudes of configuration 2, 3, 6, and 7, which have antiparallel states suddenly become large at $t = 1$. After evolving the state, the state reaches to the equilibrium distribution. Figure 10.13b shows the equilibrium distribution at $T = 10, 5, 3, 2,$ and 1. The equilibrium probability distribution at $T = 2$ looks like probability distribution at $t = 1$ (Fig. 10.13a). This is the nature of "effective temperature" mechanism.

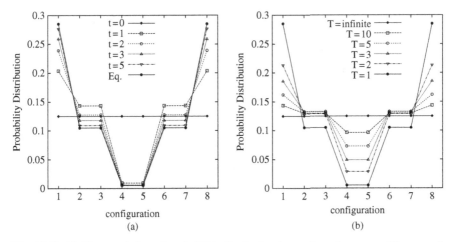

Fig. 10.13 (**a**) Time evolution of probability distribution in configuration space. The second, the third, the sixth, and the seventh configurations which have the antiparallel states, show non-monotonic relaxations ($N_d = 1$). (**b**) Equilibrium probability distribution in the configuration space ($N_d = 1$). *Lines* between points are drawn to highlight the trends

10.3.2 Quantum Dynamics

In the previous section, we considered the stochastic dynamics of the correlation function in classical system. In this section, we study the real-time dynamics of the correlation function in the decorated Ising spin system with time-dependent transverse field. Now we consider the following the time-dependent Hamiltonian:

$$\mathcal{H}(t) = \frac{t}{\tau}\mathcal{H}_c + \left(1 - \frac{t}{\tau}\right)\mathcal{H}_q, \tag{10.28}$$

$$\mathcal{H}_c = \frac{N_d J}{2}\sigma_1^z\sigma_2^z - J\sum_{i=3}^{N_d+2}\left(\sigma_1^z + \sigma_2^z\right)\sigma_i^z, \tag{10.29}$$

$$\mathcal{H}_q = -\left(\sigma_1^x + \sigma_2^x + \sum_{i=3}^{N_d+2}\sigma_i^x\right), \tag{10.30}$$

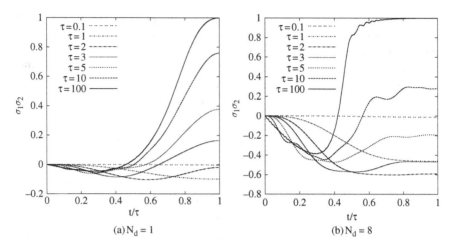

Fig. 10.14 Real-time dynamics of the correlation function. (a) $N_d = 1$, (b) $N_d = 8$

where τ corresponds to the inverse of transverse-field sweeping speed. The initial state is set to be the ground state of $\mathcal{H}(t = 0) = \mathcal{H}_q = \sum_i \sigma_i^x$, which is coherent state

$$|\psi(t = 0)\rangle = \frac{1}{\sqrt{2^N}}(|+\rangle + |-\rangle)^{\otimes N}. \tag{10.31}$$

We consider the real-time dynamics by solving the time-dependent Schrödinger equation:

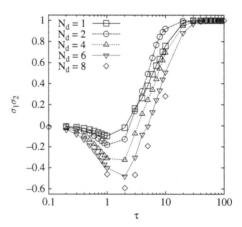

Fig. 10.15 The correlation function at the final state ($t = \tau$) as a function of τ. The *squares*, the *circles*, the *triangles*, the *inverted triangles*, and the *diamonds* denote the cases of $N_d = 1, 2, 4, 6$, and 8, respectively. *Lines* between points are drawn to highlight the trends

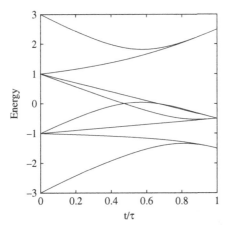

Fig. 10.16 Energy diagram of the model depicted in Fig. 10.7 in the case of $N_d = 1$

$$|\psi (t)\rangle = e^{-i\int_0^t \mathcal{H}(t')dt'} |\psi (t = 0)\rangle. \quad (10.32)$$

Figure 10.14 shows the real-time dynamics of the correlation function in the cases of (a) $N_d = 1$ and (b) $N_d = 8$.

When τ becomes large, the adiabatic quantum dynamics realizes. As τ decreases, the correlation function of the final state $|\psi (t = \tau)\rangle$ is non-monotonic as a function of τ. The correlation functions of the final state as a function of τ are shown in Fig. 10.15 for varying values of N_d.

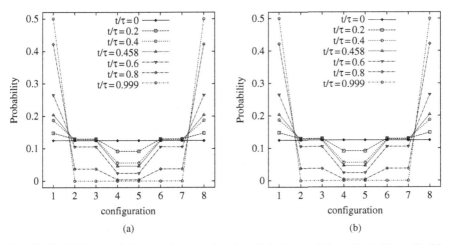

Fig. 10.17 (a) The correlation function as a function of t/τ in the adiabatic limit ($N_d = 1$). (b) The probability distribution in configuration space in the adiabatic limit ($N_d = 1$). *Lines* between points are drawn to highlight the trends

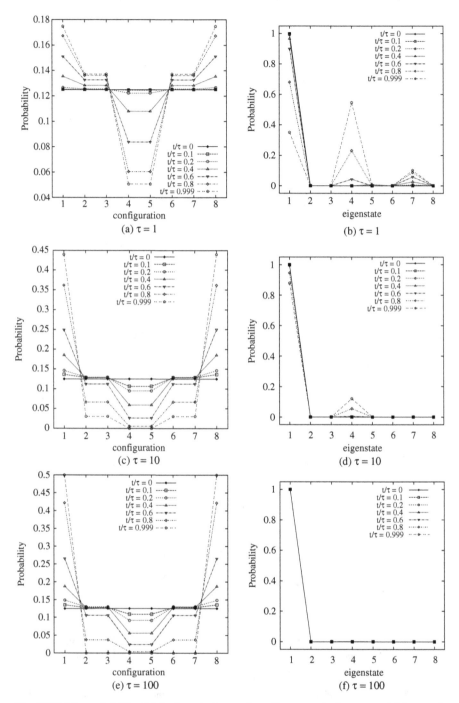

Fig. 10.18 The probability distributions in the configuration space in the cases of (**a**) $\tau = 1$, (**c**) $\tau = 10$, and (**e**) $\tau = 100$, respectively. The overlap of the wavefunction and the lth adiabatic state ($l = 1, 2, 3, 4, \cdots, 8$) for (**b**) $\tau = 1$, (**d**) $\tau = 10$, and (**f**) $\tau = 100$, respectively. *Lines* between points are drawn to highlight the trends

To consider the microscopic mechanism of this behavior, we analyze the real-time dynamics from a viewpoint of the configuration space and eigenstates. Figure 10.16 shows the energy levels as a function of t/τ.

Figure 10.17a shows the correlation function in the adiabatic limit. Now we define the adiabatic probability distribution in the configuration space as follows:

$$P^{\mathrm{ad}}(i;t/\tau) = \left| \left\langle \psi^{\mathrm{ad}}(t/\tau) \, | i \right\rangle \right|^2 , \tag{10.33}$$

where $\left| \psi^{\mathrm{ad}}(t/\tau) \right\rangle$ denotes the adiabatic wavefunction at t/τ and $|i\rangle$ corresponds to the configuration representation i as stated in the previous section. For example, $|1\rangle = |+++\rangle$, $|2\rangle = |-++\rangle$. Figure 10.17b shows the probability distribution in the configuration space.

Next we consider the probability distribution in configuration space and of eigenstates. The definition of the first one is

$$P(i;t/\tau) = |\langle \psi(t/\tau) \, |i\rangle|^2 , \tag{10.34}$$

where $|\psi(t/\tau)\rangle$ denotes the wavefunction which is obtained by solving time-dependent Schrödinger equation. Figure 10.18a, c, and e shows the probability distribution in the configuration space in the cases of $\tau = 1$, 10, and 100, respectively. As τ increases, the probability distribution approaches that of adiabatic limit as shown in Fig. 10.17b. The definition of the second one is

$$P(l;t/\tau) = \left| \left\langle \psi(t/\tau) \, | \psi_l^{\mathrm{ad}}(t/\tau) \right\rangle \right|^2 , \tag{10.35}$$

where $\left| \psi_l^{\mathrm{ad}}(t/\tau) \right\rangle$ denotes the lth eigenstate in the adiabatic limit. Figure 10.18b, d, and f shows the probability distribution of eigenstates in the cases of $\tau = 1$, 10, and 100, respectively. At small τ, the probability at higher level is large and as τ increases, the amplitude of only the first level remains.

10.4 Conclusion

We have studied the effect of quantum fluctuation on the fully frustrated Ising spin systems and compared with the effect of thermal fluctuation . In this chapter, we take the transverse field as the quantum fluctuations. We found that there are similarities and differences between the thermal and the quantum fluctuations. As a way of summary, we would like to stress some of the major points stated in this chapter.

1. *Triangular Ising antiferromagnets with transverse field:* We studied the correlation function as a function of temperature and the transverse field. The quantum-field-assisted ordering nature appears at low temperature and at small transverse field.

2. *Static properties of the decorated bond systems:* The correlation function between the system spins of the classical system behaves non-monotonically as a function of temperature. Also the correlation function in the ground state behaves non-monotonically as a function of the transverse field.

3. *Dynamical properties of the decorated bond systems:* We studied the dynamics of the correlation function between the system spins in two ways: the stochastic dynamics for classical case and the real-time dynamics for quantum case. Non-monotonic dynamics of the correlation function appears in both cases.

Acknowledgments We thank Hans de Raedt, Bernard Barbara, Eric Vincent, Hiroshi Nakagawa, Kenichi Kurihara, Hosho Katsura, Issei Sato, and Yoshiki Matsuda for fruitful discussions. This work was partially supported by Research on Priority Areas "Physics of new quantum phases in superclean materials" (Grant No. 17071011) from MEXT, and also by the Next Generation Super Computer Project, Nanoscience Program from MEXT, The authors also thank the Supercomputer Center, Institute for Solid State Physics, University of Tokyo for the use of the facilities.

References

1. G. Toulouse, Commun. Phys. **2**, 115 (1977).
2. J. Vannimenus and G. Toulouse, J. Phys. C. **10**, L537 (1977).
3. R. Liebmann, *Statistical Mechanics of Periodic Frustrated Ising Systems* (Springer-Verlag GmbH & Co. KG, Berlin and Heidelberg 1986).
4. M.F. Collins and O.A. Petrenko, Can. J. Phys. **75**, 605 (1997).
5. H. Kawamura, J. Phys. Condens. Matter, **10**, 4707 (1998).
6. H.T. Diep (eds.), *Frustrated Spin Systems* (World Scientific Pub. Co. Inc., Singapore, 2005).
7. K. Husimi and I. Syozi, Prog. Theor. Phys. **5**, 177 (1950).
8. I. Syozi, Prog. Theor. Phys. **5**, 341 (1950).
9. G. H. Wannier, Phys. Rev. **79**, 357 (1950).
10. R.M.F. Houtappel, Physica. **5**, 1009 (1964).
11. G.H. Wannier, Phys. Rev. B. **7**, 5017 (1973).
12. K. Kano and S. Naya, Prog. Theor. Phys. **10**, 158 (1953).
13. J. Villain, R. Bidaux, J.-P. Carton, and R. Conte, J. Physique. **41**, (1980) 1263.
14. C.L. Henley, J. Appl. Phys. **61**, 3962 (1987).
15. J.T. Chalker, P.C.W. Holdworth, and E.F. Shender, Phys. Rev. Lett. **68**, 855 (1992).
16. D.A. Huse and A.D. Rutenberg, Phys. Rev. B. **45**, 7536 (1992).
17. O. Nagai, S. Miyashita and T. Horiguchi, Phys. Rev. B. **47** 202 (1993).
18. J.N. Reimers and A.J. Berlinsky, Phys. Rev. B. **48**, 9539 (1993).
19. R. Moessner and J.T. Chalker, Phys. Rev. B. **58**, 12049 (1998).
20. R. Moessner, Can. J. Phys. **79**, 1283 (2001).
21. S. Tanaka and S. Miyashita, J. Phys. Condens. Matter. **19**, 145256 (2007).
22. S. Tanaka and S. Miyashita, J. Phys. Soc. Jpn. **76**, 103001 (2007).
23. A. Chubukov, Phys. Rev. Lett. **69**, 832 (1992).
24. R. Moessner and S. L. Sondhi, Phys. Rev. B. **63**, 224401 (2001).
25. N. Todoroki and S. Miyashita, J. Phys. Soc. Jpn. **74**, 2957 (2005).
26. Y. Matsuda, H. Nishimori, and H.G. Katzgraber, arXiv:0808.0365v2.
27. J. Villain, J. Phys. C. **10**, 1717 (1977).
28. G. Forgacs, Phys. Rev. B. **22**, 4473 (1980).
29. R. Moessner, S.L. Sondhi, and P. Chandra, Phys. Rev. Lett. **84**, 4457 (2000).
30. R. Moessner and S.L. Sondhi, Phys. Rev. Lett. **86**, 1881 (2001).

31. R. Moessner and S.L. Sondhi, Phys. Rev. B. **68**, 054405 (2003).
32. D. S. Rokhsar and S.A. Kivelson, Phys. Rev. Lett. **61**, 2376 (1988).
33. L.D. Landau and E.M. Lifshitz, *Statistical Physics* (Pergamon Press, London, 1959).
34. V. G. Vaks, A.I. Larkin, and Yu.N. Obchinnikov, JETP. **49**, 1180 (1965).
35. H. Nakano, Prog. Theor. Phys. **39**, 1121 (1968).
36. I. Syozi, Prog. Theor. Phys. **39**, 1367 (1968).
37. S. Miyazima and I. Syozi, Prog. Theor. Phys. **40**, 185 (1968).
38. S. Miyazima, Prog. Theor. Phys. **40**, 462 (1968).
39. I. Syozi, *Phase Transition and Critical Phenomena* **1**, Domb and Green (eds.), (Academic Press, New York, 1972).
40. E.H. Fradkin and T.P. Eggarter, Phys. Rev. A. **14**, 495 (1976).
41. S. Miyashita, Prog. Theor. Phys. **69**, 714 (1983).
42. H. Kitatani, S. Miyashita and M. Suzuki, Phys. Lett. A. **108**, 45 (1985).
43. H. Kitatani, S. Miyashita and M. Suzuki, J. Phys. Soc. Jpn. **55**, 865 (1986).
44. P. Azaria, H.T. Diep and H. Giacomini, Phys. Rev. Lett. **59**, 1629 (1987).
45. T. Yokota, Phys. Rev. B. **39**, 523 (1989).
46. P. Azaria, H.T. Diep and H. Giacomini, Europhys. Lett. **9**, 755 (1989).
47. P. Azaria, H.T. Diep and H. Giacomini, Phys. Rev. B, **39**, 740 (1989).
48. H. Asakawa and M. Suzuki, Physica A. **229**, 552 (1996).
49. S. Miyashita and E. Vincent, Eur. Phys. J. B. **22**, 203 (2001).
50. S. Tanaka and S. Miyashita, Prog. Theor. Phys. Suppl. **157**, 34 (2005).
51. S. Miyashita, S. Tanaka and M. Hirano, J. Phys. Soc. Jpn. **76**, 083001 (2007).
52. W. Känzig. *Ferroelectrics and Antiferroelectrics* (Academic Press, New York, 1957).
53. J. Burfoot. *Ferroelectrics* (Van Nostrand, London, 1965).
54. S. Tanaka and S. Miyashita, J. Phys. Soc. Jpn. **78** (8), 084002 (2009), arXiv:0711.3261.
55. S. Kirkpatrick, C. D. Gelatt Jr., and M. P. Vecchi, Science. **220**, 671 (1983).
56. S. Kirkpatrick, J. Stat. Phys. **34**, (1984) 975.
57. K. Hukushima and K. Nemoto, J. Phys. Soc. Jpn. **65**, 1604 (1996).
58. M.C. Tesi, E.J. Janse van Rensburg, E. Orlandini and S.G. Whittington, J. Statist. Phys. **82**, (1996) 155.
59. K. Hukushima, Phys. Rev. E. **60**, 3606 (1999).
60. R. Yamamoto and W. Kob, Phys. Rev. E. **61**, 5473 (2000).
61. S. Miyashita, S. Tanaka, H. de Raedt and B. Barbara, J. Phys. Conf. Ser. **143**, 012005 (2009).
62. B. Apolloni, C. Carvalho and D. de Falco, Stoc. Proc. Appl. **33**, 233 (1989).
63. A.B. Finnila, M.A. Gomez, C. Sebenik, C. Stenson and J.D. Doll, Chem. Phys. Lett. **219**, (1994) 343.
64. T. Kadowaki and H. Nishimori, Phys. Rev. E. **58** 5355 (1998).
65. J. Brooke, D. Bitko, T. F. Rosenbaum and G. Aeppli, Science. **284**, 779 (1999).
66. E. Farhi, J. Goldstone, S. Gutmann, J. Lapan, A. Lundgren and D. Preda, Science. **292**, 472 (2001).
67. G.E. Santoro, R. Martoňák, E. Tosatti and R. Car, Science. **295**, 2427 (2002).
68. A. Das and B.K. Chakrabarti, *Quantum Annealing and Related Optimization Methods*, Lect. Notes Phys. (Springer-Verlag, Berlin Heidelberg, 2005).
69. G.E. Santoro and E. Tosatti, J. Phys. A. **39**, R393 (2006).
70. A. Das and B.K. Chakrabarti, Rev. Mod. Phys. **80**, 1061 (2008).
71. S. Tanaka and S. Miyashita, J. Magn. Magn. Mater. **310**, e468 (2007).
72. K. Kurihara, S. Tanaka and S. Miyashita, Proceedings of the 25th Conference on Uncertainty in Artificial Intelligence. (2009). arXiv:0905.3424.
73. I. Sato, K. Kurihara, S. Tanaka, H. Nakagawa and S. Miyashita, Proceedings of the 25th Conference on Uncertainty in Artificial Intelligence. (2009). arXiv:0905.3425.
74. R. H. Swendsen and J. Wang, Phys. Rev. Lett. **58**, 86 (1987).
75. U. Wolff, Phys. Rev. Lett. **62**, 361 (1989).
76. S. Miyashita, J. Phys. Soc. Jpn. **63**, 2449 (1994).

77. O. Koseki and F. Matsubara, J. Phys. Soc. Jpn. **66**, 322 (1997).
78. T. Nakamura, Phys. Rev. Lett. **101**, 210602 (2008).
79. S. Morita, S. Suzuki and T. Nakamura, Phys. Rev. E. **79**, 065701 (2009).
80. K. Tanaka and T. Horiguchi, Electron. Commun. Jpn. **83**, 2117 (2000).
81. K. Tanaka and T. Horiguchi, Interdiscipl. Inform. Sci. **8**, 33 (2002).
82. D. Jaksch, C. Bruder, J.I. Cirac, C. W. Gardiner and P. Zoller, Phys. Rev. Lett. **81**, 3108 (1998).
83. M. Greiner, O. Mandel, T. Esslinger, T.W. Hänsch and I. Bloch, Nature, **415**, 39 (2001).
84. M. Greiner, O. Mandel, T. Rom, A. Altmeyer, A. Widera, T. W. Hänsch and I. Bloch, Physica B Condensed Matter **329–333**, 11 (2003).
85. I. Bloch, J. Dalibard and W. Zwerger, Rev. Mod. Phys. **80**, 885 (2008).
86. J. Stephenson, J. Math. Phys. **11**, 413 (1970).
87. S. Fujiki, K. Shutoh, Y. Abe and S. Katsura, J. Phys. Soc. Jpn. **52**, 1531 (1983).
88. H. Takayama, K. matsumoto, H. Kawahara and K. Wada, J. Phys. Soc. Jpn. **52**, 2888 (1993).
89. L. P. Landau, Phys. Rev. B. **27**, 5604 (1983).
90. S. Miyashita, H. Kitatani and Y. Kanada, J. Phys. Soc. Jpn. **60**, 523 (1991).
91. O. Nagai, M. Kang and S. Miyashita, Phys. Lett. A. **196**, 101 (1994).
92. H.F. Trotter, Proc. Am. Math. Soc. **10**, 545 (1959).
93. M. Suzuki, Prog. Theor. Phys. **56**, 1454 (1976).
94. L. Onsager, Phys. Rev. **65**, 117 (1944).

Chapter 11
Exploring Ground States of Quantum Spin Glasses by Quantum Monte Carlo Method

A.K. Chandra, A. Das, J. Inoue, and B.K. Chakrabarti

11.1 Introduction

Quantum phases in frustrated systems are being intensively investigated these days; in particular in the context of quantum spin glass and quantum ANNNI models [1]. Here we study a fully frustrated quantum antiferromagnetic model with disorder superposed on it. The finite temperature properties of sub-lattice decomposed version of this model was already considered earlier [2, 3]. The quantum phase transition and entanglement properties of the full long-range model at zero temperature was studied by Vidal et al. [4]. Here we present some results obtained by applying quantum Monte Carlo technique [6, 13] to the same full long-range model at finite temperature. We observe indications of a very unstable quantum antiferromagnetic (AF) phase (50% spin up, 50% spin down, without any sub-lattice structure) in the LRIAF model, where the antiferromagnetically ordered phase gets destabilized by both infinitesimal thermal (classical) as well as quantum fluctuations (due to tunneling or transverse field) and the system becomes disordered or goes over to the para phase [7].

The ordered phase of the long-range Ising Antiferromagnet (LRIAF) seems to be extremely volatile and loses the order (freezing of spin orientations) at any finite fluctuation level; classical or quantum. However, when a little disorder is incorporated with this pure LRIAF model, the frustration supports the spin-glass order. To

A.K. Chandra (✉)
Saha Institute of Nuclear Physics, 1/AF Bidhannagar, Kolkata 700064, India,
anjan.chandra@saha.ac.in

A. Das
The Abdus Salam International Centre for Theoretical Physics, Strada Costiera 11, 34014 Trieste, Italy, arnabdas@ictp.it

J. Inoue
Complex Systems Engineering, Graduate School of Information Science and Technology, Hokkaido University, N14-W9, Kita-ku, Sapporo 060-0814, Japan,
inoue@complex.eng.hokudai.ac.jp

B.K. Chakrabarti
Saha Institute of Nuclear Physics, 1/AF Bidhannagar, Kolkata 700064, India,
bikask.chakrabarti@saha.ac.in

Chandra, A.K. et al.: *Exploring Ground States of Quantum Spin Glasses by Quantum Monte Carlo Method*. Lect. Notes Phys. **802**, 235–249 (2010)
DOI 10.1007/978-3-642-11470-0_11   © Springer-Verlag Berlin Heidelberg 2010

check if this 'liquid'-like antiferromagnetic phase of LRIAF can get 'frozen' into spin-glass phase if a little disorder is added, we study next this LRIAF Hamiltonian with a tunable coupling with the SK spin-glass Hamiltonian and study this entire system's phase transition induced by a tunneling field. An analytic (mean field) solution suggests that an infinitesimal SK-type disorder is enough to make the system 'frozen' into a glass phase. Indeed, the stable spin-glass phase is observed for quantum fluctuations below finite threshold values [7].

Finally we study the annealing behavior of an infinite-range $\pm J$ Ising spin glass in a transverse field, using a zero-temperature quantum Monte Carlo. With this method one can simulate the low-kinetic energy ground states of the quantum Hamiltonian much more efficiently with the help of QA.

This chapter is organized in the following manner. In Sect. 11.2, we introduce the pure quantum LRIAF model and then in Sect. 11.3 discuss the (finite temperature) quantum Monte Carlo results. In Sect. 11.4, we consider the pure quantum LRIAF model with SK disorder and discuss the (analytic) mean field phase diagram. In Sect. 11.5, we focus on a zero temperature quantum Monte Carlo technique to study the annealing behavior of an infinite-range $\pm J$ Ising spin glass in a transverse field. Finally in Sect. 11.6, we present some discussions on our results.

11.2 LRIAF Without Disorder

The Hamiltonian of the infinite-range quantum Ising antiferromagnet (without any spin-glass disorder) is

$$
\begin{aligned}
H &\equiv H^{(C)} + H^{(T)} \\
&= \frac{J_0}{N} \sum_{i,j(>i)=1}^{N} \sigma_i^z \sigma_j^z - h \sum_{i=1}^{N} \sigma_i^z - \Gamma \sum_{i=1}^{N} \sigma_i^x,
\end{aligned}
\tag{11.1}
$$

where J_0 denotes the long-range antiferromagnetic ($J_0 > 0$) exchange constant; for convenience, the value J_0 has been kept 1. Here σ^x and σ^z denote the x and z components of the N Pauli spins

$$
\sigma_i^z = \begin{pmatrix} 1 & 0 \\ 0 & -1 \end{pmatrix}; \quad \sigma_i^x = \begin{pmatrix} 0 & 1 \\ 1 & 0 \end{pmatrix}; \quad i = 1, 2, \ldots, N,
$$

where h and Γ denote, respectively, the longitudinal and transverse fields. We have denoted the co-operative term of H (including the external longitudinal field term) by $H^{(C)}$ and the transverse field part as $H^{(T)}$. As such the model has a fully frustrated (infinite range or infinite dimensional) co-operative term. At zero temperature and at zero longitudinal and transverse fields, the $H^{(C)}$ would prefer the spins to orient in $\pm z$ directions only with zero net magnetization in the z-direction. This antiferromagnetically ordered state is completely frustrated and highly degenerate. Switching on the transverse field Γ would immediately induce all the spins to orient in the x-direction (losing the degeneracy), corresponding to a maximum of the kinetic energy term and this discontinuous transition to the para phase occurs at $\Gamma = 0$. However, at any finite temperature the entropy term coming from the extreme degeneracy of the antiferromagnetically ordered state and the close-by excited states does not seem to induce a stability of this phase.

11.3 Finite Temperature Quantum Monte Carlo Simulation

11.3.1 Suzuki–Trotter Mapping and Simulation

This Hamiltonian (11.1) can be mapped to a $(\infty + 1)$-dimensional classical Hamiltonian [5, 6, 13] using the Suzuki–Trotter formula. The effective Hamiltonian can be written as

$$\mathcal{H} = \frac{1}{NP} \sum_{i,j(>i)=1}^{N} \sum_{k=1}^{P} \sigma_{i,k}\sigma_{j,k} - \frac{h}{P} \sum_{i=1}^{N} \sum_{k=1}^{P} \sigma_{i,k}$$
$$- \frac{J_p}{P} \sum_{i=1}^{N} \sum_{k=1}^{P} \sigma_{i,k}\sigma_{i,k+1}, \tag{11.2}$$

where

$$J_p = -(PT/2)\ln\left(\tanh\left(\Gamma/PT\right)\right). \tag{11.3}$$

Here P is the number of Trotter replicas and k denotes the kth row in the Trotter direction. J_p denotes the nearest-neighbor interaction strength along the Trotter direction. We have studied the system for $N = 100$. Because of the diverging growth of interaction J_p for very low values of Γ and also for high values of P, and the consequent non-ergodicity (the system relaxes to different states for identical thermal and quantum parameters, due to frustrations, starting from different initial configurations), the value of P has been kept at a fixed value of 5. This choice of P value helped satisfying the ergodicity of the system up to very low values of the transverse field at different temperatures considered $T = 0.10, 0.20$, and 0.30. Starting from random initial configurations (including all up or 50–50 up–down configurations), we follow the time variations of different quantities until they relax and study the various quantities after they relax.

11.3.2 Simulation Results

We present results for three different temperatures $T = 0.10, 0.20$, and 0.30 and all the results are for $N = 100$ and 200 and $P = 5$. The following quantities were estimated after relaxation [7]:

(i) *Correlation along Trotter direction (r)*: The variation of the order parameter r was studied, where

$$r = \frac{1}{NP} \sum_{i=1}^{N} \sum_{k=1}^{P} \langle \sigma_{i,k} \sigma_{i,k+1} \rangle \qquad (11.4)$$

is the first neighbor correlation along Trotter direction. Here, $\langle \ldots \rangle$ indicates the average over initial spin configurations. This quantity r shows a smooth vanishing behavior. We consider this correlation r as the order parameter for the transition at Γ_c. A larger transverse field is needed for the vanishing of the order parameter for larger temperature. The observed values (see Fig. 11.1) of Γ_c are $\simeq 1.6, 2.2$, and 3.0 for $T = 0.1, 0.2$, and 0.3, respectively. As shown in the inset, an unique data collapse occurs when r is plotted against Γ/T and one seems to get the complete disorder immediately as the scaling does not involve any finite value T_c.

(ii) *Susceptibility (χ)*: The longitudinal susceptibility $\chi = (1/NP)\partial [\sum_{i,k} \langle \sigma_{i,k} \rangle]/\partial h$, where $h (\to 0)$ is the applied longitudinal field, has also been shown. We went up to $h = 0.1$ and estimated the χ values. As we increase the value of the transverse field Γ from a suitably chosen low value, χ initially starts with a value almost equal to unity and then gradually saturates at lower values (corresponding to the classical system where $J_p = 0$ in Eq. (11.2) as Γ is increased.

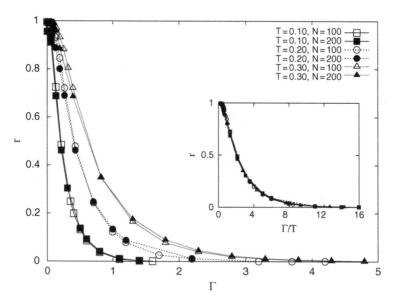

Fig. 11.1 Variation of the order parameter r (correlation in the Trotter direction) with transverse field Γ for $T = 0.10, 0.20$, and 0.30 ($h = 0$) for two different system sizes ($N = 100$ and 200). $r = 0$ for large Γ. The inset shows the plot of r against the scaled variable Γ/T

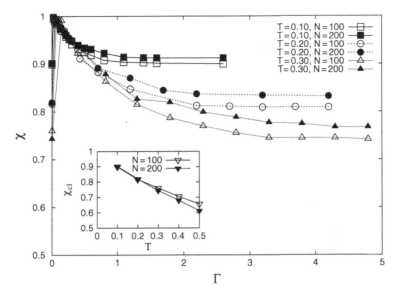

Fig. 11.2 Variation of the susceptibility χ with transverse field Γ for $T = 0.10, 0.20$, and 0.30 ($h \leq 0.1$) for two different system sizes ($N = 100$ and 200). The corresponding susceptibility χ_{cl} for various temperatures for $N = 100$ and 200 for the classical system are shown in the inset. χ converges to the classical values χ_{cl} for large Γ

Also at $\Gamma = 0$, the classical values are indicated in Fig. 11.2. This saturation value of χ decreases with temperature. Again the field at which the susceptibility saturates is the same as for the vanishing of the order parameter for each temperature.

(iii) *Average energy (E)*: We have measured the value of the co-operative energy for each Trotter index and then take its average E, i.e., $E = \langle H^{(C)} \rangle$ of Eq. (11.1) with $J_0 = 1$. It initially begins with -1.0 and after a sharp rise the average energy saturates, at large values of Γ, to values corresponding to the classical equilibrium energy (E_{cl} for $J_p = 0$ in Eq. (11.2)) at those temperatures. Again it takes larger values of Γ at higher temperatures to achieve the classical equilibrium energy. At $\Gamma = 0$, the corresponding classical values of E are plotted in Fig. 11.3. The variations of all these quantities indicate that the 'quantum order' disappears and the quantities reduce to their classical values (corresponding to $J_p = 0$ for large values of the transverse field Γ.

The continuous transition-like behavior seen from Fig. 11.1 can be justified from a mean field analysis (see Sect. 10.7.1). At finite temperature, it is the free energy that we have to minimize and the entropy term plays a crucial role. Minimization of free energy leads to an analytic variation of the total magnetization and no phase transition at any finite temperature.

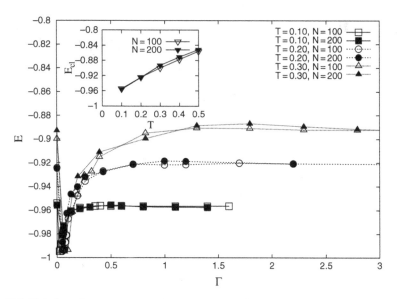

Fig. 11.3 Variation of average energy E with transverse field Γ for $T = 0.10, 0.20$, and 0.30 ($h = 0$) for two different values of $N(= 100, 200)$. The corresponding average energy E_{cl} for various temperatures for $N = 100$ and 200 are shown in the inset. E converges to the classical values E_{cl} for large Γ

11.4 LRIAF with SK Disorder: 'Liquid' Phase of the SK Spin Glasses Gets Frozen

In this section, we discuss the phase diagram for the LRIAF model with SK disorder. We find that for LRIAF the antiferromagnetic order is immediately broken when one adds an infinitesimal transverse field or thermal fluctuation to the system, whereas an infinitesimal SK-type disorder is enough to get the system 'frozen' into the glass phase.

The model we discuss here is given by the following Hamiltonian:

$$H = \frac{1}{N} \sum_{ij(j>i)} (J_0 - \tilde{J}\tau_{ij})\sigma_i^z\sigma_j^z - \Gamma \sum_i \sigma_i^x, \qquad (11.5)$$

where J_0 is a parameter which controls the strength of the antiferromagnetic bias and $\tilde{J}$ is an amplitude of the disorder τ_{ij} in each pair interaction. The Γ controls the quantum-mechanical fluctuation. When we assume that the disorder τ_{ij} obeys a Gaussian with mean zero and variance unity, the new variable $J_{ij} \equiv -J_0 + \tilde{J}\tau_{ij}$ follows the following distribution: $\Pi(J_{ij}) = \exp[-(J_{ij} + J_0)^2/2\tilde{J}^2]/\sqrt{2\pi}\tilde{J}$.

Therefore, we obtain the 'pure' antiferromagnetic Ising model with infinite-range interactions when we consider the limit $\tilde{J} \to 0$ keeping $J_0 > 0$. On the other hand, for $J_0 < 0$ with $\Gamma = 0$ is identical to the classical SK model. A phase

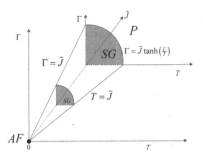

Fig. 11.4 Schematic phase diagram for the quantum long-range antiferro system. The antiferro-magnetic order exists if and only if we set $T = \Gamma = 0$. As $\tilde{J}$ decreases, the spin-glass phase gradually shrinks to zero and eventually ends up at an antiferromagnetic phase at its vertex (for $\Gamma = 0 = T = \tilde{J}$) as discussed in Sect. 2

diagram predicted by static and replica symmetric approximation (see, e.g., [8, 9] and Sect. 10.7.2) is given in Fig. 11.4. The para spin-glass boundary equations become [9]

$$\Gamma = \tilde{J} \tanh\left(\frac{\Gamma}{T}\right). \tag{11.6}$$

We are now investigating by applying quantum Monte Carlo simulation, whether introduction of disorder is capable of freezing the system (spin-glass phase) and whether it supports the predicted phase diagram (qualitatively).

11.5 Ground State of a Quantum Spin Glass Using a Zero-Temperature Quantum Monte Carlo

So far we have concentrated on finite temperature Quantum Monte Carlo method. But to determine the ground states of highly frustrated systems is a challenging task for the physicists. Due to the frustration there arise huge barriers ($O(N)$, N = system size) in the free-energy landscape of the system. In thermodynamic limit, height of such barriers occasionally goes to infinity. These barriers strongly separate different configurations of the system, so that once the system gets stuck in a deep valley in between two barriers, it practically gets trapped around that configuration for a macroscopically large time. Spin glass belongs to such frustrated systems. Several optimization techniques have been implemented to get rid of this situation. Various local search algorithms and simulated annealing have been applied. Local search algorithms which are based on certain local moves mostly end up with one local minima which are far away from the global one. Simulated annealing is based on the introduction of thermal fluctuation which helps the system to escape from local minima. However, in cases of highly frustrated systems, the energy barriers are so large that thermal fluctuations are not good enough to relax the system to the global minimum by crossing those large barriers. This problem can be overridden (in many cases) by implementing (artificially) tunable quantum fluctuations which helps tunneling through such barriers and the method is called *quantum annealing*.

Quantum annealing (QA) [6, 9], [10–17] is a method of finding the ground state (minimum energy state) of a given classical Hamiltonian $\mathcal{H}$ by tuning an external quantum fluctuation, i.e., by adding a time-dependent kinetic part $\mathcal{H}'(t)$ which does not commute with $\mathcal{H}$ and subsequently with adiabatic reduction of that (by reducing the strength of $\mathcal{H}'(t)$ from a very high initial value to zero finally). The total Hamiltonian is thus given by

$$\mathcal{H}_{\text{tot}}(t) = \mathcal{H} + \mathcal{H}'(t). \tag{11.7}$$

If the evolution is slow enough and the initial state is the ground state of $\mathcal{H}_{tot}$ (which is effectively given by the dominating kinetic part $\mathcal{H}'(t = 0)$), then, according to adiabatic theorem of quantum mechanics, the state of the system will always remain close to the ground state of the instantaneous Hamiltonian, and thus, at the end of the annealing, the system will be found in the ground state of the surviving classical part $\mathcal{H}$ with a high probability [11]. Based on this principle, algorithms can be framed to anneal complex physical systems like spin glasses as well as the objective functions of computationally hard combinatorial optimization problems (like the traveling salesman problem or TSP) mapped to glass-like Hamiltonians toward their ground (optimal) states. So far, the successful QA Monte Carlo schemes are mostly based on finite temperature Monte Carlo methods [6, 9, 18]. But to acheive the ground states of glass-like systems, implementation of some zero-temperature Quantum Monte Carlo technique is essential. This technique was first applied to explore the ground state properties of spin-1 isotropic antiferromagnetic Heisenberg spin chain [19]. The ground state properties of the transverse Ising model were also studied using this technique [20]. Annealing behavior of an infinite-range $\pm J$ classical Ising spin glass in a transverse field has also been studied using this technique [6, 21]. Not only the classical ground state was found using quantum annealing, but also the ground state of quantum spin glass, when the transverse field is low. In this section we will discuss how zero-temperature quantum annealing can be applied to simulate the ground state of a quantum spin glass.

The Hamiltonian of an infinite-range Ising spin-glass system is given by

$$\mathcal{H} = -\sum_{i,j(>i)}^{N} J_{ij}\sigma_i^z\sigma_j^z, \tag{11.8}$$

where σ_i^z is the z-component of Pauli spin, representing a classical Ising spin at site i and J_{ij}'s are random variables taking up values either $+1$ or -1 with equal probabilities. To perform the zero-temperature technique, a transverse field term $(\mathcal{H}' = \Omega(t)\sum_{i=1}^{N}\sigma_i^x)$ is added, where σ_i^x's are x-components of Pauli spins which introduce probability of tunneling between the basis states (classical configurations) and $\Omega(t)$ is the strength of the transverse field. The total Hamiltonian is thus given by

$$\mathcal{H}_{tot} = \mathcal{H} + \mathcal{H}'(t) = - \sum_{i,j(>i)}^{N} J_{ij}\sigma_i^z\sigma_j^z - \Omega(t) \sum_{i=1}^{N} \sigma_i^x. \tag{11.9}$$

In glassy systems the potential energy landscape is rugged and has valleys separated by huge energy barriers. Simulating the ground states for low kinetic energy (low value of the transverse field) using zero-temperature quantum annealing may be difficult and time-consuming. For small kinetic term, the acceptance probability may become very small the higher the potential energy states, and the system may take a long time to come out from a local minimum. To get rid of this situation one can start the annealing with a high value of kinetic energy and then reducing slowly to the low value at which the ground state is desired.

Following this annealing method we reach much closer to the exact result for the ground state of the Hamiltonian given by Eq. (11.9) (as obtained by exact diagonalization [22]) than that done directly keeping the value of Ω fixed to the low value from the onset. The results of both kinds of simulations (with and without annealing) for several random samples of the spin glass for $N = 20$ with the respective exact diagonalization results for them (see Fig. 11.5) [21].

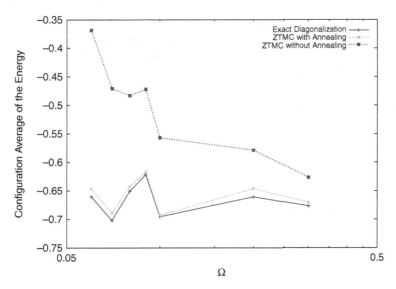

Fig. 11.5 In this figure a comparison between the results of simulation (with and without pre-annealing) of the ground state of the Hamiltonian (11.9) for different static values of Ω and the corresponding exact results obtained by numerical exact diagonalization for the same set of samples. Each data point represents an average over the same set of 40 randomly generated samples of size $N = 20$. The total number of Monte Carlo steps is 10^5 for each Monte Carlo simulation (including the annealing period for the annealed simulations). In the figure, the results of simulations with annealing are seen to be much closer to the exact diagonalization results than those without annealing for lower values of Ω

11.6 Summary

We considered here a long-range Ising antiferromagnet at a finite temperature and put in a transverse field. The antiferromagnetic order is seen to get immediately broken as soon as the thermal or quantum fluctuations are added. However, when we add the Sherrington–Kirkpatrick Hamiltonian as perturbation we find that an infinitesimal SK spin-glass disorder is enough to induce a stable glass order in this LRIAF antiferromagnet. This glass order eventually gets destroyed as the thermal or quantum fluctuations increased beyond their threshold values and the transition to para phase occurs. As shown in the phase diagram in Fig. 11.4, the antiferromagnetic phase of the LRIAF (occurring only at $\tilde{J} = 0 = \Gamma = T$) can get 'frozen' into spin-glass phase if a little SK-type disorder is added ($\tilde{J} \neq 0$); the only missing element in the LRIAF (which is fully frustrated, but lacks disorder) to induce stable order (freezing of random spin orientations) in it. As mentioned already, the degeneracy factor $e^{0.346N}$ of the ground state of the LRIAF is much larger than $e^{0.199N}$ for the SK model. Hence (because of the presence of full frustration), the LRIAF possesses a surrogate incubation property of stable spin-glass phase in it when induced by addition of a small disorder. This observation should enable eventually the study of classical and quantum spin-glass phases by using some perturbation theory with respect to the disorder.

In the case of zero-temperature quantum Monte Carlo method, since the acceptance probability for higher potential energy configurations depends on the magnitude of the kinetic term, it is hard to simulate the ground state for low values of the kinetic term. It is found that quantum annealing can be utilized in overcoming this difficulty (at least partially).

Acknowledgments We are grateful to I. Bose, A. Das, S. Dasgupta, D. Sen, P. Sen, and K. Sengupta for useful discussions and comments.

Appendix

Mean Field Analysis

First, let us consider the case of pure LRIAF model and rewrite our Hamiltonian H in Eq. (11.1) for $\tilde{J} = 0$ as

$$H = \frac{1}{2N}\left(\sum_{i=1}^{N} \sigma_i^z\right)^2 - \frac{1}{N}\sum_{i=1}^{N}(\sigma_i^z)^2 - h\sum_{i=1}^{N}\sigma_i^z - \Gamma\sum_{i=1}^{N}\sigma_i^x \qquad (11.10)$$

If we now denote the total spin by σ_{tot}, i.e., $\sigma_{\text{tot}} = \frac{1}{N}\sum_{i=1}^{N}\sigma_i$ (where $N|\sigma| = 0, 1, 2, \ldots, N$), then the Hamiltonian H can be expressed as

$$\frac{H}{N} = \frac{1}{2}(\sigma_{\text{tot}}^z)^2 - h\sigma_{\text{tot}}^z - \Gamma\sigma_{\text{tot}}^x - \frac{1}{N}. \tag{11.11}$$

Let us assume the average total spin $\langle\sigma\rangle$ to be oriented at an angle θ with the z-direction: $\langle\sigma_{\text{tot}}^z\rangle = m\cos\theta$ and $\langle\sigma_{\text{tot}}^x\rangle = m\sin\theta$. Hence the average total energy $E_{\text{tot}} = \langle H\rangle$ can be written as

$$\frac{E_{\text{tot}}}{N} = \frac{1}{2}m^2\cos^2\theta - hm\cos\theta - \Gamma m\sin\theta - \frac{1}{N}. \tag{11.12}$$

At the zero temperature and at $\Gamma = 0$, for $h = 0$, the energy E_{tot} is minimized when $\theta = 0$ and $m = 0$ (complete antiferromagnetic order in z-direction). As soon as $\Gamma \neq 0$ ($h = 0$) the minimization of E_{tot} requires $\theta = \pi/2$ and $m = 1$ (the maximum possible value) driving the system to paramagnetic phase. This discontinuous transition at $T = 0$ was also seen in [4]. As observed in our Monte Carlo study in the previous section, $\Gamma_c(T) \to 0$ as $T \to 0$. This is consistent with this exact result $\Gamma_c = 0$ at $T = 0$. For $T = 0$ (and $h = 0$), therefore, the transition from antiferromagnetic ($\theta = 0 = m$) to para ($\theta = \pi/2, m = 1$) phase, driven by the transverse field Γ, occurs at $\Gamma = 0$ itself.

One can also estimate the susceptibility χ at $\Gamma = 0 = T$. Here $E_{\text{tot}}/N = \frac{1}{2}m^2\cos^2\theta - hm\cos\theta - \frac{1}{N}$ and the minimization of this energy gives $m\cos\theta = h$ giving the (longitudinal) susceptibility $\chi = m\cos\theta/h = 1$. This is consistent with the observed behavior of χ shown in Fig. 11.2 where the extrapolated value of χ at $\Gamma = 0$ increases with decreasing T and approaches $\chi = 1$ as $T \to 0$.

At finite temperatures $T \neq 0$, for $h = 0$, we have to consider also the entropy term and minimize the free energy $\mathcal{F} = E_{\text{tot}} - TS$ rather than E_{tot} where S denotes the entropy of the state. This entropy term will also take part in fixing the value of θ and m at which the free energy $\mathcal{F}$ is minimized. As soon as the temperature T becomes non-zero, the extensive entropy of the system for antiferromagnetically ordered state with $m \simeq 0$ (around and close-by excited states with $\theta = 0$) helps stabilization near $\theta = 0$ and $m = 0$ rather than near the para phase with $\theta = \pi/2$ and $m = 1$, where the entropy drops to zero. While the transverse field tends to align the spins along x-direction (inducing $\theta = \pi/2$ and $m = 1$), the entropy factor prohibits that and the system adjusts θ and m values accordingly and they do not take the disordered or para state values ($\theta = \pi/2$ and $m = 1$) for any non-zero value of Γ (like at $T = 0$). For very large values of Γ, of course, the free energy $\mathcal{F}$ is practically dominated by the transverse field term in H and again $\theta = \pi/2$ and $m = 1$, beyond $\Gamma = \Gamma_c(T) > 0$ for $T > 0$. However, this continuous transition-like behavior may be argued (D. Sen Private Communication) to correspond to a crossover type property of the model at finite temperatures (suggesting that the observed finite values of $\Gamma_c(T)$ are only effective numerical values). In fact, for $h = 0$ one adds the entropy term $-T\ln D$, where D is the degeneracy for getting total spin $\tau = N|\sigma_{\text{tot}}|$ [4],

$$D = \frac{N!}{(N/2 + \tau)!(N/2 - \tau)!} - \frac{N!}{(N/2 + \tau + 1)!(N/2 - \tau - 1)!} \tag{11.13}$$

to E_{tot} in Eq. (10.12) to get $\mathcal{F}$ and one can then get, after minimizing the $\mathcal{F}$ with respect to m and θ, $m = \tanh(\Gamma/2T)$, which indicates an analytic variation of m and no phase transition at any finite temperature for $\tilde{J} = 0$ (antiferromagnetic phase occurs only at $\Gamma = T = 0$ and $J_0/\tilde{J} = \infty$ as shown in Fig. 11.4).

Static and Replica Symmetric Approximations

The Sherrington–Kirkpatrick (S–K) model in the presence of a non-commutating tunneling field is given by the Hamiltonian

$$\mathcal{H}_{SK} = -\sum_{ij} J_{ij}\sigma_i^z\sigma_j^z - \Gamma\sum_i \sigma_i^x \tag{11.14}$$

where J_{ij} follows the Gaussian distribution

$$\Pi(J_{ij}) = \left(\frac{N}{2\pi\tilde{J}^2}\right)^{1/2} \exp\left(\frac{-NJ_{ij}^2}{a\tilde{J}^2}\right). \tag{11.15}$$

Several analytical studies have been made to obtain the phase diagram of the transverse Ising S–K model (giving in particular the zero-temperature critical field). The problem of S–K glass in transverse field becomes a nontrivial one due to the presence of noncommuting spin operators in the Hamiltonian. This leads to a dynamical frequency-dependent (spin) self-interaction.

Mean field estimates (see, e.g., [6]):
One can study an effective spin Hamiltonian for the above quantum many body system within the mean field framework. Here we shall briefly review the replica-symmetric solution of the classical S–K model ($\Gamma = 0$) in a longitudinal field given by the Hamiltonian

$$\mathcal{H}_{SK} = -\sum_{\langle ij\rangle} J_{ij}\sigma_i^z\sigma_j^z - h\sum_i \sigma_i^z \tag{11.16}$$

where J_{ij} follows the Gaussian distribution given by Eq. (11.15). Using the replica trick, one obtains configuration averaged n-replicated partition function $\bar{Z}^n$, given by

$$\bar{Z}^n = \sum_{(\sigma_{i\alpha}=\pm1)} \int_{-\infty}^{\infty} \Pi(J_{ij})dJ_{ij} \exp\left[\beta\sum J_{ij}\sum \sigma_{i\alpha}^z\sigma_{j\alpha}^z + \beta h\sum \sigma_{i\alpha}^z.\right]$$

Performing the Gaussian integral using Hubbard–Stratonovich transformation and finally using the method of steepest descent to evaluate integrals for thermodynamically large system, one obtains free energy per site f, given by

$$-\beta f = \lim_{n \to 0} \left[\frac{\beta \tilde{J}^2}{4} \left(1 - \frac{1}{n} \sum_{\alpha,\beta} q^2_{\alpha,\beta} + \frac{1}{n} \ln Tr(\exp L) \right) \right],$$

where $L = (\beta J)^2 \sum_{\alpha,\beta} q_{\alpha\beta} \sigma^z_\alpha \sigma^z_\beta + \beta \sum^n_{\alpha=1} \sigma^z_\alpha$ and

$$q_{\alpha\beta} = N^{-1} \sum^N_{i=1} \langle \sigma^z_{i\alpha} \sigma^z_{i\beta} \rangle.$$

$q_{\alpha\beta}$ is self-consistently given by the saddle point condition $(\partial f / \partial q_{\alpha\beta}) = 0$. Considering the replica symmetric case $(q_{\alpha\beta} = q = N^{-1} \sum_i \langle \sigma^z_i \rangle^2)$, one finds

$$-\beta f = \frac{(\beta \tilde{J})^2}{2}(1 - q) + \frac{1}{\sqrt{2\pi}} \int^\infty_{-\infty} dr \ e^{-\frac{r^2}{2}} \ln [2 \cosh \{\beta h(r)\}],$$

where r is the excess static noise arising from the random interaction J_{ij} and the spin-glass order parameter q is self-consistently given by

$$q = \frac{1}{\sqrt{2\pi}} \int^\infty_{-\infty} dr \ e^{-\frac{r^2}{2}} \tanh^2 \{\beta h(r)\}$$

and $h(r) = \tilde{J}\sqrt{q}r + h$ can be interpreted as a local molecular field acting on a site. Different sites have different fields because of disorder, and the effective distribution of $h(r)$ is Gaussian with mean 0 and variance $\tilde{J}^2 q$.

At this point we can introduce quantum effect through transverse field term $-\Gamma \sum_i \sigma^x_i$ (with longitudinal field $h = 0$). The effective single particle Hamiltonian in the transverse Ising quantum glass can be written as

$$\mathcal{H}_s = -h^z(r)\sigma^z - \Gamma \sigma^x,$$

where $h^z(r)$, as mentioned earlier, is the effective field acting along the z-direction arising due to non-zero value of the the spin-glass order parameter. Treating $h^z(r)$ and σ as classical vectors in pseudo-spin space, one can write the net effective field acting on each spin as

$$h_0(r) = h^z(r)\hat{z} - \Gamma \hat{x}; \quad |h_0(r)| = \sqrt{h^z(r)^2 + \Gamma^2}.$$

One can now arrive at the mean field equation for the local magnetization, given by

$$m(r) = p(r) \tanh [\beta h_0(r)]; \quad p(r) = \frac{|h^z(r)|}{|h_0(r)|},$$

and consequently, the spin-glass order parameter can be written as

$$q = \frac{1}{\sqrt{2\pi}} \int_{-\infty}^{\infty} dr \ e^{-r^2/2} \tanh^2 \{\beta h_0(r)\} p^2(r).$$

The phase boundary can be found from the above expression by putting $q \to 0$ ($h^z(r) = \tilde{J}\sqrt{q}r$ and $h_0 = \Gamma$), when it gives

$$\frac{\Gamma}{\tilde{J}} = \tanh \left(\frac{\Gamma}{T}\right).$$

From above we get $\Gamma_c = \tilde{J}$ for $T \to \infty$.

For the Hamiltonian (14.9) with $J_0 \neq 0$ (as in above) we get, by applying the same technique as mentioned above, the saddle point equations under the static and replica symmetric approximations as follows:

$$m = \int_{-\infty}^{\infty} Dr \frac{h^z(r)}{\sqrt{(h^z(r))^2 + \Gamma^2}} \tanh \beta \sqrt{(h^z(r))^2 + \Gamma^2}, \tag{11.17}$$

$$q = \int_{-\infty}^{\infty} Dr \left\{ \frac{h^z(r)}{\sqrt{(h^z(r))^2 + \Gamma^2}} \right\}^2 \tanh^2 \beta \sqrt{(h^z(r))^2 + \Gamma^2}, \tag{11.18}$$

where $h^z(r) = \tilde{J}\sqrt{q}r - J_0 m$ and $m \equiv N^{-1} \sum_i \langle \sigma_i^z \rangle$ is the magnetization and $q \equiv N^{-1} \sum_i \langle \sigma_i^z \rangle^2$ is the spin-glass order parameter. We defined $Dr \equiv dr\,e^{-r^2/2}/\sqrt{2\pi}$. The bracket $\langle \cdots \rangle$ denotes an expectation over the density matrix: $\rho = e^{-\beta H}/\mathrm{tr}\,e^{-\beta H}$. When J_0 is negative, (Eq. (11.17)) has the only solution $m = 0$. The general phase boundaries (see Fig. 11.4) between the ferro (F), spin glass (SG), antiferro (AF), and para (P) phases can be obtained by solving the above two equations in the limit $m \to 0$; $q \neq 0$ for the F-SG boundary, $q \to 0$ for the SG-P boundary, and $m \to 0$ for the F-P boundary. In these limits, the P-SG boundary equations become

$$\Gamma = \tilde{J} \tanh \left(\frac{\Gamma}{T}\right). \tag{11.19}$$

References

1. R.N. Bhatt, *Spin Glasses and Random Fields* ed A.P. Young, (World Scientific, Singapore 1998) p. 225.
2. B.K. Chakrabarti and J. Inoue, Indian J. Phys. **80** (6), 609 (2006).
3. B.K. Chakrabarti, A. Das and J. Inoue, Eur. Phys. J. B **51**, 321 (2006).
4. J. Vidal, R. Mosseri and J. Dukelsky, Phys. Rev. A. **69**, 054101 (2004).
5. M. Suzuki, Prog. Theor. Phys. **56**, 2454 (1976).
6. B.K. Chakrabarti and A. Das, Quantum Annealing and Related Optimization Methods Das A. and B.K. Chakrabarti, Lect. Notes Phys. **679** (eds.) (Springer-Verlag, Heidelberg 2005) p. 3.
7. A.K. Chandra, J.-I. Inoue and B.K. Chakrabarti, J. Phys.: Conf. Ser. **143** 012013 (2009).
8. B.K. Chakrabarti, A. Dutta and P. Sen, *Quantum Ising Phases and Transitions in Transverse Ising Models* (Springer-Verlag, Heidelberg 1996).

9. A. Das and B.K. Chakrabarti, Rev. Mod. Phys. **80**, 1061 (2008).
10. G.E. Santoro and E. Tosatti, Nature Phys. **3**, 593 (2007).
11. E. Farhi, J. Goldstone, S. Gutmann, J. Lapan, A. Ludgren and D. Preda, Science. **292**, 472 (2001).
12. T. Kadowaki and H. Nishimori, Phys. Rev. E. **58**, 5355 (1998).
13. G.E. Santoro, R. Martoňák, E. Tosatti and R. Car, Science. **295**, 2427 (2002).
14. J. Brook, D. Bitko, T.F. Rosenbaum and G. Aeppli, Science. **284**, 779 (1999).
15. R. Martoňák, G.E. Santoro and E. Tosatti, Phys. Rev. E. **70**, 057701 (2004).
16. A. Das, B.K. Chakrabarti and R.B. Stinchcombe, Phys. Rev. E. **72**, 026701 (2005).
17. R.D. Somma, C.D. Batista and G. Ortiz, Phys. Rev. Letts. **99**, 030603 (2007).
18. G.E. Santoro and E. Tosatti, J. Phys. A. **39**, R393 (2006).
19. J.P. Neirotti and M.J. de Oliveira, Phys. Rev. B. **53**, 668 (1995).
20. M.J. de Oliveira and J.R.N. Chiappin, Physica A. **238**, 307 (1997).
21. A. Das and B.K. Chakrabarti, Phys. Rev. E. **78**, 061121 (2008).
22. J. Stoer and R. Bulirsch, *Introduction to Numerical Analysis, Text in Applied Mathematics*, **12** (Springer-Verlag, New York, 1993).

Chapter 12
Phase Transition in a Quantum Ising Model with Long-Range Interaction

A. Ganguli and S. Dasgupta

12.1 Introduction

Among the many models of Statistical Physics, Ising model [1–3] is perhaps the most ubiquitous. Till date it is being applied in varied context with reasonable success. In the "classical" version of this model, one has in general a set of lattice sites and to each site is associated a variable called spin which can assume the values $\pm\frac{1}{2}$. For a system of N sites, there can hence be 2^N possible configurations. The energy of any given configuration is given by

$$\mathcal{E} = \frac{1}{2} \sum_{i,j,i \neq j} J_{ij} s_i s_j, \tag{12.1}$$

where s_i, s_j are the spins at the sites i and j, and J_{ij} is the interaction energy between sites i and j. At a temperature T, the partition function is given by

$$Z = \sum_{\text{all configurations}} \exp\left[-\frac{\mathcal{E}}{k_B T}\right],$$

where k_B is the Boltzmann constant and the free energy is $F = -k_B T \log Z$. In general, for an arbitrary form of J_{ij} one cannot calculate the free energy without making any approximation. We shall discuss in Sect. 12.2 several combinations of lattices and interactions and show that in some cases the free energy can be calculated without any approximation, while in some other cases one cannot do without some approximation.

A. Ganguli (✉)
Department of Physics, University of Calcutta, 92 Acharya Prafulla Chandra Road, Kolkata 700009, India, anindita.ganguli@rediffmail.com

S. Dasgupta
Department of Physics, University of Calcutta, 92 Acharya Prafulla Chandra Road, Kolkata 700009, India, sdphy@caluniv.ac.in

Ganguli, A., Dasgupta S.: *Phase Transition in a Quantum Ising Model with Long-Range Interaction*. Lect. Notes Phys. **802**, 251–266 (2010)
DOI 10.1007/978-3-642-11470-0_12   © Springer-Verlag Berlin Heidelberg 2010

The energy expression Eq. (12.1) describes a *classical* Ising model in the sense that the spin variables are scalar quantities. A *quantum* Ising model is described by a Hamiltonian operator containing spin variables, which are quantum mechanical operators, rather than scalars. The spin operators in three dimensions s_i^x, s_i^y and s_i^z obey the usual angular momentum commutation relations:

$$\left[s_i^x, s_j^y\right] = i\hbar\delta_{ij} s_i^z \quad \left[s_i^y, s_j^z\right] = i\hbar\delta_{ij} s_i^x \quad \left[s_i^z, s_j^x\right] = i\hbar\delta_{ij} s_i^y. \quad (12.2)$$

For a system of size N, the matrix representation of these operators is the $2^N \times 2^N$ matrices:

$$s_i^x \equiv \frac{1}{2}\left[\underline{1} \otimes \underline{1} \otimes \cdots \otimes \sigma_x \otimes \cdots \otimes \underline{1}\right]$$

$$s_i^y \equiv \frac{1}{2}\left[\underline{1} \otimes \underline{1} \otimes \cdots \otimes \sigma_y \otimes \cdots \otimes \underline{1}\right]$$

$$s_i^z \equiv \frac{1}{2}\left[\underline{1} \otimes \underline{1} \otimes \cdots \otimes \sigma_z \otimes \cdots \otimes \underline{1}\right]$$

There are N factors within each square bracket, each factor is $\underline{1}$, except the ith one which is some σ matrix. Here $\otimes$ means direct product, $\underline{1}$ is a 2×2 unit matrix, and the σ's are Pauli spin matrices

$$\sigma_x = \begin{pmatrix} 0 & 1 \\ 1 & 0 \end{pmatrix} \quad \sigma_y = \begin{pmatrix} 0 & -i \\ i & 0 \end{pmatrix} \quad \sigma_z = \begin{pmatrix} 1 & 0 \\ 0 & -1 \end{pmatrix}.$$

One simple example of a quantum Ising Hamiltonian is

$$\mathcal{H}_{QI} = -\frac{1}{2} \sum_{i,j,i \neq j} J_{ij} s_i^z s_j^z - \Gamma \sum_j s_j^x. \quad (12.3)$$

The ground states of such models are to be found by solving the eigenproblem of this operator and the partition function is given by Trace $\{\exp[-\mathcal{H}_{QI}/(k_B T)]\}$. We shall discuss in Sect. 12.3 a Hamiltonian where one can solve for the eigenstates. We shall also discuss there a model called Lipkin–Meshkov–Glick (LMG) model which cannot be solved in closed form. In Sect. 12.4 we shall discuss a long-range antiferromagnetic quantum Ising model that has been studied recently and present in Sect. 12.5, following [4] a perturbative treatment for it along with the phase transition (Sect. 12.6) predicted therefrom. In Sect. 12.7 we present a discussion on a disordered version of this model.

12.2 Classical Ising Model

Let us start with the case of a periodic lattice in one dimension with nearest-neighbour interaction. Then the spin–spin interaction becomes

$$J_{ij} = -J \quad \text{if} \ i - j = \pm 1.$$
$$= 0 \quad \text{otherwise}$$

The energy expression Eq. (12.1) then becomes

$$\mathcal{E} = -J \sum_{j=1}^{N-1} s_j s_{j+1}. \tag{12.4}$$

Calculation of the corresponding partition function is a text book exercise [1–3]. Substitute

$$t_j = 2 s_j s_{j+1} = \pm \frac{1}{2}$$

for $j = 1, 2, \cdots, (N-1)$ and observe that each of the 2^{N-1} configurations of t_j lattice corresponds to two configurations of the s_j lattice. These two configurations have the same energy, since one of them is obtained by reversing all the spins of the other ($s_j = -s_j$, for all j). Thus,

$$Z = 2 \sum_{t_1} \sum_{t_2} \cdots \sum_{t_{N-1}} \exp \left[\beta J \left(t_1 + t_2 + \cdots t_{N-1} \right) \right] = 2 \left[2 \cosh \left(\beta J \right) \right]^{N-1},$$

where $\beta = 1/(k_B T)$. The free energy per spin for a large lattice becomes

$$\frac{F}{N} = -\frac{1}{\beta} \log \left[2 \cosh \left(\beta J \right) \right]. \tag{12.5}$$

This is an analytic function of temperature and hence one-dimensional Ising model with nearest-neighbour interaction does not show any phase transition at finite temperatures.

When the lattice is hypercubic in d-dimension ($d > 1$), the energy expression is

$$\mathcal{E} = -J \sum_{<ij>} s_i s_j,$$

where $<ij>$ denotes summation over all nearest-neighbour pairs. The calculation of partition function is now highly non-trivial even for such simple interaction. For $d = 2$, the exact solution is possible although difficult (the celebrated "Onsager solution" [5–7]) while for $d > 2$, exact solution is not possible at all. However, for large d it is reasonably accurate to take recourse to an approximation called mean field approximation.

In the mean field approximation, the system is described in terms of the magnetisation per spin $m = \frac{1}{N} \sum_j s_j$. While calculating the internal energy, one assumes that each spin is equal on the average to m. Since the number of nearest-neighbouring pairs in a d-dimensional hypercubic lattice is dN, the total interaction energy becomes $-JdNm^2$. To calculate the entropy, we note that since $\sum_j s_j = mN$, there must be $(\frac{1}{2} + m)N$ up spins and $(\frac{1}{2} - m)N$ down spins. These many up and down spins can be rearranged among themselves in

$$\frac{N!}{\left[\left(\frac{1}{2} + m\right)N\right]! \left[\left(\frac{1}{2} - m\right)N\right]!}$$

ways, and the energy is the same for all these arrangements. The expression for free energy is then

$$F_{MF} = -JdNm^2 - k_BT \log\left[\frac{N!}{\left[\left(\frac{1}{2} + m\right)N\right]! \left[\left(\frac{1}{2} - m\right)N\right]!}\right]. \tag{12.6}$$

The equilibrium value of m is obtained by minimising this expression with respect to m and using Stirling's approximation for logarithm of factorial whenever required. This gives

$$m = \frac{1}{2} \tanh(\beta dJm). \tag{12.7}$$

This equation always has a solution $m = 0$. However, plotting $\tanh(\alpha x)$ against x, it is seen that the equation $x = \tanh(\alpha x)$ has non-zero solutions for x when $\alpha > 1$. Also, $F_{MF}(m \neq 0) < F_{MF}(m = 0)$. Since $m \neq 0$ corresponds to ferromagnetic phase and $m = 0$ to paramagnetic phase, the transition between these two phases occurs at a critical temperature given by

$$T_c^{MF} = \frac{dJ}{2k_B}. \tag{12.8}$$

This predicts a phase transition for all dimensions. Such prediction is obviously *wrong* at $d = 1$ since we have already proved (without making any approximation) an absence of such a transition in this case. For $d > 1$, there is indeed a phase transition at a non-zero temperature, and the quantitative agreement of the actual critical temperature with the mean-field value Eq. (12.8) increases as the dimension increases.

Let us now consider an extreme type of spin–spin interaction when each spin interacts with *every other* spin with the same strength (i.e. the range of interaction is infinity) [8]:

$$J_{ij} = -J \text{ for all } i \neq j.$$

Since the number of nearest neighbours in a d-dimensional hypercube is $2d$, this model also corresponds to nearest-neighbour interaction on an infinite dimensional lattice. We now write the energy expression Eq. (12.1) as

$$\mathcal{E} = -\frac{J}{N} \sum_{i,j,i \neq j} s_i^z s_j^z = -\frac{J}{N} \left[\sum_j s_j^z \right]^2 \tag{12.9}$$

by introducing an extra $1/N$ factor for normalisation and dropping the constant arising from the $i = j$ terms. The internal energy is hence exactly (and not only approximately) equal to $-JNm^2$, where m is the magnetisation per site as before. Consequently, the free energy expression Eq. (12.6) is exact in this case and the critical temperature is given by

$$T_c = \frac{J}{2k_B}. \tag{12.10}$$

Thus, mean field theory becomes exact in the limit $d \to \infty$.

12.3 Quantum Ising Model

Let us start with what is called the transverse Ising model in one dimension with nearest-neighbour interaction. The Hamiltonian is

$$\mathcal{H}_{TI} = -J \sum_j s_j^z s_{j+1}^z - \Gamma \sum_j s_j^x. \tag{12.11}$$

All the eigenvalues and eigenvectors of this Hamiltonian can be solved exactly [5–7]. The ground state is ferromagnetic for $\Gamma < J$ and paramagnetic for $\Gamma > J$ and undergoes a transition at $\Gamma = J$.

We now consider a widely studied quantum Ising model called the Lipkin–Meshkov–Glick (LMG) model [9] which was originally proposed long back in 1965 as a nuclear model. Later, LMG model received a lot of attention in connection with its entanglement properties. The general form of the Hamiltonian is

$$\mathcal{H}_{LMG} = -\frac{1}{N} \left[J_x \left(S^x \right)^2 + J_y \left(S^y \right)^2 \right] - \Gamma S^z, \tag{12.12}$$

where S^x, S^y and S^z are the components of total magnetic moment defined by

$$S^x = \sum_{j=1}^N s_j^x \tag{12.13}$$

and similarly for S^y and S^z. In this model the X and Y components of each pair of spins interact with (a constant) strength J_x/N and J_y/N, respectively. LMG model has been attacked in various ways [10–12]. The simplest way is to adopt the semi-classical approximation

$$S^x = S \sin\theta \cos\phi, \quad S^y = S \sin\theta \sin\phi, \quad S^z = S \cos\theta, \qquad (12.14)$$

where S is the total spin. Obviously, such approximation is valid only when (i) S is large (so that the allowed values of it forms nearly a continuum) and (ii) the commutators of the non-commutating terms in $\mathcal{H}_{LMG}$ may be ignored. Noting from Eq. (12.2) that

$$\left[S_x^2, S_z \right] = -i(S_x S_y + S_y S_x) \qquad (12.15)$$

and that this operator has off-diagonal matrix elements $\sim N$, we conclude that the semi-classical approximation does not hold always. (To see this point explicitly, consider two spins and observe that

$$\langle \uparrow_1 \uparrow_2 | \, (S_1^x S_2^y + S_2^x S_1^y) \, | \downarrow_1 \downarrow_2 \rangle = \frac{i}{2},$$

so that for N spins such matrix elements may be $\sim N$.) Another way of solving the LMG Hamiltonian is to derive exact solutions for the eigenstates in the form of Bethe-like equations using Gaudin's Lie algebra [13]. Yet another way is to use spin-coherent-state formalism [10].

12.4 A Quantum Ising Model with Long-Range Antiferromagnetic Interaction

We now consider a model where each spin interacts with all other spins with *anti*ferromagnetic interaction of the same strength along the longitudinal direction and the system is subjected to a transverse field. The Hamiltonian is written as

$$\mathcal{H} = \frac{J}{N} \sum_{i,j,i \neq j} s_i^z s_j^z - \Gamma S^x = \frac{J}{N} \left(S^z \right)^2 - \Gamma S^x, \qquad (12.16)$$

with $J > 0$. The constant arising from the $j = k$ terms has been dropped. This model has been recently studied by Chandra, Inoue and Chakrabarti [14] and is actually a variant of the LMG model. Comparing with Eq. (12.12), we see that the LMG Hamiltonian reduces to this when either of J_x and J_y is zero and the other is negative. Two features of the Hamiltonian of Eq. (12.16) can be immediately observed. At $\Gamma = 0$, the ground state of the system has zero longitudinal (total) magnetic moment and is highly degenerate. As soon as a small Γ is switched on, the state with all spins aligned along X is chosen as the ground state. If the temperature is increased from zero now, the disorder increases and one expects an order–disorder transition. The field-induced and temperature-induced transitions were studied by Chandra, Inoue and Chakrabarti [14] using semi-classical approximation (Eq. (12.14)) and by quantum Monte Carlo method. In the remaining part of this chapter we shall study these transitions by perturbative treatment following [4]. Expressions for the ground state energy and free energy of $\mathcal{H}$ will be derived correct up to the second order in Γ, and this will lead us to conclude that there is indeed a second-order phase transition as a function of Γ at $\Gamma = 0$ and that there is also a thermal phase transition for any $\Gamma \neq 0$, although this latter transition is somewhat pathological. These results are not derivable from the existing treatments of LMG Hamiltonian discussed above as the spin-coherent-state formalism [10] excludes the parameter region considered here while the Gaudin algebra technique [13] does not yield a solution in closed form.

12.5 Perturbative Treatment of $\mathcal{H}$

In the Hamiltonian of Eq. (12.16), we shall treat the first term as the unperturbed part

$$\mathcal{H}_0 = \frac{J}{N} \left(S^z \right)^2 \tag{12.17}$$

and the second term as the perturbation

$$\mathcal{H}_p = -\Gamma S^x \tag{12.18}$$

for small values of Γ. The unperturbed eigenvalues are

$$E_M^{(0)} = \frac{J M^2}{N} \tag{12.19}$$

where

$$M = -\frac{N}{2}, -\frac{N}{2} + 1, \ldots, \frac{N}{2} - 1, \frac{N}{2} \tag{12.20}$$

is the total magnetic moment. To obtain the perturbation corrections to this eigenvalue, first note that it has degeneracy

$$\Omega_M = \frac{N!}{\left(\frac{N}{2} + M \right)! \left(\frac{N}{2} - M \right)!}. \tag{12.21}$$

Calling the set of these degenerate states as $\mathcal{M}$, note that the first-order perturbation corrections $E_M^{(1)}$ are equal to the eigenvalues of the matrix $\langle \beta \mid \mathcal{H}_p \mid \alpha \rangle$ where $\mid \alpha \rangle$, $\mid \beta \rangle \in \mathcal{M}$. Since $\mathcal{H}_p$ operating on any spin configuration flips one spin in turn, all the configurations in $\mathcal{H}_p \mid \alpha \rangle$ have a magnetic moment different from M. Hence, $\langle \beta \mid \mathcal{H}_p \mid \alpha \rangle = 0$ for all $\mid \alpha \rangle$, $\mid \beta \rangle$ and $E_M^{(1)} = 0$.

The second-order perturbation correction $E_M^{(2)}$ to the eigenvalues $E_M^{(0)}$ is the eigenvalues of a matrix $\mathbf{P}$ defined by

$$P_{\alpha\beta} = \sum_l \frac{\langle \alpha \mid \mathcal{H}_p \mid l \rangle \langle l \mid \mathcal{H}_p \mid \beta \rangle}{E_M^{(0)} - E_l^{(0)}}, \tag{12.22}$$

where $|\alpha\rangle, |\beta\rangle \in \mathcal{M}$ but $|l\rangle \notin \mathcal{M}$. (See Appendix for the working formulas of degenerate perturbation theory.) To calculate the elements of $\mathbf{P}$, first consider the diagonal elements and note that each state in $\mathcal{M}$ has $\left(\frac{N}{2} + M\right)$ up spins and $\left(\frac{N}{2} - M\right)$ down spins, so that when the operator $\mathcal{H}_p$ flips an up/down spin $E_l^{(0)}$ becomes $\frac{J}{N}(M \mp 1)^2$. Hence the diagonal elements are given by

$$P_{\alpha\alpha} = \frac{\left(\frac{N}{2} + M\right)\left(\frac{\Gamma}{2}\right)^2}{\frac{J}{N}\left[M^2 - (M-1)^2\right]} + \frac{\left(\frac{N}{2} - M\right)\left(\frac{\Gamma}{2}\right)^2}{\frac{J}{N}\left[M^2 - (M+1)^2\right]} = \frac{N\Gamma^2}{4J}\frac{4M^2 + N}{4M^2 - 1}. \tag{12.23}$$

On the other hand, the off-diagonal elements $P_{\alpha\beta}$ will be non-zero when and only when the spin distributions $|\alpha\rangle$ and $|\beta\rangle$ differ in two and only two unlike spins (since these two states must have the same magnetic moment). For example, the two spin states may be like

$$|\alpha\rangle : | \cdots + \cdots - \cdots \rangle \quad |\beta\rangle : | \cdots - \cdots + \cdots \rangle.$$

Clearly, $P_{\alpha\beta}$ will be non-zero if $| l \rangle$ is either $| \cdots + \cdots + \cdots \rangle$ with $E_l^{(0)} = \frac{J}{N}(M+1)^2$ or $| \cdots - \cdots - \cdots \rangle$ with $E_l^{(0)} = \frac{J}{N}(M-1)^2$. Hence,

$$P_{\alpha\beta} = \frac{\left(\frac{\Gamma}{2}\right)^2}{\frac{J}{N}\left[M^2 - (M-1)^2\right]} + \frac{\left(\frac{\Gamma}{2}\right)^2}{\frac{J}{N}\left[M^2 - (M+1)^2\right]} = \frac{N\Gamma^2}{2J}\frac{1}{4M^2 - 1}, \tag{12.24}$$

if $|\alpha\rangle$ and $|\beta\rangle$ differ in two unlike spins and zero otherwise.

We shall now see that the eigenproblem of $\mathbf{P}$ can be solved exactly. Let us consider the Hamiltonian

$$\mathcal{H}' = \frac{J}{N}\left(S^z\right)^2 + \frac{h}{N}\left[\left(S^x\right)^2 + \left(S^y\right)^2\right] \tag{12.25}$$

and treat the first term as the unperturbed Hamiltonian (which is incidentally, the same unperturbed Hamiltonian as before, see Eq. (12.17)) and the remaining part, namely

$$\mathcal{H}'_p = \frac{h}{N}\left[\left(S^x\right)^2 + \left(S^y\right)^2\right] \tag{12.26}$$

as perturbation. The first-order perturbation matrix $\mathbf{P}'$ can be obtained easily. The diagonal elements are

$$P'_{\alpha\alpha} \equiv \langle \alpha | \, \mathcal{H}'_p \, | \alpha \rangle$$

$$= \langle \alpha | \, \frac{h}{N} \sum_{j,k} \left(s_j^x s_k^x + s_j^y s_k^y \right) \delta_{j,k} \, | \alpha \rangle$$

$$= h/2. \tag{12.27}$$

The off-diagonal elements $P'_{\alpha\beta}$ will be h/N if $|\alpha\rangle$ and $|\beta\rangle$ differ in two (and only two) unlike spins and will be zero otherwise. Comparing the expressions of P and P', we find that they are related linearly:

$$\mathbf{P} = \frac{N\Gamma^2}{4J} \frac{4M^2 + N}{4M^2 - 1} \mathbf{1} + \frac{N^2\Gamma^2}{2Jh} \frac{1}{4M^2 - 1} \left(\mathbf{P'} - \frac{h}{2} \mathbf{1} \right), \tag{12.28}$$

where $\mathbf{1}$ is the $\Omega_M \times \Omega_M$ unit matrix. The problem of diagonalising $\mathbf{P}$ thus reduces to that of $\mathbf{P'}$ and incidentally the latter problem is easy to solve.

We rewrite $\mathcal{H}'$ as

$$\mathcal{H}' = \frac{J - h}{N} \left(S^z \right)^2 + \frac{h}{N} \left(\hat{S} \right)^2 \tag{12.29}$$

(where $\hat{S}$ is the total spin operator) and note that the eigenstates of this Hamiltonian are labelled by m and the total spin S. Using the standard relationships

$$\hat{S}^2 \, | \, S, M \rangle = S(S + 1) \, | \, S, M \rangle,$$

$$S = 0, 1, 2, \ldots, \frac{N}{2},$$

and

$$M = -S, -S + 1, \cdots, S - 1, S,$$

we obtain the eigenvalues of $\mathbf{P'}$ as

$$\frac{h}{N} [S(S + 1) - M^2]. \tag{12.30}$$

In view of Eq. (12.28) we conclude that the second-order perturbation correction $E_M^{(2)}$ is given by

$$\lambda(S, M) = \frac{N\Gamma^2}{4J(4M^2 - 1)} \left[2M^2 + 2S(S + 1) \right]. \tag{12.31}$$

We shall now calculate the degeneracy of this state (following [15]) to check whether the degeneracy has been lifted wholly or partly.

For a given S, the allowed values of M are

$$M = -S, -S + 1, \ldots, S - 1, S,$$

while for a given M (≥ 0), the allowed values of S are

$$S = M, M + 1, M + 2, \ldots, \frac{N}{2}.$$

For a given value of S, let M_1 and M_2 be two allowed values of M. Then, each of the degenerate states $|S, M_1\rangle$ must be related by raising or lowering operator to one and only one of the states $|S, M_2\rangle$. Thus, $D(S, M_1) = D(S, M_2)$ and $D(S, M)$ is independent of M. We can hence use $D(S)$ in place of $D(S, M)$ and write the total number of states with a given M as

$$\Omega_M = \sum_{S = M, M+1, \ldots, N/2} D(S).$$

Therefore, the difference $\Omega_M - \Omega_{M+1}$ gives the value of $D(S)$ for $S = M$. Using Eq. (12.21) we have then

$$D(S, M) = D(S) = \Omega_S - \Omega_{S+1}$$
$$= \frac{N!}{\left(\frac{N}{2} + S\right)! \left(\frac{N}{2} - S\right)!} - \frac{N!}{\left(\frac{N}{2} + S + 1\right)! \left(\frac{N}{2} - S - 1\right)!}. \quad (12.32)$$

Thus, the degeneracy of the unperturbed eigenstate of $\mathcal{H}_0$ is lifted only partially under second-order perturbation by $\mathcal{H}_p$.

12.6 Phase Transition Properties

We now have the eigenvalues of the Hamiltonian $\mathcal{H}$ correct up to the second order in Γ (Eq. (12.31)) and the corresponding (still remaining) degeneracy Eq. (12.32). A few conclusions can be immediately drawn.

(i) *At zero temperature,* the ground state (unperturbed) energy is $E_0^{(0)} = 0$ corresponding to $M = 0$ and the perturbation correction is

$$E_0^{(2)} = -\frac{N\Gamma^2}{2J} S(S + 1), \quad (12.33)$$

the lowest value of which is attained at $S = N/2$. The susceptibility per spin is

$$\chi \equiv \frac{1}{N} \left(\frac{d^2 E_0^{(2)}}{d\Gamma^2} \right)_{\Gamma = 0} = -\frac{N^2}{8J} \left(1 + \frac{2}{N} \right). \quad (12.34)$$

This diverges for $N \to \infty$, indicating a second-order phase transition at $\Gamma = 0$.

(ii) *At a non-zero temperature and non-zero field,* the expression for free energy is

$$F_0^{(2)} = E_0^{(2)} - k_B T \ln D(\mathcal{S}, M) = -\frac{N\Gamma^2}{2J} \mathcal{S}(\mathcal{S} + 1) - k_B T \ln D(\mathcal{S}, M). \quad (12.35)$$

By minimizing this expression with respect to $\mathcal{S}$ (and approximating the factorials in Eq. (12.32) by Stirling's formula), we find that the equilibrium value of $\mathcal{S}$ at a temperature T is given by

$$\frac{2\mathcal{S}}{N} = \tanh\left(a\frac{2\mathcal{S}}{N}\right), \quad (12.36)$$

where $a = N^2\Gamma^2/(4Jk_BT)$. Obviously, there is a mean-field type phase transition at a critical temperature

$$\frac{k_B T_c}{J} = \frac{1}{4}\left(\frac{N\Gamma}{J}\right)^2. \quad (12.37)$$

This transition is order–disorder type, since at $T < T_c$ (or $a > 1$), the solution for Eq. (12.36) is $\mathcal{S} \neq 0$ and the expectation value for the transverse magnetic moment

$$\langle \mathcal{S}, M \mid (S^x)^2 \mid \mathcal{S}, M \rangle = \frac{1}{2}(\mathcal{S}^2 + \mathcal{S} - M^2)$$

is non-zero, while for $T > T_c$ (or $a < 1$), the equilibrium value of $\mathcal{S}$ is zero and hence there is no spin ordering in the transverse direction.

It is important to note that the thermal phase transition is pathological in the sense that the expression for the critical temperature Eq. (12.37) involves the system size. For the perturbative treatment to be valid, the perturbation correction must be much less than the spacing between two adjacent eigenvalues of the unperturbed Hamiltonian. Hence, by Eq. (12.16),

$$\Gamma \ll J/N \quad (12.38)$$

so that

$$\frac{k_B T_c}{J} \ll 1. \quad (12.39)$$

Thus, the thermal phase transition of the Hamiltonian $\mathcal{H}$ will be visible *only for a finite system* at a temperature given by Eq. (12.37) provided the condition (12.38) is fulfilled. However, if we rescale the Hamiltonian as

$$\mathcal{H}_s = \left(S^z\right)^2 - bS^x \tag{12.40}$$

(where $b = \Gamma N/J$), then for $b = 0$, there is no thermal transition, but when b is *small and non-zero*, there is a transition at a temperature $T_c = b^2/4$ and this conclusion is true for all values of N.

Before we conclude this section, we point out that the study of Chandra, Inoue and Chakrabarti [14] was for a finite system, but still they did not observe any thermal transition. This is not in contradiction with our results since for their study the value of b was 100, and our treatment cannot predict anything when the condition $b \ll 1$ is not satisfied.

12.7 Effect of Disorder

The Hamiltonian under study (Eq. (12.16)) has no element of disorder. In this section we shall study what happens when a spin glass type interaction is added to the Hamiltonian of Eq. (12.16) for $\Gamma = 0$ (classical case) and $\Gamma \neq 0$ (quantum case). This aspect was studied by Chandra, Inoue and Chakrabarti [14] using replica symmetric equations (for the classical case) and quantum Monte Carlo simulation (for the quantum case). We shall use simple heuristic arguments to derive some conclusions regarding the existence of spin glass phase that agree with the results of these authors. We shall not cover the basic physics of spin glass and refer the reader to the existing reviews [16–19] for this.

Including the Sherrington–Kirkpatrick type interaction to the Hamiltonian of Eq. (12.16) one has

$$\mathcal{H}_{SG} = \frac{1}{N} \sum_{i,j} \left[J s_j^z s_k^z + \tilde{J} \tau_{ij} \right] - \Gamma S^x, \tag{12.41}$$

where $\tilde{J}$ is the strength of the disorder and the distribution function of τ_{ij} is a Gaussian of mean zero and variance unity:

$$P(\tau_{ij}) = \frac{1}{\sqrt{2\pi}} \exp\left(-\frac{1}{2}\tau_{ij}^2\right). \tag{12.42}$$

The interaction energy between the spins i and j (the J_{ij} of Eq. (12.1)) is thus

$$J_{ij} = J + \tilde{J}\tau_{ij}. \tag{12.43}$$

Let us first consider the case of $\Gamma = 0$. Now for the case of $J = 0$ and $\tilde{J} \neq 0$, it is well known [16–18] that the ground state has spin glass order with ground state energy of the order of $\tilde{J}$ and magnetisation zero. On the other hand for $J > 0$ and

Energy

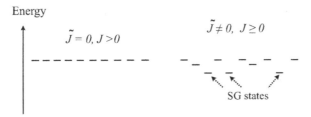

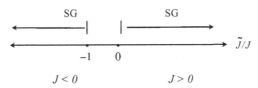

Fig. 12.1 Schematic diagram for the energy of different states. When $\tilde{J} = 0$, $J > 0$ all the states with zero magnetisation have zero energy and are ground states, while for $\tilde{J} > 0$, $J \geq 0$ some of the states with zero magnetisation have lower energy than the others and form the spin glass (SG) ground state

Fig. 12.2 Spin glass regions for different values of J and $\tilde{J}$ with the convention $\tilde{J} \geq 0$. For $J > 0$ (antiferromagnetic J), spin glass phase (SG) is present for $\tilde{J} > 0$, while for $J < 0$ (ferromagnetic J), this phase is observed for $\tilde{J} < -1$

$\tilde{J} = 0$, the ground state energy (according to Eq. (12.19)) is zero and *all* configurations having zero magnetisation are ground state implying that the spin glass states together with some non-spin glass states constitute the highly degenerate ground state. If $\tilde{J}$ is now turned on, the spin glass states have their energy preferentially lowered and the ground state shows spin glass behaviour, for all non-zero values of $\tilde{J}$ (Fig. 12.1). Thus, the ground state for $\tilde{J} \neq 0$ remains the same for $J = 0$ and $J > 0$. One must note that if J becomes ferromagnetic (negative), then the ground state energy for $\tilde{J} = 0$ is negative and of the order of J, so that when $\tilde{J}$ is turned on, there is a competition between J and $\tilde{J}$ and spin glass order prevails only when $\tilde{J} < -J$ (Fig. 12.2).

Let us now turn on the transverse field Γ keeping J antiferromagnetic. Since we have seen that for $\tilde{J} \neq 0$ the presence of J does not alter the ground state, the ground state for the Hamiltonian $\mathcal{H}_{SG}$ of Eq. (12.41) remains the same as that for $J = 0$. There exist extensive studies of the Sherrington–Kirkpatrick model under transverse field [19] which are applicable to this case.

Acknowledgments We are grateful to I. Bose, T.K. Das, G. Ortiz and D. Sen for helpful discussions and to J. Inoue for pointing out a mistake in an earlier version. One author (AG) is grateful to UGC for UPE fellowship. The work was financed by UGC-UPE Grant (Computational Group) and by CSIR project.

Appendix: Working Formulas for Degenerate Perturbation Theory

We present here the working formulas for the eigenvalues in the case of degenerate perturbation theory for ready reference.[1]

Consider a Hamiltonian

$$\mathcal{H} = \mathcal{H}_0 + V.$$

The eigenstates of $\mathcal{H}_0$ are known and we want to solve for the eigenvalues of $\mathcal{H}$ when the effect of V is small. Let $\mathcal{H}_0$ have eigenvalues $E_1^{(0)}$, $E_2^{(0)}$, ... and one of them, say $E_n^{(0)}$ be g-fold degenerate with (one choice of orthogonal) eigenvectors

$$|\psi_1^{(n)}\rangle, |\psi_2^{(n)}\rangle, \ldots, |\psi_g^{(n)}\rangle.$$

The *first-order* perturbation corrections to the eigenvalue $E_n^{(0)}$ are the eigenvalues of the $g \times g$ matrix $\mathbf{M}$ defined by

$$M_{ij} = \langle \psi_i^{(n)} | V | \psi_j^{(n)} \rangle. \tag{12.44}$$

Let us call the eigenvalues of this matrix as $\lambda_1, \lambda_2, \ldots, \lambda_g$. If no two eigenvalues of this matrix are equal, then the degeneracy is said to be completely removed in the first order. To obtain the second-order correction, one needs to calculate the eigenvectors $|e_1\rangle, |e_2\rangle, \cdots, |e_g\rangle$ of the matrix $\mathbf{M}$ and construct therefrom the set of vectors

$$|\phi_i^{(n)}\rangle = \sum_{j=1}^{g} e_i(j) |\psi_j^{(n)}\rangle, \quad i = 1, 2, \cdots, g, \tag{12.45}$$

which can be identified as the correct choice of unperturbed eigenvectors for the eigenvalue $E_n^{(0)}$. In this equation, $e_i(j)$ stands for the jth component of the vector $|e_i\rangle$. The second-order corrections to the eigenvalue $E_n^{(0)} + \lambda_i$ is now given by

$$\sum_{m, m \neq n} \frac{|\langle \phi_i^{(n)} | V | \psi^{(m)} \rangle|^2}{E_n^{(0)} - E_m^{(0)}}, \tag{12.46}$$

where $\psi^{(m)}$ is the eigenvector corresponding to an eigenvalue $E_m^{(0)}$. Note that the sum is here over all the (degenerate or non-degenerate) eigenstates of $\mathcal{H}_0$ whose eigenvalue is different from the one under consideration, namely $E_n^{(0)}$.

---

[1] All text books on Quantum Mechanics (and there is no dearth of it) cover this topic [20–22].

If all the eigenvalues of the matrix $\mathbf{M}$ are equal

$$\lambda_1 = \lambda_2 = \cdots = \lambda_g = \lambda(\text{say}),$$

then the first-order correction is still λ, but the degeneracy is not lifted (even partially). To find the second-order perturbation correction here, one needs to construct the $g \times g$ matrix $\mathbf{P}$ defined by

$$P_{ij} = \sum_{m, m \neq n} \frac{\langle \psi_i^{(n)} \mid V \mid \psi^{(m)} \rangle \langle \psi^{(m)} \mid V \mid \psi_j^{(n)} \rangle}{E_n^{(0)} - E_m^{(0)}}. \tag{12.47}$$

Here also, the sum is over all the eigenstates of $\mathcal{H}_0$ satisfying $E_m^{(0)} \neq E_n^{(0)}$. The second-order correction to $E_n^{(0)}$ are now the eigenvalues of $\mathbf{P}$.

When *some* of the eigenvalues of $\mathbf{M}$ are degenerate, the degeneracy of the eigenvalue $E_n^{(0)}$ is partially lifted, and one has to perform the degenerate perturbation theory (i.e. construct $\mathbf{P}$ and diagonalise it) over each degenerate subspace of $\mathbf{M}$.

The condition of validity of these perturbation equations are that (as mentioned above) the change in eigenstates due to the introduction of V must be small. In quantitative terms, this means

$$\text{second-order correction} \ll \text{first-order correction} \ll d,$$

where d stands for the separation of eigenvalues at $E_n^{(0)}$, that is, the smaller one of the quantities $\mid E_n^{(0)} - E_{n-1}^{(0)} \mid$ and $\mid E_n^{(0)} - E_{n+1}^{(0)} \mid$, $E_{n\pm1}^{(0)}$ being the eigenvalues of $\mathcal{H}_0$ just below and above $E_n^{(0)}$.

References

1. K. Huang, *Statistical Mechanics* Second Edition (John Wiley and Sons, New York, 1987) (see for introduction).
2. M. Plischke and B. Bergersen, *Equilibrium Statistical Physics* Third Edition (World scientific, Singapore, 2006) (see for introduction).
3. R.K. Pathria, *Statistical Mechanics* (Pergamon Press, Oxford, 1972) (see for introduction).
4. A. Ganguli and S. Dasgupta, Phys. Rev. E **80**, 031115 (2009).
5. D.C. Mattis, *The Theory of Magnetism* (Springer-Verlag, Berlin, 1985) vol. 2.
6. R.J. Baxter, *Exactly Solved Models in Statistical Mechanics* (Academic Press, London, 1982).
7. B.M. McCoy and T.T. Wu, *The Two-Dimensional Ising Model* (Harvard University Press, Cambridge, Massachusetts, 1973).
8. H.E. Stanley, *Introduction to Phase Transition and Critical Phenomena* (Oxford University Press, Oxford, 1971).
9. H.J. Lipkin, N. Meshkov and A.J. Glick, Nucl. Phys. **62**,188 (1965).
10. P. Ribeiro, J. Vidal and R. Mosseri, Phys. Rev. E **78**, 021106 (2008).
11. J. Vidal, G. Palacios and R. Mosseri, Phys. Rev. A **69**, 022107 (2004).
12. J. Vidal, R. Mosseri and J. Dukelsky, Phys. Rev. A **69**, 054101 (2004).
13. G. Ortiz, R. Somma, J. Dukelsky and S. Rombouts, Nucl. Phys. B **707**, 421 (2005).

14. A.K. Chandra, J. Inoue and B.K. Chakrabarti, J. Phys. (Conf. Ser.) **143**, 012013 (2009).
15. R.H. Dicke, Phys. Rev. **93**, 99 (1954).
16. K. Binder and A.P. Young, Rev. Mod. Phys. **58**, 801 (1986).
17. M. Mezard, G. Parisi, M. Virasoro, *Spin Glass Theory and Beyond* (World Scientific, Singapore, 1987).
18. K.H. Fischer and J.A. Hertz, *Spin Glasses* (Cambridge University Press, 1993).
19. B.K. Chakrabarti, A. Dutta and P. Sen, *Quantum Ising Phases and Transitions in Transverse Ising Models* (Springer-Verlag, Berlin, 1996) Chapter 6.
20. J.J. Sakurai, *Modern Quantum Mechanics* (Addison-Wesley, Reading, Massachusetts, 1994).
21. V.A. Fock, *Fundamentals of Quantum Mechanics* (Mir Publishers, Moscow, 1978).
22. L.D. Landau and E.M. Lifshitz, *Quantum Mechanics (Non-relativistic Theory)* (Pergamon Press, Oxford, 1977).

Chapter 13
Length Scale-Dependent Superconductor–Insulator Quantum Phase Transitions in One Dimension: Renormalization Group Theory of Mesoscopic SQUIDs Array

S. Sarkar

13.1 Introduction

The transition at zero temperature or at very low temperature belongs to the class of phase transition that is driven by quantum fluctuations of the system [1]. The quantum fluctuations are controlled by system parameters such as the charging energies and Josephson couplings of the Josephson junction array [2]. Here we present a field-theoretical renormalization group study to find the quantum dissipative phases of a lumped superconducting quantum interference device (SQUID). The quantum fluctuations of the system are controlled by the externally applied magnetic field and α, which is the ratio of quantum resistance to tunnel junction resistance. In the SQUID the total current of device is modulated by the applied magnetic flux. Therefore the total current in the SQUID is $I = 2I_c \sin(\theta) |\cos(\frac{\pi \Phi}{\Phi_0})|$, where I_c is the critical current, Φ is the magnetic flux and $\Phi_0 (= \frac{h}{2e})$ is the flux quantum, and θ is the phase of superconducting order parameter. Similar relation holds for Josephson coupling: $E_J = 2E_{J0} \cos(\theta) |\cos(\frac{\pi \Phi}{\Phi_0})|$, where E_{J0} is the bare Josephson coupling. So one may consider that a lumped SQUID system (Fig. 13.1a) can be described in terms of an array of superconducting quantum dots (SQD) but with a modulated Josephson coupling (Fig. 13.1b) and critical current [3, 4]. This effective mapping will help us to analyze (analytically) the system in detail. The experimentalists of [3, 4] have also considered the mesoscopic SQUID system as a modulated Josephson junction array, and we have been motivated by these well-accepted experimental findings [3, 4]. The plan of this manuscript is as follows. Section 13.2 contains the analytical derivations and the physical explanation for the occurrence of quantum dissipative phase in the lumped SQUID system; conclusions are presented in Sect. 13.4.

S. Sarkar (✉)

PoornaPrajna Institute of Scientific Research, 4 Sadashivanagar, Bangalore 5600 80, India,
sujit@physics.iisc.ernet.in

Sarkar, S.: *Length Scale-Dependent Superconductor–Insulator Quantum Phase Transitions in One Dimension: Renormalization Group Theory of Mesoscopic SQUIDs Array*. Lect. Notes Phys. **802**, 267–282 (2010)

DOI 10.1007/978-3-642-11470-0_13

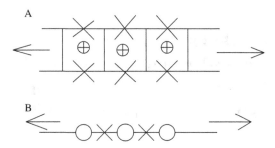

Fig. 13.1 a. Schematic diagram of a one-dimensional array of small capacitance DC SQUIDs. Each plaque is a SQUID with two Josephson junctions marked by the cross, and $\otimes$ represents the applied magnetic flux Φ. b. Equivalent representation of system a, where dots are connected through tunnel junctions and the Josephson couplings of this system are tunable due to the presence of magnetic flux Φ

13.2 Renormalization Group Study for the Quantum Dissipation Phase in Mesoscopic Lumped SQUIDs

Here we prove the appearance of quantum phase slip center (QPS) in SQD array with modulated Josephson coupling through the analysis of a minimal model. We would like to explain the physics of phase slip centers before we start our calculations: The source of dissipation at very low temperature is due to the appearance of phase slip centers. Phase slip centers are of two kinds: the first are thermally activated phase slip centers valid near the superconducting transition temperature, which are well described by LAMH theory [5–7]. The second are quantum phase slip (QPS) centers that occur at $T = 0$ K or at very low temperatures due to the quantum mechanical tunneling in different metastable states [8, 9]. The most important type of fluctuation which occurs during this QPS process is that the phase of the superconducting order parameter changes by $\pm 2\pi$ at a point in the system and the amplitude of the order parameter vanishes at one point. As the appearance of QPS in different low-dimensional superconducting systems is a common phenomenon, we have cited only those references which are related to our problem [5–9].

We consider two SQDs separated by a Josephson junction, where these two SQDs are any arbitrary SQDs of the array. The appearance of QPS is an intrinsic phenomenon (at any junction at any instant) of the system. The Hamiltonian of the system is

$$H = \sum_i \frac{n_i^2}{2C} - E_J |\cos\left(\frac{\pi\Phi}{\Phi_0}\right)| \sum_i \cos(\theta_{i+1} - \theta_i), \qquad (13.1)$$

where n_i and θ_i are, respectively, the Cooper pair density and the superconducting phase of the ith dot and C is the capacitance of the junction. The first term of the Hamiltonian presents the Coulomb charging energies between the dots and the second term is nothing but the Josephson phase term with modulated coupling due to the presence of a magnetic flux. We will see that this model is sufficient to capture the appearance of QPS in SQD systems, and therefore this Hamiltonian has sufficient merit to capture the low-temperature dissipation physics of an SQD array. In the continuum limit, the partition function of the system is given by $Z = \int D\theta(x, \tau) \, e^{-S_Q(x,\tau)}$, where $S_Q = \int d\tau \int dx \frac{E_J}{2} [(\partial_\tau \theta(x, \tau))^2 + (\partial_x \theta(x, \tau))^2]$. We derive the above action in the following way: Cooper pair density n_i is the canonical conjugate variable to the superconducting phase θ_i and they are related with each other through the commutation relation $[n_i, \theta_i] = ih$ and also with the uncertainty principle $\Delta n \Delta \theta = \hbar$. We can write the Cooper pair density as $n_i = i\frac{\partial}{\partial \theta_i}$. The partition function of the system is $Z = \int \Pi_i D\theta_i(\tau) e^{-\int_0^\beta d\tau S_1(\tau)}$, where $S_1(\tau) = \sum_i \frac{C}{2} \dot{\theta}_i^2(\tau) - \frac{E_J}{2} \sum_{ij} \cos(\theta_i(\tau) - \theta_j(\tau))$. In the continuum limit one can write $\theta_i(\tau)$ as $\theta(x, \tau)$ and $\int d\tau$ as $\int dx d\tau$. We scale x as $x = \sqrt{C/Ja^2} x$, where a is the lattice spacing. We obtain the expression of S_Q by substituting all these in consideration of continuum limit in Z and also expanding the cosine function.

This action is quadratic in the scalar field $\theta(x, \tau)$, where $\theta(x, \tau)$ is a steady and differentiable field, so one may think that no phase transition can occur for this case. This situation changes drastically in the presence of topological excitations for which $\theta(x, \tau)$ is singular at the center of the topological excitations. For this type of system, we express $\theta(x, \tau)$ into two components: $\theta(x, \tau) = \theta_0(x, \tau) + \theta_1(x, \tau)$, where $\theta_0(x, \tau)$ is the contribution from attractive interaction of the system and $\theta_1(x, \tau)$ is the singular part from topological excitations. We consider at any arbitrary time τ a topological excitation with center at $X(\tau) = (x_0(\tau), \tau_0(\tau))$. The angle measured from the center of the topological excitations between the spatial coordinate and the x-axis is $\theta_1(x, \tau) = \tan^{-1}(\frac{\tau_0 - \tau}{x_0 - x})$. The derivative of the angle is $\nabla_x \theta_1(x - X(\tau)) = \frac{1}{|x - X(\tau)|^2}[-(\tau_0 - \tau), (x - x_0)]$ which has a singularity at the center of the topological excitation. Finally we get an interesting result when we integrate along an arbitrary curve encircling the topological excitations: $\int_{C1} dx \nabla_x \theta_1(x - X(\tau)) = 2\pi$. We conclude from our analysis that when a topological excitation is present in the SQD array, the phase difference, θ, across the junction of quantum dots, jumps by an integer multiple of 2π. This topological excitation is nothing but the QPS in the (x, τ) plane. According to the phase voltage Josephson relation, $V_J = \frac{-1}{2e} \frac{d\theta}{dt}$ (t is the time), a voltage drop occurs during this phase slip, which is the source of dissipation. This analysis is valid when Φ is away from $\frac{\Phi_0}{2}$, otherwise the system is in the superconducting Coulomb blocked phase.

Now the problem reduces to finding the quantum dissipation physics of SQD array with modulated Josephson couplings and critical current. There are a few interesting studies, following the prescription of Caldeira and Legget [10], that deal with quantum dissipation physics of low-dimensional tunnel junction system [11–15]. Our starting quantum action and subsequent methods are the same as those in [11–15], because the quantum dissipative physics of SQD array system is the same as that of low-dimensional tunnel junction systems. Our prime motivation is to find out the RG equation of SQD system from the analysis of this action to explain the experimental findings of [3, 4]. We will see that the scaling analysis of RG equations derives from this action and explains the experimental findings of [3, 4].

$$S_1 = S_0 + \frac{\alpha'}{4\pi T} \sum_m \omega_m |\theta_m|^2. \tag{13.2}$$

Here, S_1 is the standard action for the system with tiled wash-board potential [11–15] to describe the dissipative physics for low-dimensional tunnel junctions, $S_0 = \int_0^\beta \frac{C}{8e^2} \left(\frac{d\theta(\tau)}{d\tau}\right)^2 + V(\theta(\tau))$, $\alpha' = \frac{R_Q}{R_s} \cos|\frac{\pi\phi}{\phi_0}|$ (the extra cosine factor which we consider in α' is entirely new in the literature to probe the effect of an external magnetic flux), the Matsubara frequency $\omega_m = \frac{2\pi}{\beta} m$, R_Q (= 6.45 $k\Omega$) is the quantum resistance, R_s is the tunnel junction resistance, and β is the inverse temperature. Here $V(\theta) = -\frac{I_c}{2e}|\cos(\frac{\pi\phi}{\phi_0})| + \frac{I\theta}{2e}$. We derive Eq. (13.2) following the prescription of Caldeira and Legget. They proposed a phenomenological idea that the interaction of the microscopic degrees of freedom θ, with the other microscopic degrees of freedom, may be expressed as a system of linearly coupled harmonic oscillators. The structure of our analytical derivations has some similarities with some interesting studies on quantum dissipation [11–15], so we mention our derivation briefly. Here we consider the Lagrangian

$$L = \frac{C}{8e^2} \left(\frac{d\theta}{dt}\right)^2 - V(\theta) + \sum_n \frac{\lambda_n x_n}{2e} \theta$$
$$+ \frac{1}{2} \sum_n \left[\left(\frac{dx_n}{dt}\right)^2 - \omega_n^2 x_n^2 - \frac{\lambda_n^2}{8m_n e^2 \omega_n^2} \theta^2 \right], \tag{13.3}$$

where x_n denotes the coordinate of the harmonic oscillator and λ_n and ω_n are the coupling constant and frequency of the harmonic oscillator, respectively. The last term of the above equation cancels out the change in the adiabatic potential of θ that arises due to coupling with a system of harmonic oscillators. We want to quantize the Lagrangian in the imaginary time formalism. One can express the partition function of the system as

$$Z = \int \prod_n Dx_n(\tau) D\theta_n(\tau) e^{-S_1(x_n, \theta_n)}, \tag{13.4}$$

where

$$S_1 = \int_0^\beta d\tau \left[\frac{C}{8e^2} \left(\frac{d\theta(\tau)}{d\tau}\right)^2 + V(\theta(\tau)) + \sum_n \frac{\lambda_n x_n(\tau)}{2e} \theta(\tau) \right.$$
$$+ \frac{m_n}{2} \sum_n \left[\left(\frac{dx_n(\tau)}{d\tau}\right)^2 - \omega_n^2 x_n^2(\tau) \right.$$
$$\left. + \sum_n \frac{\lambda_n^2}{8m_n e^2 \omega_n^2} \theta^2(\tau) \right]. \tag{13.5}$$

One can write the action S_1 in an effective action, after performing the Gaussian integral over $x_n(\tau)$.

$$S_1 = S_0 + \sum_{\omega_m} K(i\omega_m)|\theta(i\omega_m)|^2, \qquad (13.6)$$

where

$$S_0 = \int_0^\beta d\tau \left[\frac{C}{8e^2}\left(\frac{d\theta(\tau)}{d\tau}\right)^2 + V(\theta(\tau)) \right] \qquad (13.7)$$

is the non-interacting part of the action and

$$K(i\omega_m) = \sum_n \frac{\lambda_n^2}{2m_n} \left(\frac{1}{\omega_n^2} - \frac{1}{\omega_n^2 + \omega_m^2} \right). \qquad (13.8)$$

One can also express this action in imaginary time formalism:

$$S_1 = S_0 + \int_0^\beta d\tau \int_0^\beta d\tau' K(\tau - \tau')\theta(\tau)\theta(\tau)'. \qquad (13.9)$$

$K(|\tau - \tau'|) = (\frac{1}{4\pi R_s})\frac{1}{(|\tau - \tau'|)^2}$. After some mathematical manipulation in Eq. (13.9) and using the periodicity of θ, we can write the action in the following form:

$$S_1 = S_0 + \int_{-\infty}^{+\infty} d\tau \int_0^\beta d\tau' K(|\tau - \tau'|)(\theta(\tau) - \theta(\tau'))^2. \qquad (13.10)$$

The final expression of the action becomes

$$S_1 = \int_0^\beta d\tau \left[\frac{C}{2}(h/2e)^2 \left(\frac{d\theta(\tau)}{d\tau}\right)^2 + V(\theta(\tau)) \right.$$
$$\left. + \frac{1}{16\pi e^2 R_s}\left(\frac{\theta(\tau) - \theta(\tau)'}{(|\tau - \tau'|)} \right)^2 \right] \qquad (13.11)$$

One can also write the above expression as

$$S_1 = S_0 + \frac{\alpha'}{4\pi T} \sum_m \omega_m |\theta_m|^2. \qquad (13.12)$$

We would like to exploit the renormalization group (RG) calculation for weak potentials without loss of generality, so we perform analysis for $I = 0$ case as finite I only inclined the potential profile. Since we are interested in the low-energy excitations, we can ignore the contribution of $|\omega_m|^2$ compared to $|\omega_m|$. So in S_0, we only consider the second term:

$$S_0 = \frac{-I_c}{2e} |\cos\left(\frac{\pi\Phi}{\Phi_0}\right)| \int_0^\beta \cos(\theta)d\tau = V_1 \int_0^\beta \cos(\theta)d\tau. \qquad (13.13)$$

We can write the final action as

$$S_1 = \frac{\alpha'}{4\pi T} \sum_m |\theta_m|^2 + V_1 \int_0^\beta \cos(\theta)d\tau. \qquad (13.14)$$

The RG equation of the above action applicable to study of low-energy excitations is derived following the method given in [16]:

$$\frac{dV_1}{d\ln b} = \left(1 - \frac{1}{\alpha'}\right)V_1, \qquad (13.15)$$

where $b = \frac{d\Lambda}{\Lambda}$. In the RG process we integrate out the higher energy modes of the system; Λ is the higher energy cutoff and $d\Lambda$ is the infinitesimal change of energy around Λ. The time evolution of the coupling constant is $V_1(t) = V_1(0)e^{(1-1/\alpha')t}$. We are mainly interested in the low-energy theory of the system and if we consider the low-energy frequency as ω_m then the corresponding time is $t = \ln\left(\frac{\Lambda}{\omega_m}\right)$.

The coupling constant at maximum time reduces to $V_1 = V_1(0)\left(\frac{\Lambda}{\omega_m}\right)^{1-\frac{1}{\alpha'}}$ and we consider the lowest Matsubara quantization frequency, i.e., $\omega_m = 2\pi T$, so $V_1(T) \propto T^{\frac{1}{\alpha'}-1}$. When we consider a finite system of length, L ($L > 1$), we might argue that the mode $\theta(k)$ is quantized with $\omega_1 = \frac{\pi v}{L}$, where ω_1 is the lowest frequency and v is the velocity of low-energy excitations. So the effective value of the coupling at the lowest frequency is $V_1(L) \propto L^{1-\frac{1}{\alpha'}}$. We observe that the potential V_1 increases for $\alpha' > 1$ and decreases for $\alpha' < 1$, so $\alpha' = 1$ is the phase boundary. When $\alpha' > 1$ owing to the strong dissipation effects, particle comes to rest at one of the minima of the potential (local or particle-like character), i.e., the system is confined in one of the metastable current carrying states, hence the system is in the superconducting phase. It is interesting to observe that dissipation acts to stabilize the superconducting phase, where $\alpha' < 1$ implies weak dissipation has no effect on the potential. The phase fluctuation is large around the dots (nonlocal or wave-like process). As a result, there is no phase-coherent state in the system and therefore there is no superconducting phase. When Φ is zero or an integer multiple of flux quantum, the flux has no effect on the dissipation physics. Larger values of Φ (small α') increase the quantum fluctuations in the system, thereby destroying the phase coherence of states such that the higher magnetic field drives the system from the superconducting phase to the insulating phase, which is consistent with the experimental findings [3, 4]. The analytical structure of our derived RG equation Eq. (13.15) has some similarity with the RG equation of single impurity

Luttinger liquid [17, 18]. But the initial Hamiltonians of these two problems are quite different.

In the strong potential, tunneling between the minima of the potential is very small. In the imaginary time path integral formalism tunneling effect in the strong coupling limit can be described in terms of instanton physics, so we expect that the strong coupling physics of our system can be described in terms of tunneling physics. This view of appearance of instanton for strong coupling phase of low-dimensional correlated condensed matter system is supported by many works [17–22]. In the imaginary time path integral formalism, the potential is inverted and therefore the particle cannot reside at the maximum of the potential for a long time and rolls down to one of the potential minima. It is convenient to characterize the profile of θ in terms of its time derivative,

$$\frac{d\theta(\tau)}{d\tau} = \sum_i e_i h(\tau - \tau_i), \tag{13.16}$$

where $h(\tau - \tau_i)$ is the time derivative at time τ of one instanton configuration. τ_i is the location of the ith instanton, $e_i = 1$ and -1 is the topological charge of instanton and anti-instanton, respectively. Integrating the function h from $-\infty$ to ∞, $\int_{-\infty}^{\infty} d\tau h(\tau) = \theta(\infty) - \theta(-\infty) = 2\pi$. It is well known to us that the instanton (anti-instanton) is almost universally constant except for a very small region of time variation. In the QPS process the amplitude of the superconducting order parameter is zero only in a very small region of space as a function of time and the phase changes by $\pm 2\pi$. In our case width of QPS center in time scale is quite small because the barrier is high. The time interval between the two QPS process is much larger than the width of individual QPS. We have also seen from the analysis of Eq. (13.7) that a phase difference, $\pm 2\pi$, occurs in θ profile when instanton (anti-instanton) occurs in the system. So we can conclude that physics of QPS process is nothing but the instanton (anti-instanton) physics. Larkin et al. [19] also support the idea of QPS as appearance of instanton (anti-instanton). So our system reduces to a neutral system consisting of equal numbers of instanton and anti-instanton. Now our prime task is to present the partition function of the system. After a few steps, we will come to that stage. One can find the expression for $\theta(\omega)$, after Fourier transform to both sides of Eq. (13.16) and that yields

$$\theta(\omega) = \frac{i}{\omega} \sum_i e_i h(i\omega) e^{i\omega \tau_1}. \tag{13.17}$$

Now we substitute this expression for $\theta(\omega)$ in the second term of Eq. (13.2) and finally we get this term as $\sum_{ij} F(\tau_i - \tau_j) e_i e_j$, where $F(\tau_i - \tau_j) = \frac{\pi \alpha}{T} \sum_m \frac{1}{|\omega_m|} \simeq ln(\tau_i - \tau_j)$. We obtain this expression for very small values of ω ($\rightarrow 0$). So effectively $F(\tau_i - \tau_j)$ represents the Coulomb interaction between the instanton and the anti-instanton. This term is the main source of dissipation physics of the system. Following the standard prescription of imaginary time path integral formalism, we can write the partition function of the system [17–22] as

$$Z = \sum_{n=0}^{\infty} z^n \sum_{e_i} \int_0^{\beta} d\tau_n \int_0^{\tau_{n-1}} d\tau_{n-1} \dots \int_0^{\tau_2} d\tau_1 e^{-F(\tau_i - \tau_j)e_i e_j}, \quad (13.18)$$

where z^n is the contribution from instanton , $z = e^{-S_{inst}}$ ($S_{inst} \simeq \sqrt{V_1 C}$). We may also write this expression as

$$Z = \sum_{n=0}^{\infty} z^n \sum_{e_i} \int_0^{\beta} d\tau_n \int_0^{\tau_{n-1}} d\tau_{n-1} \dots \int_0^{\tau_2} d\tau_1$$

$$e^{\sum_m \frac{|\omega_m||q(\omega_m)|^2}{(4\pi\alpha' T)} + i \sum_i e_i q_i(\tau_i)}. \quad (13.19)$$

Here q_τ is the auxiliary field which arises during the functional integral. After extensive calculation, we get the partition function

$$Z = \int Dq(\tau) e^{-\frac{1}{4\pi\alpha' T} \sum_m |\omega||q_m|^2 + 2z \int_0^{\beta} d\tau \cos(q(\tau))}. \quad (13.20)$$

$q(\tau)$ is the auxiliary field. Following the method of previous paragraphs, we finally obtain the RG equation:

$$\frac{dz}{d\ln b} = (1 - \alpha')z. \quad (13.21)$$

The analytical structure of the quantum action and the RG equation is the same with weak potential, which implies the following mappings $\alpha' \leftrightarrow \frac{1}{\alpha'}$ and $V_1 \leftrightarrow z$. Hence there is a duality in this problem between the weak and strong potential. One can also find the scaling expression of z as a function of temperature and length as we derive for weak coupling case, the only change being α' replaced by $1/\alpha'$. Fugacity depends on temperature and length scale as $z(T) \propto T^{\alpha'-1}$, $z(L) \propto L^{1-\alpha'}$.

In this complicated system, we estimate the behavior of resistance from the behavior of fugacity. It is expected to scale as z^2 (the major contribution of voltage/resistance occurs from the second-order expansion of partition function, i.e., from the square of fugacity). In our study, the resistance is evolving due to dissipation effect at very low temperature (less than the superconducting Coulomb blocked temperature). According to our calculations, for large dissipation ($\alpha' > 1$), $R(T) \propto R_Q T^{\beta_1}$ and $\beta_1 > 0$. Therefore at very low temperature, the system shows superconducting behavior. When $\alpha' < 1$, the resistance of the system $R(T) \propto R_Q T^{-\beta_2}$ and $\beta_2 > 0$. So at very low temperature, the resistance of the system shows Kondo-like divergence behavior and the system is in the insulating phase. According to our calculations, for large dissipation ($\alpha' > 1$), $R(T) \propto R_Q L^{-\gamma_1}$ and $\gamma_1 > 0$. Therefore the longer array system shows the less resistive state than shorter array in the superconducting phase. When $\alpha' < 1$, the resistance of system $R(T) \propto R_Q L^{\gamma_2}$, where $\gamma_2 > 0$ ($\beta_1, \beta_2, \gamma_1,$ and γ_2 are independent numbers). So the resistance in the insulating state is larger for longer array system than shorter one. We find the dual behavior of the resistance (voltage) for lower and higher values of magnetic field. When $\alpha' = 1$, i.e., $\Phi = (\Phi_0/\pi)\cos^{-1}(\frac{R_s}{R_Q})$, the system has no length scale-dependent superconductor–insulator transition at very low temperature. This is the critical behavior of the system for a specific value of magnetic field. These theoretical findings are consistent with the experimental observations [3, 4]. Figure 13.2 shows the variation of resistance with temperature. At higher temperature, larger than T_c, SQD array system is in the normal phase and the tunneling between the dots is the sequential tunneling (one after another, tunneling of Cooper pairs). We have shown in [23], that superconductivity occurs due to the co-tunneling effect (higher order tunneling of Cooper pairs, virtual process). The presence of finite resistance at the superconducting phase (between T_1 and T_2) is due to the dissipation effect and also the presence of finite tunneling conductance (i.e., the finite resistance).

The low resistance superconducting phase or insulating phase occurs at very low temperature, a few millikelvin, which is even smaller than the superconducting Coulomb blocked temperature. Experimental measurements have been made down to 50 mK, so they have not found the decaying tendency of resistance at very low temperature. We have not considered the classical phase ($E_J >> E_C$) of the system. In this phase, one can also obtain the dissipative phase with phase slip centers, but one can lose the informations of intermediate quantum phases for the system [23]. Chakarvarty et al. [12, 13] and Larkin et al. [19] have studied QPS for the classical phase. Currently Fistual et al. [24] have done some interesting work on collective Cooper pair transport in the insulating state of one- and two-dimensional Josephson junction array. They have studied the current–voltage characteristics revealing thermally activated conductivity at small voltages and threshold voltage depinning.

Now we explain the effect of impurities in superconducting dots or in the tunnel barrier. Nonmagnetic impurities were found not to affect the Josephson supercurrent . This is in agreement with Anderson's theorem [25] of dirty superconductor. In the presence of paramagnetic impurities, there is an exchange interaction between the spin of conduction electrons with the magnetic impurity spin. The spin

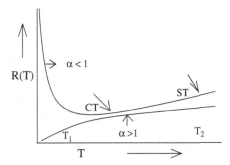

Fig. 13.2 Schematic phase diagram showing the variation of resistance with temperature. ST (see the text) is sequential tunneling regime and CT (see the text) is the Co-tunneling regime. T_2 ($\sim$ few Kelvin) is higher than superconducting transition temperature and T_1 ($\sim$ few millikelvin) is the temperature region much less than the superconducting Coulomb blocked transition temperature

of the magnetic impurity polarizes the spin of electrons and interferes with their tendency for pair formation in the singlet state. As one increases the probability (Γ) of scattering with spin flip, for higher values of Γ, the system enters into the gapless region, and above a critical value ($\Gamma_c = \frac{\Delta(0,0)}{2}$, $\Delta(0,0)$ being the order parameter for the superconductor at zero temperature in the absence of impurity) the superconductivity is destroyed [26]. The Josephson supercurrent is also zero for a clean dot when the applied magnetic flux of the system is a half-integer multiple of flux quantum. At those values of magnetic flux, our system is in the Coulomb blockade insulating phase.

13.3 Quantum Field-Theoretical Study of Model Hamiltonian of the System and Derivation of Tunneling Resistance at the Quantum Critical Point

In this section we define the model Hamiltonian of the system which explains the basic physics of mesoscopic SQUIDs array. Based on this model Hamiltonian, we calculate the tunneling resistance of the system at the quantum critical point. Before we proceed further, we would like to discuss the classical physics of the system. This discussions also present the merit of our work over the existing result in the literature [3].

We have already discussed in the previous section (Fig. 13.1) that one-dimensional array of mesoscopic SQUIDs can be described as an array of Josephson junction with modulated coupling due to the presence of external magnetic flux. It is well known in the literature that a d-dimensional quantum system can be mapped to a $(d+1)$-dimensional classical system [2]. Therefore, the one-dimensional Josephson junction array can be mapped to the two-dimensional XY model. The quantum phase slip centers, which occur across the one-dimensional array at zero or very low temperature, correspond to the appearance of free vortices in two-dimensional XY model. The temperature appears in this system as an extra imaginary time dimension. The authors of [3] have found the temperature-independent flat tail in the resistance measurement at a very low temperature. This regime of temperature is so low that imaginary time dimension is larger and the finite size in real space determines the energy for free vortex formation. They have expressed the zero bias resistance at the minimum temperature as $R_0(T_{min}) \sim N^{2-\pi J}$, where $J = \sqrt{\frac{E_J(B)}{\Lambda E_{C0}}}$, Λ is a dimensional number at $J = J^*$, $R_0(T_{min})$ is independent of temperature. For $J < J^*$, R_0 increases as N (size) of the system increases and for $J > J^*$, R_0 decreases with N [3]. The analytical expression and the physical explanation for this size-dependent zero bias resistance at the quantum critical point is not explicit but rather qualitative due to the lack of rigorous theoretical foundations. In our theoretical study, we calculate the tunneling resistance of the junction by using the quantum field-theoretical calculations in the following paragraphs. To our best knowledge this is the first study in the literature to calculate the tunneling resistance of an array of mesoscopic SQUIDs array at the quantum critical point.

At first, we write the model Hamiltonian of SQD with nearest-neighbor (NN) Josephson coupling and also with the presence of the on-site, NN, and NNN charging energy between the SQD,

$$H = H_{J1} + H_{EC0} + H_{EC1} + H_{EC2}. \qquad (13.22)$$

We recast the different parts of the Hamiltonian in Quantum Phase model

$$H_{J1} = -E_{J1} \sum_i \cos(\phi_{i+1} - \phi_i),$$

where the Hamiltonian H_{J1} is the Josephson tunneling Hamiltonian for NN tunneling between the SQDs. We have

$$H_{EC0} = E_{C0} \sum_i n_i(n_i - 1),$$

$$H_{EC1} = E_{Z1} \sum_i n_i \, n_{i+1},$$

$$H_{EC2} = E_{Z2} \sum_i n_i \, n_{i+2},$$

where H_{EC0}, H_{EC1}, and H_{EC2} are the Hamiltonians for on-site, NN and next-nearest-neighbor (NNN) charging energies of the SQD, respectively. Furthermore n_i is the number of Cooper pairs at the ith site and Hamiltonian H_{EC0} accounts for the on-site charging energy. When the ratio $\frac{E_{J1}}{E_{C0}} \to 0$, the SQD array is in the insulating state having a gap of the width $\sim E_{C0}$, since it costs an energy $\sim E_{C0}$ to change the number of pairs at any dot. Now we would like to recast our basic Hamiltonians in the spin language. This is valid when $E_{C0} >> E_J$, i.e., the Coulomb charging energy dominates over the Josephson tunneling. It is also observed from the experiments that the quantum critical point exists for the larger values of the magnetic field, when the magnetic field-induced Coulomb blockade phase is more prominent than the Josephson tunneling-induced superconducting phase. So our theoretical model is consistent with the experimental findings. During this mapping process we follow [27]:

$$H_{J1} = -2 E_{J1} \sum_i (S_i^\dagger S_{i+1}^- + h.c),$$

$$H_{EC0} = E_{C0} \sum_i S_i^Z,$$

$$H_{EC1} = 4E_{Z1} \sum_i S_i^Z \, S_{i+1}^Z,$$

$$H_{EC2} = 4E_{Z2} \sum_i S_i^Z \, S_{i+2}^Z.$$

One can express spin chain systems to spinless fermion systems through the application of Jordan–Wigner transformation. In Jordan–Wigner transformation the relations between the spin and the electron creation and annihilation operators are [20]

$$S_i^z = \psi_i^\dagger \psi_i - 1/2, S_i^- = (-1)^i \psi_i \exp\left[i\pi \sum_{j=-\infty}^{i-1} n_j\right], S_i^+ = (S_i^-)^\dagger, \quad (13.23)$$

where $n_j = \psi_j^\dagger \psi_j$ is the fermion number at site j. Therefore,

$$H_{J1} = -2E_{J1} \sum_i (\psi_{i+1}^\dagger \psi_i + \psi_i^\dagger \psi_{i+1}), \quad (13.24)$$

$$H_{EC0} = -2hE_{C0} \sum_i (\psi_i^\dagger \psi_i - 1/2), \quad (13.25)$$

$$H_{EC1} = 4E_{Z1} \sum_i (\psi_i^\dagger \psi_i - 1/2)(\psi_{i+1}^\dagger \psi_{i+1} - 1/2), \quad (13.26)$$

$$H_{EC2} = 4E_{Z2} \sum_i (\psi_i^\dagger \psi_i - 1/2)(\psi_{i+2}^\dagger \psi_{i+2} - 1/2). \quad (13.27)$$

We have realized during our calculations that to get an attractive interaction between the Jordan and Wigner (spinless) fermions, we shall have to consider the higher order expansion in $\frac{E_{J1}}{E_{C0}}$. The higher order expansion leads to the virtual state with energies exceeding E_{C0}. This type of incident is also occurring in the co-tunneling process in quantum dot. In this second-order process, the effective Hamiltonian reduces to the subspace of charges 0 and 2 and takes the form [23, 27, 28],

$$H_C = -\frac{3E_{J1}^2}{4E_{C0}} \sum_i S_i^z S_{i+1}^z - \frac{E_{J1}^2}{E_{C0}} \sum_i (S_{i+2}^\dagger S_i^- + h.c). \quad (13.28)$$

With this corrections H_{C1} become

$$H_{EC1} \simeq \left(4E_{Z1} - \frac{3E_{J1}^2}{4E_{C0}}\right) \sum_i (\psi_i^\dagger \psi_i - 1/2)(\psi_{i+1}^\dagger \psi_{i+1} - 1/2).$$

In order to study the continuum field theory of these Hamiltonians, we recast the spinless fermion operators in terms of field operators by the relation [20]

$$\psi(x) = [e^{ik_F x} \psi_R(x) + e^{-ik_F x} \psi_L(x)], \quad (13.29)$$

where $\psi_R(x)$ and $\psi_L(x)$ describe the second quantized fields of right- and the left-moving fermions, respectively. We want to express the fermionic fields in terms of bosonic field by the relation

$$\psi_r(x) = \frac{U_r}{\sqrt{2\pi\alpha}} \, e^{-i\,(r\phi(x) - \theta(x))}, \tag{13.30}$$

r denotes the chirality of the fermionic fields, right (1) or left movers (-1). The operators U_r commute with the bosonic field. U_r of different species commute and U_r of the same species anti-commute. ϕ field corresponds to the quantum fluctuations (bosonic) of spin and θ is the dual field of ϕ. They are related by the relations $\phi_R = \theta - \phi$ and $\phi_L = \theta + \phi$. Our model Hamiltonian consists of two parts, H_0 and H_1. H_0 is the non-interacting part of the Hamiltonian Eq. (13.23),

$$H_0 = \left(\frac{v}{2\pi} + \frac{8E_{C0}}{\pi^2} - 4E_{J2} - \frac{2E_{J1}^2}{E_{C0}}\right) \int dx \, [:(\partial_x\theta)^2: + :(\partial_x\phi)^2:]$$

$$+ \left(16E_{C0} - 8E_{J2} - 4\frac{E_{J1}^2}{E_{C0}}\right) \int dx :(\partial_x\theta - \partial_x\phi)(\partial_x\theta + \partial_x\phi): \tag{13.31}$$

$$H_1 = +\frac{4E_{Z12}}{(2\pi\alpha)^2} \int dx : \cos\left(4\sqrt{K}\phi(x)\right):$$

$$+ \frac{E_{C0}}{\pi\alpha} \int (\partial_x\phi(x)) \, dx, \tag{13.32}$$

v is the velocity of low-energy excitations. It is one of the Luttinger liquid (LL) parameter and the other is K. We have noticed that the co-tunneling effect Eq. (13.29) has two terms: one is the interaction from the z component of the spins between the NN sites and the other introduces the NNN hopping between SQDs.

At first we would like to consider the first term of the co-tunneling process; in this limit system can be described as the Heisenberg XXZ spin chain. The Luttinger liquid parameter of this system is

$$K_1 = \sqrt{\left[\frac{\pi}{\pi + 2\sin^{-1}\left(\frac{2E_{Z1}}{E_{J1}} - \frac{3E_{J1}}{8E_{C0}}\right)}\right]}. \tag{13.33}$$

When we consider the total effect of co-tunneling process then the system reduces to Heisenberg spin chain with NN and NNN interactions. In this limit the LL parameter of the system is

$$K_2 = \sqrt{\left[\frac{E_{J1} - \frac{32}{\pi}E_{Z2}}{E_{J1} + \frac{4}{\pi}\left(4E_{Z1} - \frac{3E_{J1}^2}{4E_{C0}}\right)}\right]}. \tag{13.34}$$

We have already proved in the previous sections of this chapter that the dissipative physics of this system can be described from the analysis of RG equations derived from the Caldeira–Leggett prescription of dissipative physics. The prescription of

Caldeira–Leggett is a simple formalism but it covers many interesting problems in physics, one of them is the array of the SQUIDs [10–13, 29]. In the previous section, during the derivation of RG, we have considered the source of dissipative energy as the tunneling between the two metastable states of tiled wash-board potential. Effectively one can think of it as a tunneling between the two minima of the tiled wash-board potential. It is already shown in the literature from the spectral function analysis of an Ohmic dissipation phase of the system that the dissipative strength (α) is the same as the LL parameter for the superconducting tunnel junction system [29]. We have already shown in the previous section that at $\alpha = 1$, the system shows the quantum phase transition from superconducting phase to the insulating phase. Now our prime interest is to derive the analytical expression of tunneling resistance at the quantum critical point where the zero bias resistance of all mesoscopic SQUIDs array is the same for a particular value of the external magnetic flux [3]. These experimental findings also imply that the tunneling resistance for that particular value of magnetic flux is the same for all the SQUIDs array. For $\alpha = 1$, it implies from the expression for K_1 that

$$E_J^2 = \frac{16E_{Z1}E_{C0}}{3\cos^2|\frac{\pi\Phi}{\Phi_0}|}. \tag{13.35}$$

From this equation we get finally

$$\Phi = \frac{\Phi_0}{\pi}\cos^{-1}\left(\sqrt{16E_{Z1}E_{C0}/3E_J^2}\right). \tag{13.36}$$

If we compare this equation with

$$\Phi = \frac{\Phi_0}{\pi}\cos^{-1}\left(\frac{R_Q}{R_s}\right), \tag{13.37}$$

we derive the analytical expression for tunneling resistance as

$$R_S^{(1)} = \frac{4R_Q}{E_J}\sqrt{\frac{E_{Z1}E_{C0}}{3}}. \tag{13.38}$$

Now we consider the total co-tunneling effect of the system. In this limit our system reduces to a Heisenberg spin chain with NN and NNN interactions. In this limit, the LL parameter of the system is K_2. At the quantum critical point

$$E_J^2 = \frac{64E_{C0}}{3\pi\cos^2|\frac{\pi\Phi}{\Phi_0}|}(E_{z1} + E_{z2}), \tag{13.39}$$

and finally we obtain

$$R_S^{(2)} = \frac{8R_Q}{E_J}\sqrt{\frac{E_{C0}}{3\pi}}(E_{Z1} + E_{Z2}).$$ (13.40)

The ratio of the two tunneling resistances, $\left(\frac{R_s^{(2)}}{R_s^{(1)}}\right)$, is larger than one. From this calculation, it implies that the NNN correction in the co-tunneling process has appreciable effect at the quantum critical point. For different kinds of systems, the strength of on-site, NN, and NNN Coulomb charging energies is different. Therefore the tunneling resistance at the quantum critical point will be different.

13.4 Conclusions

We have observed the length scale-dependent superconductor–insulator quantum phase transition and the dual behavior of the system for smaller and larger values of the magnetic field in a one-dimensional mesoscopic lumped SQUID system. We have also observed a critical behavior where the resistance is independent of length for some critical magnetic field. We find that weak and strong coupling results are self-dual. Our theoretical findings have experimental relevance for lumped SQUID system [3, 4]. We have also done the quantum field-theoretical study of the model Hamiltonian of the system and have calculated the tunneling resistance of the system at the quantum critical point. We have concluded that the tunneling resistance at the quantum critical point will be different due to the different strengths of interactions.

Acknowledgments The author would like to thank the conveners of 'Quantum Phase Transition and Dynamics: Quenching, Annealing and Quantum Computation (Centre for Applied Mathematics and Computational Science, Saha Institute of Nuclear Physics, Kolkata)' for inviting him to participate in this international workshop. The author would like to thank CMTU of the Physics Department of IISC for the facilities extended. Finally the author would like to thank Prof. N. Behera for reading the manuscript critically.

References

1. S. Sachdev, *Quantum Phase Transition* (Cambridge University Press, Cambridge, 1998)
2. S.L. Sondhi., et al., Rev. Mod. Phys. **69**, 315–327 (1997).
3. E. Chow, P. Delsing and D.B. Haviland, Phys. Rev. Lett, **81**, 204–207 (1998).
4. M. Watanbe and D.B. Haviland, Josephson Structures and Superconducting Electronics, Vol. 6 A.V. Narlikar (ed.) (Nova Science Publishers, New York, 2002).
5. M. Tinkham, *Introduction to Superconductivity* (McGRAW-HILL, New York 1996).
6. J.S. Langer and V. Ambegaokar, Phys. Rev, **164**, 498–512 (1967).
7. D.E. MaCumber and B.I. Halperin, Phys. Rev. B, **1**, 1054–1064 (1970).
8. D.S. Golubev and A.D. Zaikin, Phys. Rev. B, **64** 014504–014520 (2001).
9. A.D. Zaikin et al., Phys. Rev. Lett., **78**, 1552–1555 (1997).
10. A.O. Caldeira and A.J. Leggett, Phys. Rev. Lett, **46**, 211–214 (1981).
11. A. Schmid, Phys. Rev. Lett. **51**, 1506–1509 (1983).
12. S. Chakravarty et al., Phys. Rev. B **37**, 3238–3248 (1988).
13. S. Chakravarty, Phys. Rev. Lett. **49**, 681–684 (1082).

14. S. Schon and A.D. Zaikin, Phys. Reports **198**, 237–307 (1990).
15. R. Fazio and H. van der Zant, Physics Report **355**, 235–297 (2001).
16. R. Shankar, Rev. Mod. Phys **66**, 129–191 (1994).
17. C. Kane and M.P.A Fisher, Phys. Rev. B **46**, 15233–15265 (1992).
18. C. Kane and M.P.A Fisher, Phys. Rev. Lett **68**, 1220–1223 (1992).
19. K.A. Matveev, A.I. Larkin and L.I. Glazman, Phys. Rev. Lett. **89**, 0968902–0968905 (2002).
20. T. Giamarchi, *Quantum Physics in One Dimension* (Clarendon Press, Oxford, 2004).
21. A. Furusaki and N. Nagaosa, Phys. Rev. B **47**, 4631–4647 (1993).
22. A. Furusaki and N. Nagaosa, Phys. Rev. B **47**, 3827–3832 (1993).
23. S. Sarkar, Europhys. Lett, **76**, 1172–1178 (2006).
24. M.V. Fistul, V.M. Vinokur and T.I. Batrunia, cond-mat/0708.2334.
25. P.W. Anderson, J. Phys. Chem. Solids **11**, 26–31 (1959).
26. A. Barone and G. Paterno, *Physics and Application of the Josephson Effect* (John Wiley and Sons, USA, 1982).
27. L.I. Glazman and A.I. Larkin, Phys. Rev. Lett. **79** 3736 (1997).
28. S. Sarkar, Phys. Rev. B, **75**, 014528–014534 (2007).
29. A.O. Gogolin, A.A. Nersesyan and A.M. Tsvelik, *Bosonization and Strongly Correlated System* (Cambridge University Press, Cambridge, 1998).

Chapter 14
Quantum-Mechanical Variant of the Thouless–Anderson–Palmer Equation for Error-Correcting Codes

J. Inoue, Y. Saika, and M. Okada

14.1 Introduction

Statistical mechanics of information has been applied to problems in various research topics of information science and technology [1, 2]. Among those research topics, error-correcting code is one of the most developed subjects. In the research field of error-correcting codes, Nicolas Sourlas showed that the so-called convolutional codes can be constructed by spin glass with infinite range p-body interactions and the decoded message should be corresponded to the ground state of the Hamiltonian [3]. Ruján pointed out that the bit error can be suppressed if one uses finite temperature equilibrium states as the decoding result, instead of the ground state [4], and the so-called Bayes-optimal decoding at some specific condition was proved by Nishimori [5] and Nishimori and Wong [6]. Kabashima and Saad succeeded in constructing more practical codes, namely low-density parity check (LDPC) codes by using the infinite range spin glass model with finite connectivities [7]. They used the so-called TAP (Thouless–Anderson–Palmer) equations to decode the original message for a given parity check.

As we shall see later on, an essential key point to obtain the Bayes-optimal solution is controlling the 'thermal fluctuation' in order to satisfy the condition on the Nishimori line (the so-called Nishimori–Wong condition [6]).

J. Inoue (✉)
Complex Systems Engineering, Graduate School of Information Science and Technology, Hokkaido University, N14-W9, Kita-ku, Sapporo 060-0814, Japan, j_inoue@complex.eng.hokudai.ac.jp

Y. Saika
Department of Electrical and Computer Engineering, Wakayama National College of Technology, Nada-cho, Noshima 77, Gobo-shi, Wakayama 644-0023, Japan, saika@wakayama-nct.ac.jp

M. Okada
Division of Transdisciplinary Science, Graduate School of Frontier Science, The University of Tokyo, 5-1-5 Kashiwanoha, Kashiwa-shi, Chiba 277-8561, Japan, okada@k.u-tokyo.ac.jp

Inoue, J. et al.: *Quantum-Mechanical Variant of the Thouless–Anderson–Palmer Equation for Error-Correcting Codes*. Lect. Notes Phys. **802**, 283–295 (2010)
DOI 10.1007/978-3-642-11470-0_14
© Springer-Verlag Berlin Heidelberg 2010

Then, a simple question inspired by quantum mechanics is arisen, namely

- Is it possible to obtain the Bayes-optimal solution by means of the 'quantum fluctuation' induced by tunneling effects?
- What is condition for the optimal control of the fluctuation?

To answer these essential questions, Tanaka and Horiguchi introduced a quantum-mechanical fluctuation into the meanfield annealing algorithm and showed that performance of image restoration is improved by controlling the quantum fluctuation appropriately during its annealing process [8, 9]. The average-case performance is evaluated analytically by one of the present authors [10]. However, there are few studies concerning such a quantum meanfield algorithm for information processing described by spin glasses .

In this chapter, we examine a quantum version of TAP-like meanfield algorithm for the problem of error-correcting codes. For a class of the so-called Sourlas error-correcting codes, we check the usefulness to retrieve the original message with a finite length. The decoding dynamics is derived explicitly and we evaluate the average-case performance numerically through the bit-error rate (BER). We find that TAP-like meanfield approach examined here is useful to decode the original message with a low BER for a relatively large signal-to-noise ratio.

This chapter is organized as follows. In the next section, we explain our model system and comment on the Shannon's bound, namely we mention the generic properties of our error-correcting codes systems. In Sect. 14.3, the Bayesian approach to the problem is introduced. It is shown that the maximum likelihood estimate for each bit can be constructed for both classical and quantum systems. The quantum-mechanical variant of the maximum likelihood estimate for each bit and the corresponding Hamiltonian that gives the estimate in the framework of Bayesian statistics are referred to as *quantum Sourlas codes*. Then, the quantum Sourlas codes and the average-case performance for the case of $p \to \infty$ (p is the number of bit products in the parity check) are investigated by the replica theory. In Sect. 14.4, we construct the TAP-like meanfield decoding algorithm for the Sourlas codes with finite p and examine the performance numerically. The last section is a concluding remark.

14.2 The Model System and the Generic Properties

In this section, we introduce our model system of error-correcting codes and mention the Shannon's bound . In our error-correcting codes, in order to transmit the original message $\{\xi\} \equiv (\xi_1, \ldots, \xi_N), \xi_i \in \{-1, 1\}$ through some noisy channel, we send all possible combinations $_NC_p$ of the products of p-components in the N-dimensional vector $\{\xi\}$ such as

$$J^0_{i1 i2 \cdots ip} = \xi_{i1} \xi_{i2} \cdots \xi_{ip} \tag{14.1}$$

as 'parity.' Therefore, the rate R, which is defined as the ratio between the number of the original N-bits in the message and the number of redundant information to be sent in order to send the original N-bits, is now evaluated by

$$R = \frac{N}{{}_N C_p} \simeq \frac{p!}{N^{p-1}} \tag{14.2}$$

in the limit of $N \to \infty$ keeping the p finite. Obviously, the above rate R goes to zero as N becomes infinity.

On the other hand, when we assume that each parity $J^0_{i1 i2 \cdots ip}$ is sent through the additive white Gaussian noise (AWGN) channel with mean $\mathcal{M} \equiv (J_0 p!/N^{p-1})$ $J^0_{i1 i2 \cdots ip}$ and variance $\Sigma^2 \equiv \{J\sqrt{p!/2N^{p-1}}\}^2$, the output of the channel $J_{i1 i2 \cdots ip}$ is given by

$$J_{i1 i2 \cdots ip} = \left(\frac{J_0 p!}{N^{p-1}}\right) J^0_{i1 i2 \cdots ip} + J\sqrt{\frac{p!}{2N^{p-1}}}\, \eta, \quad \eta = \mathcal{N}(0, 1). \tag{14.3}$$

The factors $p!/N^{p-1}$ or $\sqrt{p!/2N^{p-1}}$ appearing in Eq. (14.3) are needed to take a proper thermodynamic limit (to make the energy of order 1 object) as will be explained in the next section.

We should keep in mind that one can rewrite the output of the channel (14.3) in terms of the conditional probability:

$$P(J_{i1 i2 \cdots ip}|J^0_{i1 i2 \cdots ip}) = \frac{1}{\sqrt{2\pi}}\, e^{-\frac{\eta^2}{2}} = \frac{\exp\left[-\frac{N^{p-1}}{2J^2 p!}\left(J_{i1 i2 \cdots ip} - \frac{J_0 p!}{N^{p-1}} J^0_{i1 i2 \cdots ip}\right)^2\right]}{(J^2 \pi p!/N^{p-1})^{1/2}}$$

$$= \frac{\exp\left[-\frac{N^{p-1}}{2J^2 p!}\left(J_{i1 i2 \cdots ip} - \frac{J_0 p!}{N^{p-1}} \xi_{i1} \xi_{i2} \cdots \xi_{ip}\right)^2\right]}{(J^2 \pi p!/N^{p-1})^{1/2}}$$

$$= P(J_{i1 i2 \cdots ip}|\xi_{i1} \xi_{i2} \cdots \xi_{ip}). \tag{14.4}$$

If each parity $J^0_{i1 i2 \cdots ip}$ is degraded independently, we obtain

$$P(\{J\}|\{\xi\}) = \prod_{i1 i2 \cdots ip} P(J_{i1 i2 \cdots ip}|\xi_{i1} \xi_{i2} \cdots \xi_{ip})$$

$$= \frac{\exp\left[-\frac{N^{p-1}}{2J^2 p!} \sum_{i1 i2 \cdots ip}\left(J_{i1 i2 \cdots ip} - \frac{J_0 p!}{N^{p-1}} \xi_{i1} \xi_{i2} \cdots \xi_{ip}\right)^2\right]}{(J^2 \pi p!/N^{p-1})^{1/2}}, \tag{14.5}$$

where $\prod_{i1 i2 \cdots ip}(\cdots)$ or $\sum_{i1 i2 \cdots ip}(\cdots)$ is taken for all possible combinations of the p-components in the N-dimensional vector $\{\xi\}$. Then, the channel capacity C of the above AWGN channel [11] leads to

$$C = \frac{1}{2}\log_2\left(1 + \frac{\mathcal{M}^2}{\Sigma^2}\right) = \frac{1}{2}\log_2\left(1 + \frac{\{(J_0 p!/N^{p-1})J^0_{i1 \cdots ip}\}^2}{J^2 p!/2N^{p-1}}\right) \simeq \frac{J_0^2 p!}{J^2 N^{p-1}\log 2} \tag{14.6}$$

in the same limit as in derivation (14.2) (we also used the fact $(J_{i1\cdots ip})^2 = \xi_{i1}^2 \xi_{i2}^2 \cdots \xi_{ip}^2 = 1$).

Then, the channel coding theorem tells us that zero-error transmission is achieved if the condition $R \leq C$ is satisfied. For the above case, we have $R/C = (J/J_0)^2 \log 2 \leq 1$, that is,

$$\frac{J_0}{J} \geq \sqrt{\log 2} \equiv \left(\frac{J_0}{J}\right)_c. \tag{14.7}$$

The above inequality means that if the signal-to-noise ratio J_0/J is greater than or equal to $\sqrt{\log 2}$, the error probability of decoding behaves as $P_e \simeq 2^{-N(C-R)} \to 0$ in the thermodynamic limit $N \to \infty$. In this sense, we might say that the zero-error transmission is achieved asymptotically in the limit $N \to \infty, C, R \to 0$ keeping $R/C = \mathcal{O}(1) \leq 1$ for the above *Sourlas codes*.

These properties mentioned above are generic for the Sourlas codes. However, it is very important for us to consider the concrete algorithm to estimate the original messages for a given set of the noisy parity $\{J\}$. To answer the question, we use the Bayesian approach in the next section.

14.3 The Bayesian Approach

For the error-correcting codes mentioned in the previous section, Sourlas pointed out that there exists a close relationship between the error-correcting codes and an Ising spin glass model with infinite range p-body interactions [3]. In this section, we briefly show the relationship for the classical system and then we shall extend the system to the quantum version.

14.3.1 Decoding for Classical System

To decode the original message $\{\xi\}$, we construct the posterior distribution:

$$
\begin{aligned}
P(\{\sigma\}|\{J\}) &\propto P(\{J\}|\{\sigma\})P(\{\sigma\}) \\
&= \{\text{A model of channel } P(\{J\}|\{\xi\})\} \times \{\text{A model of message } P(\{\xi\})\} \\
&= \frac{\exp\left[-\frac{N^{p-1}}{2a^2 p!} \sum_{i1i2\cdots ip}\left(J_{i1\cdots ip} - \frac{a_0 p!}{N^{p-1}}\sigma_{i1}\cdots\sigma_{ip}\right)^2\right]}{(a^2\pi p!/N^{p-1})^{1/2}} \times \frac{1}{2^N},
\end{aligned} \tag{14.8}
$$

where $\{\sigma\} = (\sigma_1, \ldots, \sigma_N)$ denotes an estimate of the original message $\{\xi\}$ and a and a_0 are the so-called hyperparameters corresponding to J_0 and J, respectively. It should be noted that we assumed that the prior $P(\{\sigma\})$ is uniform such that

$P(\{\sigma\}) = 2^{-N}$. For the above posterior distribution, the MAP (maximum a posterior) estimate is obtained as the ground state of the following Hamiltonian:

$$H(\{\sigma\}|\{J\}) = -\log P(\{\sigma\}|\{J\}) \simeq -\log P(\{J\}|\{\sigma\})$$

$$= \frac{N^{p-1}}{2a^2 p!} \sum_{i1 i2 \cdots ip} \left(J_{i1 \cdots ip} - \frac{a_0 p!}{N^{p-1}} \sigma_{i1} \cdots \sigma_{ip} \right)^2, \qquad (14.9)$$

where a constant term $N \log 2$ was neglected for simplicity. Therefore, finding the minimum energy state corresponds to maximizing the log-likelihood $\log P(\{J\}|\{\sigma\})$. In this sense, the MAP estimate here is exactly same as a maximum likelihood estimate, namely

$$\{\bar{\xi}\} = \arg\max_{\{\sigma\}} P(\{\sigma\}|\{J\}). \qquad (14.10)$$

On the other hand, from the viewpoint of statistical mechanics of disordered spin systems, it is obvious that the system $\{\sigma\}$ described by the above Hamiltonian is an Ising spin glass with infinite range p-body interactions. Therefore, the decoding is achieved by finding the ground state of Eq. (14.9) via, for instance, simulated annealing.

In the context of the MPM (maximizar of the posterior marginal) estimate instead of the MAP, the Bayes-optimal solution is obtained for each bit as a simple majority vote:

$$\bar{\xi}_i = P(\sigma_i = +1|\{J\}) - P(\sigma_i = -1|\{J\}) = \text{sgn}\left(\sum_{\sigma_i = \pm 1} \sigma_i P(\sigma_i|\{J\}) \right) \equiv \text{sgn}(\langle \sigma_i \rangle),$$

$$(14.11)$$

where $P(\sigma_i|\{J\})$ is a posterior marginal calculated as

$$P(\sigma_i|\{J\}) = \text{tr}_{\{\sigma\} \neq \sigma_i} P(\{\sigma\}|\{J\}). \qquad (14.12)$$

From the viewpoint of statistics, the above estimate is corresponding to the maximum likelihood estimate *for each bit* because Eq. (14.11) is rewritten as

$$\bar{\xi}_i = \arg\max_{\sigma_i} P(\sigma_i|\{J\}) \qquad (14.13)$$

in contrast with the maximum likelihood estimate *for configuration* (14.10).

It might be convenient for physicists to rewrite the above estimate $\bar{\xi}_i$ in terms of the local magnetization of the system described by the Hamiltonian (14.9) as

$$\bar{\xi}_i = \text{sgn}\left(\frac{\text{tr}_{\{\sigma\}} \sigma_i \exp[-H(\{\sigma\}|\{J\})]}{\text{tr}_{\{\sigma\}} \exp[-H(\{\sigma\}|\{J\})]} \right). \qquad (14.14)$$

In the classical system specified by a given finite temperature $T = 1$, the Bayes-optimal solution $\bar{\xi}_i = \text{sgn}(\langle \sigma_i \rangle)$ minimizes the following BER:

$$p_B = \frac{1}{2}\left(1 - \frac{1}{N}\sum_i \xi_i \bar{\xi}_i\right) = \frac{1}{2}(1 - [\xi\bar{\xi}]_{\{\xi\},\{J\}}) \tag{14.15}$$

$$[\cdots]_{\{\xi\},\{J\}} \equiv \text{tr}_{\{\xi\}}\text{tr}_{\{J\}}(\cdots)P(\{J\}|\{\xi\})P(\{\xi\}) \tag{14.16}$$

on the Nishimori line $a_0/a^2 = J_0/J^2$ [6]. These are all about the classical decoding for the Sourlas codes. In the next section, we extend the classical decoding to the quantum-mechanical decoding.

14.3.2 Decoding for Quantum System

Apparently, essential key point to obtain the Bayes-optimal solution is controlling the 'thermal fluctuation' in order to satisfy the condition on the Nishimori line $T = 1, a_0/a^2 = J_0/J^2$. Then, a simple question arises, namely Is it possible to obtain the Bayes-optimal solution by means of the 'quantum fluctuation' induced by tunneling effects? or What is the condition for the optimal control of the fluctuation? However, in the corresponding quantum system, the condition is not yet clarified. In our preliminary study [12], we considered the quantum version of the posterior by modifying the Hamiltonian as

$$\hat{H}(\{\sigma\}|\{J\}) = \frac{N^{p-1}}{a^2 p!}\sum_{i1 i2\cdots ip}\left(J_{i1\cdots ip} - \frac{a_0 p!}{N^{p-1}}\hat{\sigma}_{i1}^z \cdots \hat{\sigma}_{ip}^z\right)^2 - \gamma\sum_i \hat{\sigma}_i^x$$

$$\hat{\sigma}_i^{z,x} \equiv I_{(1)} \otimes \cdots \otimes \sigma_{(i)}^{z,x} \otimes \cdots \otimes I_{(N)}$$

$$I = \begin{pmatrix} 1 & 0 \\ 0 & 1 \end{pmatrix}, \quad \sigma^z = \begin{pmatrix} 1 & 0 \\ 0 & -1 \end{pmatrix}, \quad \sigma^x = \begin{pmatrix} 0 & 1 \\ 1 & 0 \end{pmatrix} = |+\rangle\langle-| + |-\rangle\langle+|,$$

where the subscript such as $\{\cdots\}_{(i)}$ of each matrix denotes the order in the tensor product. Then, a single bit flip $|+\rangle \equiv {}^t(1,0) \to |-\rangle \equiv {}^t(0,1)$ or $|-\rangle \to |+\rangle$ is caused due to the existence of the second term in the Hamiltonian $\hat{H}$. As the result, the Bayes-optimal solution

$$\bar{\xi}_i^{(\text{Quantum})} = \lim_{\beta,\gamma\to\infty} \text{sgn}[\text{tr}(\hat{\sigma}_i^z \hat{\rho}_\beta)] \quad \text{(keeping } \gamma/\beta \equiv \Gamma = \mathcal{O}(1)) \tag{14.17}$$

with the density matrix

$$\hat{\rho}_\beta \equiv \frac{\exp[-\beta\hat{H}(\{\sigma\}|\{J\})]}{\text{tr}\exp[-\beta\hat{H}(\{\sigma\}|\{J\})]} \tag{14.18}$$

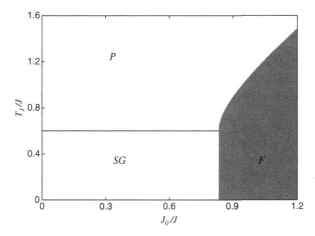

Fig. 14.1 Phase diagram of the Sourlas codes for $p \to \infty$. In the shaded area (F), zero-error transmission is achieved. The area P denotes the para-magnetic phase and the area SG is the spin glass phase. For instance, at the ground state, the critical signal-to-noise ratio is $(J_0/J)_c = \sqrt{\log 2} = 0.8326$. We set $T_J \equiv \beta_J^{-1}$

could be constructed by the quantum fluctuation (which is controlled by the scaled amplitude Γ) even at zero temperature $\beta \to \infty$ because we also take the limit $\gamma \to \infty$ at the same time. With the assistance of the replica method combining the static approximation in the Suzuki–Trottter formula [13], the phase diagram for the case of $p \to \infty$ is easily obtained within one step replica symmetry breaking scheme as shown in Fig. 14.1. At the ground state, the ferromagnet–SpinGlass transition takes place at the critical signal-to-noise ratio $(J_0/J)_c = \sqrt{\log 2} \simeq 0.8326$. As the result, we find that $R \leq C$, namely zero-error transmission $p_B = 0$ is achieved beyond $(J_0/J)_c$. It should be noted that the critical behavior is independent of the scaled amplitude Γ.

However, for finite p, the minimum BER state is dependent on Γ [14] and we should control it when we construct the algorithm based on the TAP-like meanfield approximation.

14.4 Meanfield Decoding via the TAP-Like Equation

In the previous section, we explained the generic properties of the average-case performance of the classical and quantum-mechanical variants for the maximum likelihood estimate for each bit. To evaluate the performance, we assumed that thermally equilibrium states can be obtained by means of both classical and quantum Monte Carlo method. However, it is of course difficult to realize and we should examine the performance for much more realistic scenario. In practice, for a given set of the parity $\{J\}$, we should calculate the estimate

$$\bar{\xi}_i^{(Quantum)} = \lim_{\beta \to \infty} \text{sgn}[\text{tr}(\hat{\sigma}_i^z \hat{\rho}_\beta)], \quad \hat{\rho}_\beta = \frac{\exp[-\beta \hat{H}]}{\text{tr} \exp[-\beta \hat{H}]} \quad (14.19)$$

for each bit. As we mentioned, to calculate the trace effectively by sampling, the quantum Monte Carlo method (QMCM) might be applicable and useful [13]. However, unfortunately, the QMCM approach encounters several crucial difficulties. First, it takes quite long time for us to simulate the quantum states for large number of the Trotter slices. Second, in general, it is technically quite hard to simulate the quantum states at zero temperature. Thus, we are now stuck for the computational cost problem.

Nevertheless, as an alternative to decode the original message practically, we here examine a TAP (Thouless–Anderson–Palmer)-like meanfield algorithm which has a lot of the variants applying to various information processing [15, 16]. In this chapter, we shall provide a simple attempt to apply the meanfield equations to the Sourlas error-correcting codes for the case of $p = 2$. In the following, the derivation of the equations is briefly explained.

We shall start the Hamiltonian:

$$\hat{H} = - \sum_{ij} J_{ij} \hat{\sigma}_i^z \hat{\sigma}_j^z - \Gamma \sum_i \hat{\sigma}_i^x \tag{14.20}$$

$$J_{ij} = \left(\frac{2J_0}{N}\right) J_{ij}^0 + \frac{J}{\sqrt{N}} \eta, \quad J_{ij}^0 = \xi_i \xi_j, \quad \eta = \mathcal{N}(0, 1). \tag{14.21}$$

Then, we rewrite the above Hamiltonian as follows:

$$\hat{H} = - \sum_i (\Gamma \hat{\sigma}_i^x + h_i \hat{\sigma}_i^z) + \sum_{ij} J_{ij}(m_i \hat{I}_i)(m_j \hat{I}_j) - \sum_{ij} J_{ij}(\hat{\sigma}_i^z - m_i \hat{I}_i)(\hat{\sigma}_j^z - m_j \hat{I}_j) \equiv \hat{H}^{(0)} + \hat{V},$$
$$\tag{14.22}$$

$$\hat{H}^{(0)} \equiv - \sum_i (\Gamma \hat{\sigma}_i^x + h_i \hat{\sigma}_i^z) + \sum_{ij} J_{ij}(m_i \hat{I}_i)(m_j \hat{I}_j), \tag{14.23}$$

$$\hat{V} \equiv - \sum_{ij} J_{ij}(\hat{\sigma}_i^z - m_i \hat{I}_i)(\hat{\sigma}_j^z - m_j \hat{I}_j), \quad h_i \equiv 2 \sum_j J_{ij} m_j, \tag{14.24}$$

where we defined the $2^N \times 2^N$ identity matrix $\hat{I}_i$, which is formally defined by $\hat{I}_i \equiv I_{(1)} \otimes \cdots \otimes I_{(i)} \otimes \cdots \otimes I_{(N)}$. m_i, as the local magnetization for the system described by the meanfield Hamiltonian $\hat{H}^{(0)}$, that is,

$$m_i \equiv m_i^z = \lim_{\beta \to \infty} \mathrm{tr}(\hat{\sigma}_i^z \hat{\rho}_\beta^{(0)}), \quad \hat{\rho}_\beta^{(0)} \equiv \frac{\exp(-\beta \hat{H}^{(0)})}{\mathrm{tr} \exp(-\beta \hat{H}^{(0)})}. \tag{14.25}$$

Shortly, we derive closed equations to determine m_i. It is a very difficult problem for us to diagonalize the $2^N \times 2^N$ matrix $\hat{H}$, whereas it is rather easy to diagonalize the meanfield Hamiltonian $\hat{H}^{(0)}$. Actually, we immediately obtain the ground state internal energy as

$$E^{(0)} = -\sum_i E_i + \frac{1}{2}\sum_i h_i m_i, \quad E_i \equiv \sqrt{\Gamma^2 + h_i^2}. \tag{14.26}$$

Then, taking the derivative of the $E^{(0)}$ with respect to m_i and setting it to zero, namely $\partial E^{(0)}/\partial m_i = \sum_k (\partial h_k/\partial m_i)\{h_k/\sqrt{\Gamma^2 + h_k^2} - m_k\} = 0$, we have

$$(\forall_i) \quad m_i = \frac{h_i}{\sqrt{\Gamma^2 + h_i^2}}, \quad h_i = 2\sum_j J_{ij} m_j. \tag{14.27}$$

The above equations are nothing but the so-called *naive meanfield equations* for the Ising spin glass (the Sherrington–Kirkpatrick model [17]) in a transverse field. It should be noted that the equations are reduced to $m_i = h_i/|h_i| = \mathrm{sgn}(h_i) = \lim_{\beta\to\infty} \tanh(\beta h_i)$ ($\forall_i$) which is naive meanfield equations at the ground state for the corresponding classical system.

To improve the approximation, according to [18, 19], we introduce the reaction term R_i for each pixel i and rewrite the local field h_i such that $2\sum_j J_{ij} m_j - R_i$. Then, the naive meanfield equations (14.27) are rewritten as

$$(\forall_i) \quad m_i = \frac{2\sum_j J_{ij} m_j - R_i}{\sqrt{\Gamma^2 + (2\sum_j J_{ij} m_j - R_i)^2}} \simeq \frac{h_i}{(\Gamma^2 + h_i^2)^{3/2}}\left[1 - \frac{\Gamma^2}{\Gamma^2 + h_i^2}\left(\frac{R_i}{h_i}\right)\right].$$

$$\tag{14.28}$$

In the last line of the above equation, we expanded the equation with respect to R_i up to the first order. We next evaluate the expectation of the Hamiltonian $\hat{H}$ by using the eigenvector that diagonalizes the meanfield Hamiltonian $\hat{H}^{(0)} = -\sum_i (\Gamma\hat{\sigma}_i^x + h_i\hat{\sigma}_i^z) + \sum_{ij} J_{ij}(m_i\hat{I}_i)(m_j\hat{I}_j)$. We obtain

$$E_g = E^{(0)} - \Gamma^4 \sum_{ij}\left(\frac{J_{ij}^2}{2E_i^2 E_j^2(E_i + E_j)}\right). \tag{14.29}$$

Then, $(\partial E_g/\partial m_i) = 0$ gives

$$m_i = \frac{h_i}{(\Gamma^2 + h_i^2)^{3/2}}$$
$$\times \left[1 - \frac{\Gamma^2}{\Gamma^2 + h_i^2}\left(\frac{1}{h_i}\right)\sum_j \frac{J_{ij}^2 m_i[2(1 - m_i^2)(1 - m_j^2)^{\frac{3}{2}} + 3(1 - m_i^2)^{\frac{1}{2}}(1 - m_j^2)^2]}{2\Gamma[(1 - m_i^2)^{\frac{1}{2}} + (1 - m_j^2)^{\frac{1}{2}}]^2}\right].$$

$$\tag{14.30}$$

By comparing Eqs. (14.28) and (14.30), we might choose the reaction term R_i for each bit i consistently as

$$R_i = \sum_j \frac{J_{ij}^2 m_i [2(1 - m_i^2)(1 - m_j^2)^{\frac{3}{2}} + 3(1 - m_i^2)^{\frac{1}{2}}(1 - m_j^2)^2]}{2\Gamma[(1 - m_i^2)^{\frac{1}{2}} + (1 - m_j^2)^{\frac{1}{2}}]^2}. \quad (14.31)$$

Therefore, we now have a decoding dynamics described by the following coupled equations:

$$m_i(t+1) = \frac{2\sum_j J_{ij} m_j(t) - R_i(t)}{\sqrt{\Gamma^2 + \{2\sum_j J_{ij} m_j(t) - R_i(t)\}^2}}, \quad (14.32)$$

$$R_i(t) = \sum_j \frac{J_{ij}^2 m_i(t)[2(1 - m_i(t)^2)(1 - m_j(t)^2)^{\frac{3}{2}} + 3(1 - m_i(t)^2)^{\frac{1}{2}}(1 - m_j(t)^2)^2]}{2\Gamma[(1 - m_i(t)^2)^{\frac{1}{2}} + (1 - m_j(t)^2)^{\frac{1}{2}}]^2} \quad (14.33)$$

for each bit i. Then, the MPM estimate is given as a function of time t as $\bar{\xi}_i^{\text{(Quantum)}}(t) = \text{sgn}[m_i(t)]$ and the BER is evaluated at each time step through the following expression:

$$p_B(t) = \frac{1}{2}\left(1 - \frac{1}{N}\sum_i \xi_i \bar{\xi}_i^{\text{(Quantum)}}(t)\right). \quad (14.34)$$

We should notice that the naive meanfield equations are always retrieved by setting $R_i = 0$ for all i. The naive meanfield equations were applied to image restoration by Tanaka and Horiguchi [8]. In the next section, we numerically examine the above TAP-like meanfield equations for a finite length N of the message.

14.4.1 Numerical Results

We plot several results in Fig. 14.2. In the left panel of this figure, we plot the dynamics of meanfield decoding. We plot them for several cases of the signal-to-noise ratio. During the decoding dynamics, we control the Γ by means of

$$\Gamma(t) = \Gamma_0\left(1 + \frac{c}{t+1}\right), \quad (14.35)$$

where t denotes the number of time steps in the TAP-like update described by Eqs. (14.32) and (14.33). In Fig. 14.2, we set $\Gamma_0 = 0.5$. From this figure, we find that the BER drops monotonically as the number of iterations increases. We also find in the right panel that the BER drops beyond the SN ratio $J_0/J \simeq 1$. Although the above results are still at preliminary level, however, from these limited results, we might confirm that TAP-like meanfield approach examined here is useful to decode

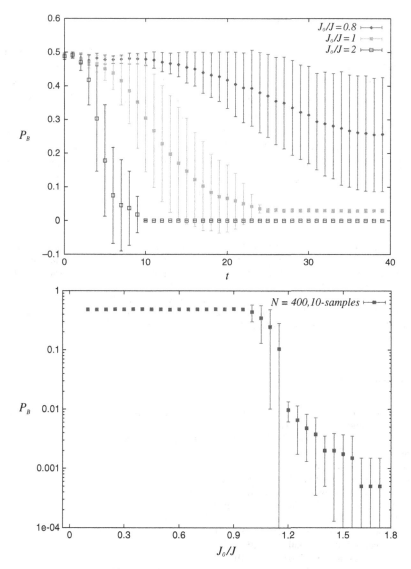

Fig. 14.2 The dynamics of the TAP-like meanfield decoding (*left*, $N = 1000$, the error bars are evaluated by independent 10 samples). We set $p = 2, \Gamma_0 = 0.5$, and $J/J_0 = 0.8, 2$, and 1. The horizontal axis t in the left panel denotes the number of time step in the TAP-like update described by Eqs. (14.32) and (14.33). The right panel shows the signal-to-noise ratio dependence of the BER. We set $p = 2$ and $\Gamma_0 = 0.5$

the original message with a low BER for relatively large SN ratio. It might be important for us to consider the relationship between the performance of the TAP-like meanfield algorithm and the averaged case performance predicted by the replica symmetry theory under the static approximation. However, to clarify this issue, we need more careful and extended numerical studies.

14.5 Concluding Remark

We examined a quantum-mechanical variant of the TAP (Thouless–Anderson–Palmer)-like meanfield algorithm at zero temperature for the problem of error-correcting codes . The algorithm seems to work effectively for our decoding problem. Much more extended studies are now ongoing. For instance, we have problems to be clarified such as the structure of basin (the initial condition dependence of the decoding dynamics), studies for the case of $p \geq 3$, a comparison of the results with those obtained by the QMCM , the relationship between the convergence of the algorithm and the Almeida–Thouless instability which was investigated for the case of the LDPC (low-density parity check) codes [20]. Some of these issues will be investigated extensively in our future studies.

Acknowledgments One of the authors (J.I.) acknowledges Kazutaka Takahashi for useful comments on the analysis of the infinite range transverse Ising spin glass model. We thank Arnab Das, Anjan K. Chandra, and Bikas K. Chakrabarti for organizing the international workshop *Quantum Phase Transition and Dynamics: Quenching, Annealing and Quantum Computation* in Kolkata. We were financially supported by *Grant-in-Aid Scientific Research on Priority Areas "Deepening and Expansion of Statistical Mechanical Informatics (DEX-SMI)" of The Ministry of Education, Culture, Sports, Science and Technology (MEXT)* No. 18079001.

References

1. H. Nishimori, *Statistical Physics of Spin Glasses and Information Processing: An Introduction* (Oxford University Press, Oxford, 2001).
2. M. Mézard, G. Parisi and M.A. Virasoro, *Spin Glass Theory and Beyond* (World Scientific, Singapore, 1987).
3. N. Sourlas, Nature **339**, 693 (1989).
4. P. Ruján, Phys. Rev. Lett., **70**, 2968 (1993).
5. H. Nishimori, J. Phys. Soc. Japan, **62**, 2973 (1993).
6. H. Nishimori and K.Y.M Wong, Phys. Rev. E, **60**, 132 (1999).
7. Y. Kabashima and D. Saad, Europhys. Lett. **45**, 98 (1999).
8. K. Tanaka and T. Horiguchi, IEICE **J80-A-12** 2217 (in Japanese) (1997).
9. K. Tanaka, J. Phys. A: Math. Gen. **35**, R81 (2002).
10. J. Inoue, Phys. Rev. E **63**, 046114 (2001).
11. T.M. Cover and J.A. Thomas, *Elements of Information Theory* (Wiley Series in Telecommunications and Signal Processing, New York) (1991).
12. J. Inoue, Quantum Spin Glasses, Quantum Annealing, and Probabilistic Information Processing, in *Quantum Annealing and Related Optimization Methods*. Lect. Notes Phys. **679**, Das A. and Chakrabarti B.K (eds.) (Springer, Berlin Heidelberg, 2005), p. 259.
13. M. Suzuki, Prog. Theor. Phys. **56**, 1454 (1976).
14. J. Inoue, Y. Saika and M. Okada, *Journal of Physics: Conference Series* 012019, *Proceedings of International Workshop on Statistical-Mechanical Informatics 2008 (IW-SMI 2008)*, Hayashi, M., Inoue, J., Kabashima, Y. and Tanaka, K. (eds.) (IOP Publishing) (2009).
15. M. Opper and D. Saad, *Advanced Mean Field Methods: Theory and Practice*, (The MIT Press, Massachusetts, 2001).

16. M.I. Jordan, *Learning in Graphical Models* (The MIT Press, Massachusetts, 1998).
17. D. Sherrington and S. Kirkpatrick, Phys. Rev. Lett. **35**, 1792 (1975).
18. H. Ishii and T. Yamamoto, J. Phys. C **18**, 6225 (1985).
19. T. Yamamoto, J. Phys. C **21**, 4377 (1988).
20. Y. Kabashima, J. Phys. Soc. Japan **72**, 1645 (2003).

Chapter 15
Probabilistic Model of Fault Detection in Quantum Circuits

A. Banerjee and A. Pathak

15.1 Introduction

Since the introduction of quantum computation, several protocols (such as quantum cryptography, quantum algorithm, quantum teleportation) have established quantum computing as a superior future technology. Each of these processes involves quantum circuits, which are prone to different kinds of faults. Consequently, it is important to verify whether the circuit hardware is defective or not. The systematic procedure to do so is known as fault testing. Normally testing is done by providing a set of valid input states and measuring the corresponding output states and comparing the output states with the expected output states of the perfect (fault less) circuit. This particular set of input vectors are known as test set [6]. If there exists a fault then the next step would be to find the exact location and nature of the defect. This is known as fault localization. A model that explains the logical or functional faults in the circuit is a fault model. Conventional fault models include (i) stuck at faults, (ii) bridge faults, and (iii) delay faults. These fault models have been rigorously studied for conventional irreversible circuit. But with the advent of reversible classical computing and quantum computing it has become important to enlarge the domain of the study on test vectors. In the recent past people have realized this fact and have tried to provide good reversible fault models [5, 7, 10] which are independent of specific technology. The existing reversible fault models are as follows:

1. Single missing gate fault (SMF) where a single gate is missing in the circuit.
2. Multiple missing gate fault (MMGF) where many gates are missing in the circuit.
3. Repeated gate fault (RGF) where the same gate is repeated consecutively many times.
4. Partial missing gate fault (PGF) which can be understood as a defective gate.

A. Banerjee (✉)
J. I. I. T University, A-10, Sector-62, Noida, UP-201307, India, anindita.phd@gmail.com

A. Pathak
J. I. I. T University, A-10, Sector-62, Noida, UP-201307, India, anirban.pathak@jiit.ac.in

Banerjee, A., Pathak, A.: *Probabilistic Model of Fault Detection in Quantum Circuits*. Lect. Notes Phys. **802**, 297–304 (2010)
DOI 10.1007/978-3-642-11470-0_15

5. Cross point fault [10] where the control points disappear from a gate or unwanted control points appear on other gate.
6. Stuck at fault model which includes single stuck at fault (SSF) and multiple stuck at fault (MSF) for zero and one, respectively.

These works are concentrated on circuits composed of gates from NCT,[1] and Generalized Toffoli gate libraries which are part of Maslov's benchmark [4] but it all work in the domain of classical reversible circuit. So most of the existing fault testing protocols [1, 3, 5, 8], except [6] have deterministic nature and are valid in the domain of classical fault testing only. But the advantage of quantum computing becomes prominent only when we use superposition gates like Hadamard gate whereas hardly any effort has been made so far to include these gates in the fault testing protocols. Further it is shown by Ito et al. [9] that given a reversible circuit C, it is NP-hard to generate a minimum complete test set for stuck at faults on the set of wires of C. These facts have motivated us to aim to obtain an efficient algorithm for fault testing and generation of test set for quantum circuits. Our effort in that direction yields several interesting characteristics of quantum fault. As expected the distinguishing nature of quantum fault is only when we consider probabilistic gates and superposition states in qubit line. The fault testing models [5, 7, 10] studied so far are deterministic (D) but in quantum circuits containing superposition gates the notion of determinism fails. To be precise if probabilistic gate like Hadamard gate is taken then the classical notion of test vector fails. This has recently been realized by Perkowski et al. [6] and a new notion of probabilistic test generation has been introduced by them [6]. Present work follows independent approach and reports several new and distinguishing features of quantum fault and provides a general methodology for detection of quantum fault.

To understand the basic nature of quantum fault, let us consider a Hadamard gate which maps $|0\rangle$ to $\frac{1}{\sqrt{2}}|0\rangle + \frac{1}{\sqrt{2}}|1\rangle$ and $|1\rangle$ to $\frac{1}{\sqrt{2}}|0\rangle - \frac{1}{\sqrt{2}}|1\rangle$. Now if we consider $|x\rangle$ as test vector and get $|\bar{x}\rangle$ in the output then we know that the gate exists but there is 50% probability of getting $|x\rangle$ in the output and in that case we shall not be able to conclude anything about the missing gate fault. Increase in number of trials will increase the probability of detecting missing gate fault and after n trials the probability of getting a missing Hadamard fault will be $1 - \frac{1}{2^n}$. As the probability reduces to unity if and only if $n = \infty$, therefore we can conclude that it is impossible to design a test vector which can always deterministically identify a missing Hadamard gate. The conclusion remains valid for every superposition gate and it is even valid if we do not work in the computational basis (i.e., even if you change the measurement basis). In the case of Hadamard gate $|0\rangle$ and $|1\rangle$ are equally good test vectors but if we consider a 1 qubit gate $G_1 = \begin{pmatrix} a_{11} & a_{21} \\ a_{31} & a_{41} \end{pmatrix}$ in general then the probability of success of detecting a missing gate fault after n measurements by using $|0\rangle$ as test vector is $1 - |a_{11}|^{2n}$ and the same with $|1\rangle$ as test vector is $1 - |a_{31}|^{2n}$.

---

[1] This gate library has NOT, CNOT, and Toffoli gates. All these gates can also be achieved in the domain of reversible classical computing.

Consequently, both $|0\rangle$ and $|1\rangle$ can work as test vectors but if $|a_{11}| > |a_{31}|$ then $|1\rangle$ is a better test vector and if $|a_{31}| > |a_{11}|$ then $|0\rangle$ is a better test vector. In general if we consider a generalized n qubit quantum gate G_n which maps states $|i\rangle$ (where i varies from 0 to $2^n - 1$) as $G_n|i\rangle = \sum_{j=0}^{2^n-1} g_{ij}|j\rangle$ then we have to compare all $2^n - 1$ values g_{ii} and find out the lowest value among that. The state $|i\rangle$ corresponding to the lowest value of g_{ii} will provide the best probabilistic test vector. Further we would like to note that in contrary to the classical stuck at fault,[2] the number of possible stuck at fault in quantum circuit is infinite as a qubit line can stuck at $\alpha|0\rangle + \beta|1\rangle$ $\forall \alpha, \beta, \varepsilon \mathbb{C}: |\alpha^2| + |\beta^2| = 1$. Practically, in a finite size circuit we do not need to consider all such stuck at faults but the number of stuck at fault models will not remain restricted to two.

In the next section we have provided an algorithm to generate test vectors for quantum circuit. We have also considered some specific circuits as examples to find different test vectors for the same and in the end we have obtained the time complexity of the algorithm presented here. Section 15.3 is dedicated for conclusions.

15.2 Methodology for Detection of Fault

The proposed fault testing algorithm can be logically divided into two parts. In the first part we find the total circuit matrix and in the second part we find the test vector for quantum circuit. It comprised of two qubit gates and one qubit gate and n qubit lines and is at present computable for m gates.

1. First let us consider an arbitrary quantum circuit having n qubit lines and m gates. The matrix of each gate is expressed by a $2^n \times 2^n$ matrix. This is easy because we just need to take tensor product with identity operator for all those qubit lines which are not addressed by a particular gate. For example, if we consider a quantum circuit having three qubit lines and if the first gate (a NOT gate) is in second qubit line then the matrix of this gate will be $I \otimes NOT \otimes I$. After this simple matrix product of all m matrices (corresponding to m gates) are obtained in sequence to obtain a $2^n \times 2^n$ matrix which is equivalent to the total circuit.

2. In the second part we find the matrices with all possible single and multiple missing gate faults and relevant stuck at faults and compare them with the circuit matrix with no faults. If any row of faulty circuit does not match with the resultant circuit matrix then the corresponding input vector becomes the test vector. Test vectors are not unique; in case we obtain more than one test vector we can follow the approach discussed in Sect. 15.1 in relation with the missing gate fault of a generalized n qubit gate G_n. Further we would like to note that if a particular test vector appears for all faults then it will comprise the test set. Otherwise, an

---

[2] which are of only two types, namely stuck at 0 and stuck at 1

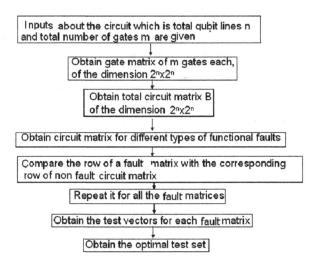

Fig. 15.1 Flowchart showing an algorithm for detection of fault in a quantum circuit

optimized set of test vectors will comprise the test set. But in this optimization process more importance should be given on the success rate than the order of the test set. To be precise, one should choose all the test vectors which have highest possible success rate. In case if two vectors have the same success rate (which is higher than all other possible test vectors) in detecting a fault then both are initially kept in the list and then it is found how many other faults each of these vectors can detect. The one which detects more faults get selected and become an element of the test set.

The algorithm discussed above can be clearly visualized through Fig. 15.1.

15.2.1 Specific Examples

Consider an EPR circuit of two qubit lines consisting of a NOT gate in first qubit line and a CNOT gate with target in second qubit line. The circuit is shown in Fig. 15.2a. This circuit is a key component in teleportation and all the other cases where entanglement generation is required. The total circuit matrix is given by B

$$B = B_0.B_1.B_2 = \begin{pmatrix} 0\,0\,0\,1 \\ 0\,0\,1\,0 \\ 1\,0\,0\,0 \\ 0\,1\,0\,0 \end{pmatrix}$$

where B_0 is the identity matrix, B_1 is the matrix of first gate tensor product with identity, and B_2 is the matrix of second gate.

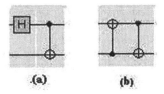

Fig. 15.2 Circuits for finding test vectors

If first gate is missing, then the faulty circuit matrix will be equivalent to $B_2 = \begin{pmatrix} 1\,0\,0\,0 \\ 0\,1\,0\,0 \\ 0\,0\,0\,1 \\ 0\,0\,1\,0 \end{pmatrix}$ and we note that all the rows of B_2 matrix and B are different, consequently all input vectors can detect the fault and if second gate is missing then the faulty circuit matrix will be equivalent to $B_1 = \begin{pmatrix} 0\,0\,1\,0 \\ 0\,0\,0\,1 \\ 1\,0\,0\,0 \\ 0\,1\,0\,0 \end{pmatrix}$ and we note that the last two rows are identical and thus $|00\rangle$ and $|01\rangle$ can only detect the fault. Now if all the gates are missing then again all the rows of $B_0 = \begin{pmatrix} 1\,0\,0\,0 \\ 0\,1\,0\,0 \\ 0\,0\,1\,0 \\ 0\,0\,0\,1 \end{pmatrix}$ matrix and B are not identical and all input vectors can detect the fault. Thus we find that $|00\rangle$ and $|01\rangle$ can detect all the missing gate faults.

Consider another circuit of two qubit line as shown in Fig. 15.2b consisting of a CNOT gate with target in second qubit line and another CNOT gate with its target on first qubit line.

$$B_0 = I \otimes I = \begin{pmatrix} 1\,0\,0\,0 \\ 0\,1\,0\,0 \\ 0\,0\,1\,0 \\ 0\,0\,0\,1 \end{pmatrix}, B_1 = \begin{pmatrix} 1\,0\,0\,0 \\ 0\,1\,0\,0 \\ 0\,0\,0\,1 \\ 0\,0\,1\,0 \end{pmatrix}, B_2 = \begin{pmatrix} 1\,0\,0\,0 \\ 0\,0\,0\,1 \\ 0\,0\,1\,0 \\ 0\,1\,0\,0 \end{pmatrix}$$

Thus total circuit matrix

$$B = B_0 . B_1 . B_2 = \begin{pmatrix} 1\,0\,0\,0 \\ 0\,0\,0\,1 \\ 0\,1\,0\,0 \\ 0\,0\,1\,0 \end{pmatrix}$$

If first gate is missing, then the faulty circuit matrix will be equivalent to B_2 and we note that first and second rows of B_2 matrix and B are identical and thus $|10\rangle$ and $|11\rangle$ input vectors can detect the fault and if second gate is missing then the faulty circuit matrix will be equivalent to B_1 and we note that the second and third

Circuit	Test vectors	Detected fault	Nature of fault
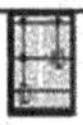 Half Adder circuit	$\lvert110\rangle\,\lvert111\rangle$	S.M.F & M.M.F	D
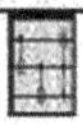 Substractor circuit	$\lvert100\rangle$	S.M.F & M.M.F	D
 Quantum circuit	$\lvert00\rangle\,\lvert01\rangle$	S.M.F & M.M.F	P
 Quantum circuit	$\lvert10\rangle\,\lvert11\rangle$	S.M.F & M.M.F	P
 Quantum circuit	$\lvert10\rangle\,\lvert11\rangle$	S.M.F & M.M.F	P
 EPR circuit	$\lvert00\rangle\,\lvert01\rangle\,\lvert10\rangle\,\lvert11\rangle$	S.M.F & M.M.F	P
 Inverse EPR circuit	$\lvert00\rangle\,\lvert01\rangle\,\lvert10\rangle\,\lvert11\rangle$	S.M.F & M.M.F	P
 Teleportation circuit	$\lvert100\rangle\,\lvert101\rangle\,\lvert110\rangle\,\lvert111\rangle$	S.M.F & M.M.F	P
 Shor's code	$\lvert010\rangle\,\lvert011\rangle\,\lvert111\rangle$	S.M.F & M.M.F	P

Fig. 15.3 Test vectors for different quantum circuits

rows are identical and thus $\lvert01\rangle$ and $\lvert10\rangle$ can detect the fault. Now if all the gates are missing then as we can see that only first row is identical with circuit matrix and thus $\lvert01\rangle$, $\lvert10\rangle$, and $\lvert11\rangle$ input vectors can detect the fault. Thus we find that $\lvert10\rangle$ can detect all the missing gate faults. A set of other examples for single missing gate fault (SMF) and multiple missing gate fault (MMF) with their test vectors and the nature of fault which is probabilistic (P) or deterministic (D) are given in Fig. 15.3. For large circuits as in [2] we divide it into smaller sub-circuits and find test vectors for each sub-circuit. This is shown in Fig. 15.5.

15.2.2 Time Complexity of the Fault Detection Algorithm

The first step requires $m2^{2n}$ multiplications to obtain m matrices of $2^n \times 2^n$ dimension which correspond to m gates. The next step to obtain the resultant matrix requires $(m-1)2^{3n}$ multiplications and $(m-1)2^{2n}$ additions. Thus it requires $\mathcal{O}\left(m2^{3n}\right)$ steps

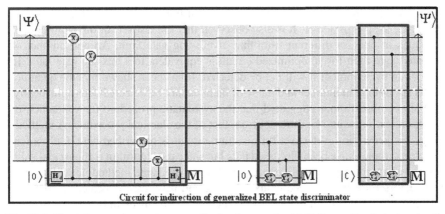

Fig. 15.4 Circuit for non-destructive generalized orthonormal qubit Bell state discriminator circuit

Circuit	Test vector	Detected fault	Nature of fault
	$|110\rangle|111\rangle$,	S.M.F & M.M.F	P
	$|001\rangle|011\rangle$ $|101\rangle|111\rangle$	S.M.F & M.M.F	P
	$|1000\rangle$ to $|1111\rangle$	S.M.F & M.M.F	P
	$|0000\rangle$ to $|1111\rangle$	S.M.F & M.M.F	P

Fig. 15.5 Test vectors for non-destructive generalized orthonormal qubit Bell state discriminator circuit

to obtain the equivalent matrix of the circuit; similarly, it requires the same number of steps to find fault matrices corresponding to each fault. So if we wish to check p faults it requires $\mathcal{O}\left(pm2^{3n}\right)$ steps to construct all the matrices. Now in order to compare the matrices we require $p \times 2^{2n}$ steps in the worst case. Thus the total time complexity is $\mathcal{O}\left(pm2^{3n}\right)$. As it has a linear relation with the number of faults to be considered and the total number of stuck at fault is infinite, we cannot detect all faults in finite time but the methodology will work for all practical purposes where the number of faults of practical interest is finite because of the physical restrictions.

15.3 Conclusions

We have followed an independent approach for generation of test set for quantum circuits and have reported several new and distinguishing features of quantum fault. We have seen that for a quantum gate the classical notion of test vector fails and theoretically it is impossible to determine a test set for Hadamard gate. Further we have observed that in contrary to the classical stuck at fault the number of possible stuck at fault in quantum circuit is infinite as the qubit line can be stuck at $\alpha \, |0\rangle + \beta \, |1\rangle$ $\forall \alpha, \beta, \varepsilon \mathbb{C}$: $|\alpha^2| + |\beta^2| = 1$. It is also observed that the test set for quantum circuit, for stuck at fault, is different from that of missing gate fault. An odd number of repeated gate fault for an optimized circuit is equivalent to a missing gate fault. In case of an even number it will not be detected. It has been shown that the quantum faults are infinite in number and many of them cannot be detected deterministically. The above observations suggested that the systematic procedure for generation of quantum test set would be different from that of the classical procedure. A methodology for generation of test set for quantum circuit is prescribed here. Only a few simple examples have been discussed here but since the algorithm is robust and valid for any quantum or reversible circuits, following the same methodology, in future test vectors for other useful circuits (existing and new) can be presented. Further, attempts to reduce the complexity of the method can be made in future.

References

1. A. Chakraborty, Synthesis of reversible circuits for testing with universal test set and C testability of reversible iterative logic arrays, Proceedings VLSI Design 249–254 (2005).
2. M. Gupta, A. Pathak, R. Srikanth and P.K. Panigrahi, Int. J. of Quantum Inform. **5**, 627–640 (2007).
3. J.P. Hayes, I. Polian and B. Becker, Proceedings of 13th ATS. 100–105 (2004).
4. D. Maslov, G. Duek and N. Scott, Reversible logic synthesis benchmarks page, http://www.cs.uvic.ca/~dmaslov/ (2004).
5. K.N. Patel, J.P. Hayes and I.L. Markov, Proceedings VLSI test Symposium. 410–416 (2003).
6. M. Perkowski, J. Biamonte and M. Lucak, Proceedings of the 35th International Symposium MVL. 62–68 (2005).
7. I. Polian, J.P. Hayes, F. Thomas and B. Becker, Proceedings of the 14th ATS. **2005** 422–427 (2005).
8. H. Rahaman, K.K. Dipak, D.K. Das and B.B. Bhattacharya, *Detection of bridging faults in a reversible circuit*, (Elite, New Delhi, 2006).
9. I. Shigeru, I. Yusuke, T. Satoshi, U. Shuichi, Proceedings of the IEICE General Conference. **2005** 1349–1369 (2005).
10. J. Zhong and J.C. Muzio, Analyzing fault models for reversible logic circuits, IEEE Congress on Evolutionary Computation, (Vancouver, Canada 2422–2427 2006).

Subject Index

Chandra, A.K. et al. (eds.): *Subject Index*. Lect. Notes Phys. **802**, 305–307 (2010)
DOI 10.1007/978-3-642-11470-0   © Springer-Verlag Berlin Heidelberg 2010